ACHIEVEMENT AND ASSURANCE
OF SAFETY

Related titles:

Directions in Safety-critical Systems
Proceedings of the First Safety-critical Systems Symposium, Bristol 1993
Redmill and Anderson (eds)
3-540-19817-2

Technology and Assessment of Safety-critical Systems
Proceedings of the Second Safety-critical Systems Symposium, Birmingham 1994
Redmill and Anderson (eds)
3-540-19859-8

SAFECOMP '93
Proceedings of the 12th International Conference on Computer Safety, Reliability and
Security, Poznań-Kiekrz, Poland 1993
Górski (ed.)
3-540-19838-5

ACHIEVEMENT AND ASSURANCE OF SAFETY

Proceedings of the Third Safety-critical Systems Symposium

Brighton, UK
7–9 February 1995

Edited by
FELIX REDMILL and TOM ANDERSON

Springer
London Berlin Heidelberg New York
Paris Tokyo Hong Kong
Barcelona Budapest

Felix Redmill
Redmill Consultancy
22 Onslow Gardens
London N10 3JU, UK

Tom Anderson
Centre for Software Reliability
University of Newcastle-upon-Tyne
Newcastle-upon-Tyne NE1 7RU, UK

ISBN-13:978-3-540-19922-9   e-ISBN-13:978-1-4471-3003-1
DOI: 10.1007/978-1-4471-3003-1

British Library Cataloguing in Publication Data
A catalogue record for this book is available from the British Library

Typesetting: Camera ready by authors

34/3830-543210 Printed on acid-free paper

PREFACE

Each year there are improvements in safety-critical system technology. These arise both from developments in the contributing technologies, such as safety engineering, software engineering, human factors and risk assessment, and from the adoption or adaptation of appropriate techniques from other domains, such as security.

For these improvements to be of real benefit, they need to be applied during the appropriate stage in the life cycle of the system, whether it be development, assessment, or operation. For this to occur, they must be communicated and explained.

Each year the Safety-critical Systems Symposium offers a distinguished forum for the presentation of papers on such developments, and also for papers from industry on the lessons learned from the use of technologies and methods. The results of many collaborative research projects, with components from both industry and academia, are reported in a universally understandable form.

In 1995 the Symposium was held in Brighton, a venue calculated to stimulate not just the presenters of papers, but all the delegates. Yet, this book of Proceedings is intended not only for the delegates but also for readers not able to attend the event itself. We welcome both categories of reader. Delegates have the benefit of attending the presentations and the opportunity to participate in the discussions; those who take up this book after the event can peruse it at their leisure and, perhaps, on account of it will resolve to attend subsequent symposia. We hope that all who read it will find the chapters both readily comprehensible and informative - they have been commissioned to provide a broad view of what is occurring in the field of safety-critical systems.

The opening chapter offers a view of the work of the Health and Safety Executive - a body which, through its standards, guidelines and other publications, has been influential in the safety-critical domain, both in the UK and internationally. Thereafter, the volume covers a wide range of topics, culminating with chapters on the applications of new technologies in the field of safety-critical systems.

Exploration of the common ground between safety and security, and the lessons which the safety-critical systems field can learn from proven security technology (and vice-versa!) is overdue. It is therefore gratifying to have two chapters which report on work in this area. Chapters on verification and validation provide insights into good practice in Germany and new practice in the UK. The growing awareness of the importance of risk analysis, hazard identification, and safety assessment is reflected and, we hope, will be stimulated, by chapters on these topics. Then there are chapters on Programmable Logic Controllers, and the use of languages in safety-critical software development. Finally, new technologies are represented by chapters on neural networks, artificial intelligence, formal methods and robotics. Several of the chapters report on the goals and results of collaborative projects and, thus, on technologies which are being prepared for use in the immediate future.

Contriving the receipt of all papers in time for publication is never an easy task, and we would like to thank the authors for their part in making this Proceedings a full record of the Symposium. For considerable effort and dedication in organising the event, special appreciation goes to Joan Atkinson.

Felix Redmill and Tom Anderson
October 1994.

Contents

The Safety-Critical Systems Club
sponsor and organiser
of the
Safety-critical Systems Symposium

The Club was inaugurated in 1991 under the sponsorship of the Department of Trade and Industry and the Science and Engineering Research Council. Responsibility for running the club was contracted to the British Computer Society (BCS) and the Institution of Electrical Engineers (IEE) jointly. The Centre for Software Reliability (CSR) was contracted to organise and operate the club.

The Club's purpose is to facilitate the transfer of information, technology, and current and emerging practices and standards. It seeks to involve both technical and managerial personnel within all sectors of the safety-critical community. By doing so, it can facilitate communication among researchers, the transfer of technology from researchers to users, feedback from users, and the communication of experience between users. It also facilitates the union of industry and academia for collaborative projects and provides the means for their publicity and the reporting of their results. The benefits are more effective research, a more rapid and effective transfer and use of technology, the identification of best practice, the definition of requirements for education and training, and the dissemination of information on legal developments, social views and new ideas in the safety-critical domain.

The Club is a non-profit organisation set up to provide a service to all. It sets out to cooperate with all bodies involved or interested in safety-critical systems.

The principal means of achieving the club's goals are organising an annual symposium, holding at least four other events per year, and publishing three issues of a newsletter annually.

The events are of one or two days duration, offering the opportunity to meet others with common interests, to learn from experts, and to participate in both formal and informal discussion.

Membership is open, and for the first three years was free. Now that the period of sponsorship has expired, it is necessary to request an annual subscription - in order to cover planning, mailing and other infrastructure costs. This can be paid at the first meeting attended. Members pay a reduced fee for attendance at Club events and receive, free of charge, all issues of the newsletter, information on forthcoming events, and any special-interest mail shots. Further, by participating in club activities, they can help to determine the club's future direction.

To join or enquire about the Club or its activities, please contact Mrs Joan Atkinson at: CSR, Bedson Building, University of Newcastle upon Tyne, NE1 7RU; Telephone: 0191 221 2222; Fax: 0191 222 7995.

Achieving Safety in Complex Control Systems

Dr Adrian F Ellis
Health and Safety Executive
Technology & Health Sciences Division
Stanley Precinct, Bootle
Merseyside L20 3QZ, UK

Introduction

The power of modern computers has enabled engineers to create and develop equipment which is so complex that only through computers are we able to control them. This presents an enormous challenge to all involved, including the Health and Safety Executive (HSE) as a regulatory authority.

In this paper I will set out HSE aims and describe our approach to achieving safety in complex control systems. There is a considerable range of activities in which we are involved as we attempt to address any contribution to achieving a safer workplace. I will also review our position at the present time and consider some future activities.

Aims of the Health & Safety Executive (HSE)

HSE is responsible for the enforcement of health and safety legislation, primarily the Health & Safety at Work etc. Act 1974 [1], in the UK at over 650,000 workplaces mainly in the industrial sector. To do this HSE employs 4500 staff, including over 1500 inspectors and some 600 specialist and scientific staff [2].

The aims of HSE are to protect the health, safety and welfare of employees and to safeguard others, principally the public who may be exposed to risks from industrial activity. Within HSE, the Technology & Health Sciences Division (THSD) develops technical policy and exists to be the centre of specialist expertise in safety technology and health sciences. THSD initiates, develops and promotes expertise and seeks to influence the actions of employers, employees and others.

Some key activities of HSE are :

- Setting standards by proposing and reforming legislation, issuing guidance and co-operating with other standards setting bodies.

- Enforcement through inspections and the giving of advice.

- Carrying out, publishing and promoting research both generically and for forensic reasons.

- Promoting open debate on safety issues by providing advice and information to Government, by making as much information as is practicable available to the public, by co-operating with other regulatory bodies within UK, EC and worldwide.

Risk Based Approach and ALARP Principle

In general, HSE adopts a risk based approach to its activities.

The words " so far as is reasonably practicable " in the Health & Safety at Work etc. Act 1974 have implied a risk based approach in achieving safety where the safety achieved is weighed against the relevant social, technical and financial factors. More recently it has been introduced formally into UK Regulations. For example Regulation 3 of " The Management of Health and Safety at Work Regulations 1992" [3] requires that:-

"Every employer shall make a suitable and sufficient assessment of:-

- the risks to the health and safety of his employees to which they are exposed whilst they are at work; and

- the risks to the health and safety of persons not in his employment arising out of or in connection with the conduct by him of his undertaking ....".

HSE has also published a short guide entitled " 5 Steps to Risk Assessment "[4].

The risk based approach followed by HSE is based on the principle of ALARP (As Low As Reasonably Practicable)[5]. This principle requires that any risk must be reduced so far as is reasonably practicable, or to a level which is " as low as reasonably practicable ". If a risk falls between the two extremes (i.e. " unacceptable region " and " broadly acceptable region ") and the ALARP principle has been applied, then the resulting risk is the tolerable risk for that specific application. This "three zone" approach is shown in Figure 1.The required level of safety is met when the tolerable risk level is achieved.

The main tests that are applied in regulating industrial risks are similar to those we apply to daily life. They involve determining whether:

♦ the risk is so great that it must be refused altogether; or

♦ the risk is, or has been made, so small as to be insignificant; or

♦ the risk falls between these two states specified above and that it has been reduced to the lowest level practicable, bearing in mind the benefits flowing from its acceptance and taking into account the costs of any further reduction.

Legally speaking, this means that unless the expense undertaken is in gross disproportion to the risk, the employer must undertake that expense. This principle also means that employers are entitled to take into account how much it is going to cost them to take a safety precaution, and that there is some point beyond which the regulator (HSE) should not press them to go; but that they must err on the side of safety.

The principle is easy to explain but it is often very difficult in practice to agree and establish the required level of safety since, achieving the tolerable risk, implies that there is always a residual risk which may still not be tolerable to a minority interest. The problems associated with the public perception of, and attitudes to, risk have to be considered by the regulator [6]. These perceptions vary considerably as any comparison between the nuclear and aviation industries and road traffic accidents would show.

Increasing Awareness

In the 14 years or so HSE has been actively working in this field we have become aware of a wide variation of knowledge and understanding of the problems arising for achieving safety in complex control systems. This variation occurs between practitioners in a single industry and between industry sectors. We have constantly strived to obtain good overall standards of safety by acting as the catalyst for the introduction of good practices.

We have attempted to stimulate debate through regular contact with industry, academia and in the relevant professional institutions. One such contact is our membership of the Safety Critical Systems Committee of the Institution of Electrical Engineers (IEE). The institutions have actioned a number of initiatives to improve awareness of safety issues in complex control systems. Of particular note is the booklet: 'Safety Related Systems: Guidance for Engineers' [7] - previously known as the 'Professional Brief' and originally developed within the IEE but now adapted for a much wider range of engineers working in this field. We welcome and support this co-operative initiative which was developed by a number of professional institutions to highlight the responsibilities and problems facing those involved in the safe use of complex systems.

Within Government we support and now convene the Working Group on Safety-Related Systems. This group is attended by Government Departments who encounter complex control systems in their industry sectors and aims to be a focal point for the exchange of information and co-ordinates approaches in international standards work.

During the 1980's computer based systems (generically referred to as programmable electronic systems (PESs)) were increasingly being used to carry out safety functions. The driving force was improved functionality and economic benefits (particularly when viewed on a total lifecycle basis). Also, the viability of certain designs could only be realised when computer technology was used. The adoption of PESs for safety purposes had, potentially, many safety advantages, but it was recognised this would only be realised if appropriate design and assessment methodologies were used. Many of the features of PESs do not enable the safety integrity (that is, the safety performance of the systems carrying out the required safety functions) to be predicted with the same degree of confidence that has traditionally been available for less complex hardware-based ("hardwired") systems. It was recognised that whilst testing was necessary for complex systems it was not sufficient on its own. This meant that even if the PES was implementing relatively simple safety functions the level of complexity of the programmable electronics was significantly greater than the hardwired systems that had traditionally been used. This rise in complexity meant that the design and assessment methodologies had to be given much more consideration than previously was the case and the level of personal competence required to achieve adequate levels of performance of the safety-related systems was subsequently greater.

In order to tackle these problems several bodies in Europe and the USA published or began developing guidelines to enable the safe exploitation of PES technology. In the UK, the Health and Safety Executive (HSE) developed the PES guidelines [8] in 1987. These guidelines tackled the difficult area of how to handle the design and assessment methodologies needed for complex computer control systems and did much in the UK to encourage industry to address the safety of complex systems in a comprehensive manner. A particularly important, and successful, objective of the guidelines was to stimulate the development of guidelines for individual industry sectors.

Other countries have also produced guidance and since the late 1980's a number of standardisation initiatives [9] have taken place; to which HSE is making a significant contribution - more on this later.

After the PES guidelines were launched, HSE continued to take an active interest in developments in the technology and had a number of concerns with respect to chemical plant automation (particularly when computers were being used to implement safety functions). Concerns have previously been described [10,11] and are summarised below:

◆ The introduction of computer control was poorly thought out and planned;

◆ Inadequate safety requirements specification;

◆ Inadequate procedures with respect to the validation of software;

◆ Evidence of poor workmanship with respect to the standard of plant installation;

5

- Inadequate documentation with respect to what was actually on the plant (as distinct from what was thought to be on the plant);

- Less than fully effective operation and maintenance procedures;

- Concerns over the competence of persons to perform the duties required of them.

HSE will consider revising the PES guidelines when the work on the emerging international standards has stabilised.

More recently, HSE is in the process of publishing the booklet - 'Out of Control' [12] which considers some of the concerns in further detail. HSE examined 34 incidents involving control systems and many useful lessons are demonstrated; most significantly that over 60% of all causes of failure had been 'built-in' before the control system had been taken into use.

One example in the booklet concerns the control system of a commercial microwave oven which failed in service, allowing the microwave generators to operate whilst the oven door was open. The fault was traced to welded contacts of the unit controlling power to the oven, but the underlying failure was that the designers had been given inadequate information. As a result the wrong unit was specified for the job, with premature failure to a dangerous state guaranteed. In another example it was shown how human factors added to the problems of a process operator who was deceived into operating the wrong valve by the design of a "page" on a Visual Display Unit. In this case the designers of the control system allowed a single operator error to cause many tonnes of acrylic acid to pass to drain.

The booklet also highlights the need for a comprehensive classification scheme for failure data from accidents and incidents and suggests a scheme based on the Safety Lifecycle [13]. The primary causes are classified into:-

- Inadequate specification;

- Inadequate design and implementation;

- Inadequate installation and commissioning;

- Inadequate operation and maintenance;

6

◆ Inadequate change control after commissioning.

The analysis is shown graphically in Figure 2.

Increasing Personal Competence

Whilst increasing awareness alerts more people to the potential dangers and to the difficulties in achieving safety, this clearly is not sufficient. Wider awareness needs to be developed into understanding and personal competence. HSE has approached this aspect of safety by encouraging the provision of training. We are now seeking ways of developing competencies based on work experience as well as through academic training.

We have been keen for the safety principles associated with safety-related systems to be included as part of undergraduate courses but we recognise the difficulties created by the already heavy pressures on degree course syllabi. We will continue our efforts in this direction. It is a longer term aim of HSE that many more courses dealing with computing and control should include the safety aspects as an integrated topic.

HSE has co-operated with IEE on the syllabus requirements of a postgraduate course on Safety Related Systems. The aims of the course are:

◆ to acquaint students with the theories, practices and standards associated with safety-related systems and to make them aware of the appropriate interactions between these;

◆ to instil a responsible and well-balanced attitude to safety matters;

◆ to identify the requirements that need to be met by all professional engineers (and those for whom they are responsible) working in the area of safety-related systems;

◆ to instil public confidence that safety issues are being handled in a responsible manner by professional engineers.

The syllabus has been designed on a modular basis to meet the differing requirements of participants and their employers. This format will enable courses of study to be set up where Part 1 of the syllabus may be offered as a course leading to a Certificate and where study of both Parts 1 & 2 of the

7

syllabus will lead to a Diploma. Successful completion of Parts 1, 2 and a project will lead to the award of a Master's degree.

Over the past few years HSE has co-operated with the IEE in the planning and presentation of its Vacation School on Safety Critical Systems. This is a successful venture which is proving to be popular with a significant number of engineers.

The role of National Vocational Qualifications (NVQ) in our system of education and training is well established but its principles have not been fully extended into the field of safety-related systems. Whilst HSE is still considering the contribution of NVQ's in various safety-related disciplines we are acutely aware of the need for engineers and others to have the necessary competencies to do the tasks with which they have been assigned. Frequently, during forensic investigations, HSE inspectors discover that people involved in accident situations did not fully understand the implications of their actions which, in some cases, had contributed to the accident. HSE has been in contact with several industry lead bodies to examine how competencies for safety can be included with, or added to, existing competencies in the fields of computing and control.

We are currently setting up a study with the Institution of Electrical Engineers and the British Computer Society which it is hoped will define competencies for main stream practitioners, dealing with both hardware and software for safety-related systems. Even concentrating on main stream competencies this work will be complex and diverse since it is important that the results of the study fit in with and compliment current activities in this area and does not result in unacceptable bureaucracy. The defining of competencies has not previously been set out and we expect that the published results of the study will provide industry with useful guidance. We view the successful completion of this study as a fundamental building block for the development of competencies.

Standards Activities

HSE views the successful completion of the current standardisation activities as particularly important in enabling the technology to be used in a harmonised way and a significant milestone for achieving safety. To that end we are actively involved in the standards making process at both a European and International level [9]. From an HSE perspective the most important current standardisation activity in the area of control systems is the proposed international standard being developed under the auspices of the International Electrotechnical Commission (IEC) entitled 'Functional safety: Safety-related systems'. The proposed Parts of this standard are: 1: General requirements, 2: Requirements for electrical/electronic/programmable electronic systems and 3: Software requirements. This draft IEC standard provides a systematic and coherent approach to the functional safety of safety-related systems employing electrical technology. The standard addresses all relevant Safety Lifecycle phases (from initial concept through to decommissioning) and adopts a risk based approach to the determination of the safety integrity requirements for the safety functions. A major objective of the proposed standard is to facilitate the development of standards for different sectors.

One standard for a specific sector is the draft European standard prEN954-1 'Safety of machinery - Safety related parts of control systems - Part 1: General principles for design' which is being drafted by a joint working group of the European Committee for Standardisation (CEN) and the European Committee for Electrotechnical Standardisation (CENELEC). This standard is intended to become a harmonised standard to support the 'Essential Safety Requirements' (ESR's) of the Machinery Directive. The purpose of harmonised standards in this European sense is to provide people with a way of meeting the ESR's. Complying with a harmonised standard means that one has been 'deemed to comply' with the relevant ESR's.

This European standard is just one example of the need to harmonise between European and International standards. Also, HSE views the harmonisation between generic standards and application specific standards as very important but with contributors from all around the world this is not easy to achieve.

Rapid developments in the technology create an ever changing backdrop of experiences which influence opinion. Therefore, it is essential that key safety principles are embedded in the generic standards in such a way so that adequate safety can be achieved without inhibiting technological development. This also meets HSE's aims of achieving adequate safety through a risk based, systematically managed approach.

Research Activities

HSE is a major research generator and 1992/1993 our total research budget was £38million, 80% of which was spent extermurally. Within THSD it is an important part of our activities. We use this HSE sponsored research for the systematic investigation into, and study of , technologies for the purpose of:

- developing techniques to improve safety in the workplace,

- mitigating the harmful effects arising from the use of these technologies, and

- for forensic evaluation following accidents and incidents.

The use of advanced computing techniques to control complex safety-related systems is viewed in these ways. Computing and control is a field of rapid development in which we have a number of research activities. The following are recent examples:

- We have carried out a preliminary survey [14] of the use of knowledge based systems in the chemical process sector. Use of these type of systems appears likely to increase in the future and so there is need for HSE to assess the implications of their use.

- There has been a growing concern that the introduction of computer based control into process plants has created a supervisory role for the operator which is not adequately catered for in the current approaches to the design of control room operator interfaces. As an initial step, in dealing with this as a potential problem, HSE commissioned a review of the current status of human factors techniques, standards and guidelines [15].

Examples of current research activities are:

- Investigation of the safety implications knowledge based systems, neural networks and fuzzy logic.

- Safety Integrity requirements for motor drive systems.

- Safety of interactive robot systems.

- Development of electromagnetic compatibility safety criteria for safety-related systems.

- Hardware modelling of complex safety-related system architectures.

It can be seen that the breath of technology in the field of complex control systems is very wide. Therefore it is important that the research funding which HSE uses in this area is highly focussed if the safety objectives of HSE are to be met.

Way Forward

Since the early 1980's we have made significant strides in the safe use of complex safety-related systems but we are only at the very early stages of the full technical development of such systems. Future technical innovations will bring, I am sure, new problems but they will also bring new safety benefits for us to harness. To do this effectively we must approach the technology systematically and ensure that the measures we are developing are applied to all safety lifecycle phases and that proper attention is paid to the safety management and the personal competencies issues.

HSE intends to play its part in the future safe development of these systems and is committed to a continuing collaborative approach with others working in this field.

References

1. Health and Safety at Work etc. Act 1974 (1974 c.37).

2. Enforcing Health and Safety Legislation in the Workplace. Report by the Comptroller and Auditor General. Available from HMSO Publication Centre. PO Box 276, London SW8 5DT.

3. The Management of Health and Safety at Work Regulation 1992 (SI 1992 No. 2051) : also " Guidance on Regulations : The Management of Health and Safety at Work Regulations 1992 ". ISBN 0 11 8863304. Available from HSE Books, PO Box 1999, Sudbury, Suffolk CO10 6FS. UK - Tel : 0787 881165 Fax : 0787 313995.

4. 5 Steps to Risk Assessment. HSE Leaflet IND(G)163L. Available from HSE Books, PO Box 1999, Sudbury, Suffolk, CO10 6FS. UK - Tel : 0787 881165 Fax : 0787 313995.

5. 'The tolerability of risk from nuclear power stations'. ISBN 0 11 886368 1. Available from HSE Books, PO Box 1999, Sudbury, Suffolk, CO10 6FS. UK - Tel : 0787 881165 Fax : 0787 313995.

6. Coping with Technological Risk : A 21st Century Problem. J D Rimington. Engineering and Society : The 1993 CSE Lecture to the Royal Academy of Engineering.

7. Safety Related Systems: Guidance for Engineers. To be published late 1994 by the Hazards Forum.

8. Programmable electronic systems in safety-related applications: 1. 'An introductory guide'. ISBN 011 8839062.   2. 'General technical guidelines'. ISBN 011 8839063. Both documents available from HSE Books, PO Box 1999, Sudbury, Suffolk CO10 6FS. UK - Tel : 0787 881165 Fax : 0787 313995.

9. Activities to harmonise worldwide safety interlock standards. V Maggioli and R Bell. Proceeding of International Symposium and Workshop on Safe Chemical Process Automation: Houston, Texas, USA - September 27-29 1994.

10. Safety overview of computer control for chemical plant; P G Jones; in Hazards X: Process Safety in Fire and Speciality Chemical Plants Including Developments in Computer Control of Plants, I Chem E Symposium Series No. 115. Available from Davis Building, 165-189 Railway Terrace, Rugby, CV21 3HQ, UK.

11. Computer in chemical plant: A need for safety awareness; P G Jones; in Hazards XI; I Chem E Symposium Series No. 124. Available from Davis Building, 165-189 Railway Terrace, Rugby, CV21 3HQ, UK.

12. 'Out of Control' (Control systems: Why things went wrong, and how they could have been prevented). To be published soon by HSE Books, PO Box 1999, Sudbury, Suffolk, CO10 6FS. UK - Tel : 0787 881165 Fax : 0787 313995.

13. ' Out of control' - The anatomy of control system failure. G R Ward and A R Jeffs. Proceeding of Safety and Reliability Society conference, Altringham, Cheshire, UK - October 12/13 1994.

14. 'The use of knowledge based systems in the chemical process industry - a summary report'. Editors E M Borrill, J A Davies, A N Lelland, J Brazendale, E Fergus. HSE SRD 2170441 93 Issue 1. To be published soon by HSE Books, PO Box 1999, Sudbury, Suffolk, CO10 6FS. UK - Tel : 0787 881165 Fax : 0787 313995.

15. 'Safety management of process faults: A position paper on Human Factors approaches for the design of operator interfaces to computer-based process control systems'. M S Carey. ISBN 0 7176 0409 8. HSE Contract Research Report No 60/1993.

Figures

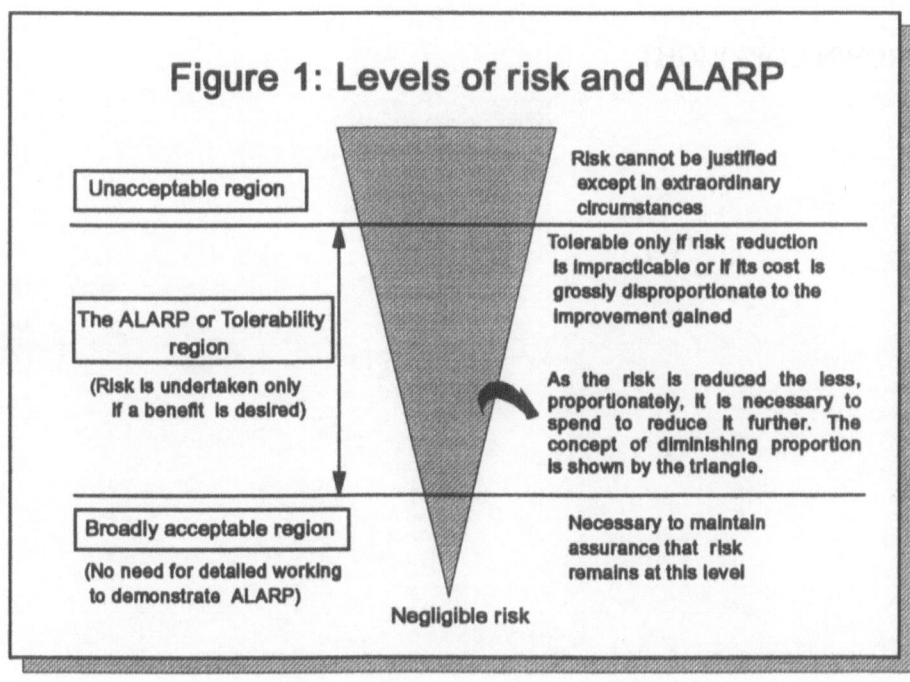

Figure 1: Levels of risk and ALARP

Unacceptable region

Risk cannot be justified except in extraordinary circumstances

The ALARP or Tolerability region

(Risk is undertaken only if a benefit is desired)

Tolerable only if risk reduction is impracticable or if its cost is grossly disproportionate to the improvement gained

As the risk is reduced the less, proportionately, it is necessary to spend to reduce it further. The concept of diminishing proportion is shown by the triangle.

Broadly acceptable region

(No need for detailed working to demonstrate ALARP)

Necessary to maintain assurance that risk remains at this level

Negligible risk

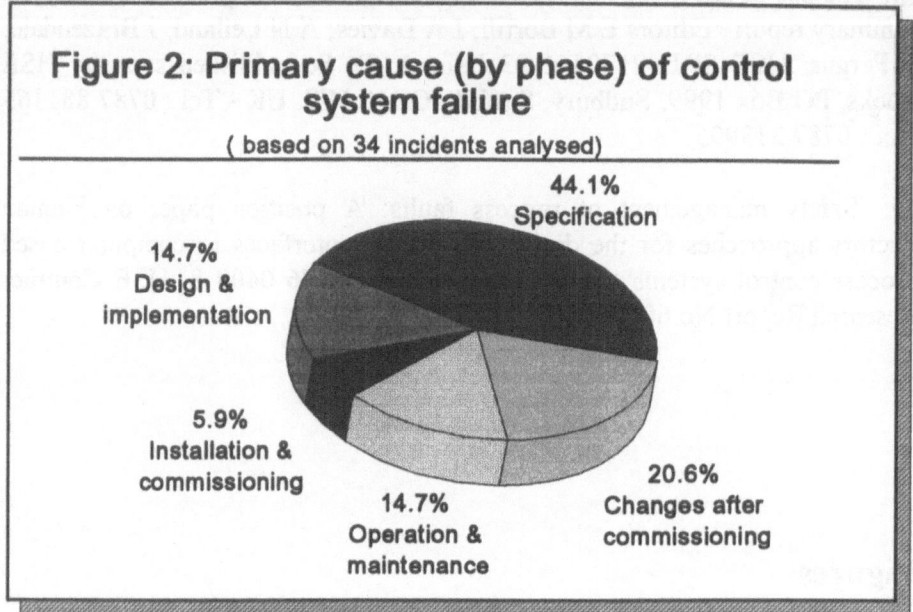

Figure 2: Primary cause (by phase) of control system failure
(based on 34 incidents analysed)

44.1%
Specification

14.7%
Design &
implementation

5.9%
Installation &
commissioning

14.7%
Operation &
maintenance

20.6%
Changes after
commissioning

MEASURING THE BENEFITS OF TRANSPORT SAFETY*

M.W. Jones-Lee
G. Loomes

University of Newcastle upon Tyne
University of York

1   Introduction

Two inescapable facts confront those concerned to
determine the appropriate level of provision of safety
on any particular transport mode. First, safety is
typically not a free good, and second, society's
resources are not limitless. This means that a
responsible decision for or against any proposed
transport safety improvement will require a judgement as
to whether or not the reduction in risk afforded by the
improvement is large enough to justify its cost of
provision. This then raises the question of *how* such a
judgement might be made. At the intuitive level, most
of us would no doubt agree that a safety improvement
which would cost just a few thousand pounds, and which
could be expected to prevent several fatalities, would
be well-warranted. Equally, most people would have
little hesitation in deciding *against* a proposal that
would cost several millions of pounds but which would,
at best, prevent only one or two minor injuries. In
less extreme cases, however, matters are not quite so
straightforward.

Clearly, were it possible to obtain an acceptable
measure of the *monetary value* of safety, then this would
go a long way towards resolving the dilemma inherent in
most safety expenditure decisions. In particular, given
such a measure, it would be possible to weigh the
benefits of safety improvement explicitly against other
costs and benefits - such as capital costs and time
savings - in the course of transport project appraisal.
But how are monetary values of safety to be defined and
estimated? And can we expect the value for, say, the
prevention of a fatality to remain uniform across

----

*   This paper is based in part on Jones-Lee (1990),
Jones-Lee et al (1993) and Jones-Lee and Loomes (1994a,
1994b).

different transport modes such as roads, rail and air, or will the value vary from mode to mode, reflecting differing degrees of voluntariness and control in relation to risk or differences in the likelihood of large-scale "catastrophic" accidents? These are some of the questions that this paper will seek to address.

2. Valuing Transport Safety : Issues of Principle

While a number of different approaches have been proposed for the definition and estimation of values of safety, only two would seem to deserve serious consideration, namely the "gross output" (or "human capital") approach and the "willingness-to-pay" approach. Under the gross output approach - until recently employed by most of those countries that associate explicit monetary values with transport safety - the major component of the cost of an accident involving one fatality, for example, is the discounted present value of the victim's future output (or income) extinguished as a result of his or her premature demise. In the case of individuals whose services are not marketed, such as housewives, imputations are typically made for such services. An allowance is then added for various other economic effects such as vehicle damage, police and medical costs and so on. As such, the gross output approach can be viewed as an attempt to measure the impact of death or injury on current and future levels of national output, broadly construed to include various non-marketed services. In some countries a further more or less arbitrary allowance for the "pain, grief and suffering" of the victim (or the latter's dependents, relatives and friends) is also incorporated in the gross output measure. Values of accident prevention are, in turn, defined in terms of the costs avoided. To give the reader an idea of the magnitude of the costs and values that emerge under the gross output approach, the U.K. Department of Transport's most recent gross output based cost of a fatality was £180,330 in 1985 prices, of which about 28% was an allowance for pain, grief and suffering[1]

The major objection to the gross output approach - raised by one of the authors and others for some years now - is that most of us almost certainly value safety principally because of our aversion to the prospect of our own and others' death and injury *per se*, rather than because of a concern to preserve current and future levels of output and income. If this is so, then values of safety ought ideally to be defined so as to reflect people's "pure" preferences for safety, as such, rather than in terms of effects on output and income, as in the gross output approach. In order to define and estimate values of safety in this way, we clearly require some means of measuring people's preferences for safety and, more particularly their *strength* of preference. Now a natural measure of the extent of a person's preference for anything is the maximum amount that he or she would be willing to pay for it. This amount reflects not only the person's valuation of the desired good or service relative to other potential objects of expenditure, but also the individual's *ability* to pay - which is itself a manifestation of society's overall resource constraint. Thus, under what has not surprisingly come to be known as the "willingness-to-pay" approach to the valuation of safety, one attempts first to determine the amounts that those affected would individually be willing to pay for (typically small) improvements in their own and others' safety. These amounts are then simply aggregated - possibly with distributional weights[2] - across all individuals to arrive at an overall value for the safety improvement concerned. The resultant figure is a clear reflection of what the safety improvement is "worth" to the affected group, relative to the alternative ways in which each individual might have spent his or her limited income. Furthermore, defining values of safety in this way effectively "mimics" the operation of market forces - in circumstances in which markets do not exist - insofar as such forces can be seen as vehicles for allowing individual preferences to interact with relative scarcities and production possibilities in determining the allocation of a society's scarce resources.

In order to standardize values of safety that emerge from the willingness-to-pay approach and render them comparable with values derived under other approaches (such as gross output) the concept of avoidance of a "statistical" death or injury is employed. To illustrate this concept, suppose that a group of 100,000 people enjoy a safety improvement that reduces the probability of death during a forthcoming period by 1 in 100,000 for each and every member of the group. The expected number of deaths within the group during the forthcoming period will thereby be reduced by precisely one and the safety improvement is thus described as involving the avoidance of one "statistical" death.

Now suppose that individuals within this group are, on average, willing to pay £v for the 1 in 100,000 reduction in the probability of death afforded by the safety improvement. Aggregate willingness to pay will then be given by £v x 100,000. Notice that this, in turn, is equal to the average willingness to pay, £v, *divided* by the individual risk reduction of 1 in 100,000. Since an individual's willingness to pay divided by the reduction in his risk of death is simply his *marginal rate of substitution* of wealth for risk, it is clear that under the willingness-to-pay approach, the value of avoiding one statistical death (or, more succinctly, "the value of statistical life") is given by the *mean, over the affected group, of individual marginal rates of substitution of wealth for risk*. In fact, the value of statistical life does not have to be defined with reference to a population or group of specific size (such as 100,000) each afforded the same probability reduction (such as 1 in 100,000). As it turns out, provided that (a) a safety improvement entails a reduction in the expected number of fatalities of precisely one, and (b) individual probability reductions are small, and (c) the affected group of individuals is not significantly atypical in terms of income, age, attitudes to risk and so on, then the value of statistical life for such a safety improvement will be effectively independent of the precise size of the affected group and the precise pattern of individual risk reductions.[3]

Before proceeding to consider the various ways in which researchers have attempted to obtain empirical estimates of values of safety using the willingness-to-pay approach, two further refinements should be noted. First, so far only passing reference has been made to people's concern - and hence willingness to pay - for others', as well as their own safety. To the extent that people do display such "altruistic" concern then *prima facie* one would expect that it would be appropriate to augment values of statistical life to reflect the amounts that people would be willing to pay for an improvement in others' safety. Indeed, this is precisely the position advocated by - amongst others - Mishan (1971), Needleman (1976), Jones-Lee (1976) and, more recently, by Viscusi et al (1988). Empirical evidence suggests that augmenting values of statistical life in this way will lead to an increase in the region of 40-50% (see Needleman (1976) and Jones-Lee et al (1985)).

However, it turns out that under quite plausible assumptions regarding the nature of people's altruistic concern for others' safety on the one hand and their material wellbeing on the other (the latter being reflected by their wealth or consumption), augmenting values of statistical life to reflect willingness to pay for others' safety would involve a form of double-counting and would therefore ultimately be unwarranted. Thus, the issue of whether and how people's concern for others' safety ought to be accommodated within the willingness-to-pay approach hinges on the essentially empirical question of the relationship between such concern and concern for others' wealth or consumption.

A second refinement of the willingness-to-pay approach involves recognition of the fact that safety improvements also have "direct" economic effects, such as avoidance of net output losses, (4) material damage, medical and police costs and so on. To the extent that people appear in the main not to take account of such factors in assessing their willingness to pay for improved safety (and there is some evidence that they tend not to - see Jones-Lee et al (1985)) then an allowance for these factors should clearly be added to values of statistical life and safety. However, such additions tend to be very small in relation to the typical magnitude of aggregate willingness to pay for safety *per se*, at least in the case of risks of death or serious injury.

3.   Empirical Estimates of Values of Road Safety

Having examined the basic principles of the willingness-to-pay approach, let us now turn to the question of how one might, in practice, obtain empirical estimates of values of avoidance of statistical fatalities and non-fatal injuries.

For risks in general, and road risks in particular, basically two types of estimation procedure have been employed. These are known respectively as the "revealed-preference" (or "implicit value") and the "contingent valuation" (or "questionnaire") approaches. Essentially, the revealed preference approach involves the identification of situations in which people actually do trade off income or wealth for physical risk, typically - though by no means invariably - in labour markets where riskier jobs can be expected to command clearly identifiable wage premia. By contrast, the contingent valuation approach involves asking a sample of people more or less directly about their willingness to pay for improved safety, or willingness to accept compensation for increased risk. While the revealed preference approach has the advantage of dealing with actual rather than hypothetical choices, it suffers from the disadvantage that wage rates can be expected to depend on many factors besides occupational risk, so that it is necessary to disentangle those other effects in estimating the relationship between wage rates and job risk. It is also the case that observed wage-risk relationships may owe as much to the intervention of regulatory bodies, such as the U.K. Health and Safety Executive, as they do to the interaction of workers' preferences and employers' profit motives. Thus, what one may be picking up in wage-risk studies is the rate at which *regulatory bodies* are willing to allow employers to trade workers' wealth for risk, rather than the marginal rate which is, in fact, acceptable to workers. By contrast with the

19

relative complexity of the revealed preference approach, the contingent valuation approach allows the researcher to go directly and unambiguously to the required wealth/risk trade-off, at least in principle.

By now there are numerous examples of both types of study and to report the details of all of them would be beyond the scope of a paper such as this. The remainder of this section will therefore focus upon the more important examples of those studies that have dealt principally with *road* safety, the main results of which are drawn together below in Table 1. However, for the sake of comparison the results of other studies will also be briefly summarised.[5]

(i)   Blomquist (1979)

The earliest large-scale empirical study aimed explicitly at obtaining a willingness-to-pay based value of statistical life for road safety was a revealed preference exercise for the USA reported in Blomquist (1979).[6]   Basically, the study focused upon people's willingness to trade time and inconvenience for safety in their decisions to wear (or not to wear) car seat-belts in the absence of compulsory seat-belt legislation.

Blomquist obtained data concerning seat-belt use and other personal characteristics of drivers and passengers from *A Panel Study of Income Dynamics 1968-1974* (Survey Research Center, 1972, 1973, 1974) which gives data on over 5,500 households in the USA over a seven-year period. Estimates of the effectiveness of seat-belt use in reducing fatal and non-fatal accident risks were obtained from various safety-engineering studies. The average cost of a non-fatal road injury (which was required in order to isolate willingness to pay for the reduction of fatality risks) was a U.S. Department of Transportation gross output-based figure, which included an arbitrary "pain, grief and suffering" component. Finally, an estimate of the average time taken to put on, adjust and take off a seat belt was obtained from a simple time-and-motion study conducted by Blomquist himself.

Given that those who choose to wear seat-belts must place a value on safety improvement that is at least equal to the time and inconvenience costs of wearing them, Blomquist estimated the value of statistical life to be, at a minimum, a figure in a range of $250,000 to $280,000 in 1978 prices, with a "preferred" estimate in excess of $368,000. This converts to about £530,000 in sterling 1992 prices. Finally, Blomquist estimated the elasticity of the value of statistical life with respect to lifetime earnings to be about 0.3 - that is, a doubling of average annual income would increase aggregate willingness to pay for risk-reduction by about 30%. It would therefore appear on the basis of Blomquist's findings that safety can be regarded as a normal, "non-luxury" good. [7]

(ii) Jones-Lee et al (1985)

The next major development in the empirical estimation of values of road safety was a contingent valuation study reported in Jones-Lee et al (1985). This study was commissioned by the U.K. Department of Transport, following a one year feasibility exercise. [8] The study involved a nationally representative sample survey carried out in 1982 by National Opinion Polls Ltd (NOP) using a questionnaire designed and piloted by the authors of the study and refined by NOP. The stratified random sample of 1718 was drawn from England, Scotland and Wales and, with a response rate of 67%, produced 1103 full and 47 partial interviews. Finally, a follow-up survey of some 200 of the original respondents was conducted to test for the temporal consistency of the responses. The questionnaire was administered in respondents' own homes by professional NOP interviewers and typically took about 45 minutes to complete. Broadly speaking, questions fell into three categories, namely :

Contingent Valuation questions - designed to provide estimates of relevant marginal rates of substitution, relative valuation of reductions in the risk of fatal and non fatal accidents, and so on. For example, one question asked how much extra the respondent would be willing to pay - when buying a new car - for an additional safety feature that would reduce driver and passenger risks by various specified amounts. Other contingent valuation questions asked about willingness to pay for coach safety on a foreign trip, local road safety and, for the sake of comparison, willingness to pay to reduce deaths from cancer and heart disease.

Perception/consistency questions - designed to test the quality of respondents' perceptions of transport risks and ability to deal with probability concepts. Questions designed to test the veracity and stability of valuation responses were also included.

Factual and other questions - concerning vehicle ownership, annual mileage and so on, as well as the usual questions concerning age, income etc.

The main findings of the study were as follows :

21

i)   People were, in the main, willing and able to provide answers to the questions posed.

ii) While some respondents gave apparently inconsistent answers to some questions and experienced difficulty with the concepts involved, it was the authors' considered opinion that there was sufficient evidence that perception of transport risks, ability to process risk-information and truthfulness of responses were adequate to justify the inference of at least a range of values of statistical life for transport risks.

iii) The values of "own" statistical life implied by the mean responses to questions concerning willingness-to-pay for car and coach safety ranged from £1,200,000 to £2,200,000 in 1982 prices, while values based on medians[9] ranged from £500,000 to £1,200,000. Drivers' value of "passenger" statistical life implied by mean responses was about £500,000, though, as noted above, the appropriateness of adding such a figure to the value for "own" statistical life has been shown to be open to question. Based on these results the authors argue for a value of "own" statistical life for transport risks of at *least* £500,000 and more probably in the region of about £1,500,000 in 1982 prices. In 1992 prices these figures would be equivalent to £850,000 and £2,600,000 respectively.

iv) Regression analysis indicated that individual willingness to pay for road safety is most significantly affected by income and age. In particular, the implied income elasticity of the value of statistical life is much the same as that obtained by Blomquist, namely 0.3, and the coefficients of linear and quadratic age variables imply that individual valuation of safety follows an "inverted-u" life cycle, peaking at an age of a little over forty.

v)   As noted earlier, it is possible that considerations of equity and distributive justice might lead one to wish to apply distributional weights to individual willingness to pay in arriving at overall values of safety. In order to explore the implications of such weighting, values of statistical life were computed with distributional weights that were (i) inversely related to household income, raised to various integer powers and (ii) scaled so as to have a mean of one (the latter was done so as to avoid arbitrary inflation or deflation of values of

22

statistical life). Roughly speaking, with weights inversely related to income, values of statistical life fell by, at most, about 20% while with weights inversely related to the square of income values fell by, at most, about 40%.

vi)   While the study did not produce specific estimates of values of avoidance of statistical non-fatal injuries, it did provide evidence indicating that such values could be expected to be at least one hundredth of the value of statistical life for the case of serious non-fatal injuries.

As far as the author is aware, the only three other major studies of valuation of road safety to have been undertaken to date are all contingent valuation exercises based in part upon the Jones-Lee et al study. The first, reported in Persson (1989), was conducted in Sweden, the second, reported in Maier et al (1989), in Austria and the third, by Miller and Guria (1991), in New Zealand.

(iii) Persson (1989)

While the underlying methodology and approach employed in Persson's study parallel those employed in the Jones-Lee et al study, the survey instrument embodied some important modifications and refinements. Amongst the most significant of these was that in many cases individual marginal rates of substitution of wealth for risk were estimated on the basis of respondent's *subjective perceptions* of risk levels, as opposed to the "objective" counterparts to these risk levels employed in the Jones-Lee et al study. These subjective perceptions were elicited essentially by providing respondents with "baseline" risk levels - such as the annual risk of a motorcycle or bus fatality - and then asking for respondents' estimates of the corresponding risks to car drivers and passengers. A second difference between Persson's study and that conducted by Jones-Lee et al was that questions asking directly about willingness to pay for reductions in the risk of non-fatal injury were also included.

The survey, conducted in Sweden during the period from Autumn 1986 to Spring 1987, was based on a random sample of 1000 individuals. Respondents were not interviewed directly, but instead were asked to complete a postal questionnaire. There was, however, extensive telephone contact with respondents to resolve misunderstandings and to encourage respondents to complete and return the questionnaire. The response rate to the survey was 50.6%, producing 506 usable questionnaires. In general, the survey appears to have been very successful, judged by various relevant criteria and to have produced

23

results that are, with one or two exceptions, gratifyingly similar to the findings of the Jones-Lee et al study. In particular the values of "own" statistical life implied by mean responses from the subsample of individuals whose subjective perceptions of the risk of being killed as a driver or passenger corresponded to the "objective" level ranged from SEK 14,500,000 to SEK 17,600,000 in 1986 prices, while those implied by median responses ranged from SEK 4,000,000 to SEK 8,000,000. In 1992 prices these convert to ranges of about £1,800,000 to £2,200,000 and £500,000 to £1,000,000. Income elasticities of the value of statistical life were also very similar to those found in the Blomquist and Jones-Lee et al studies. Indeed, the only major difference between Persson's findings and those of other questionnaire based studies concerns the relationship between individual valuation of safety and age. For road safety, Persson's regression results imply that individual valuation of safety is either monotonically *declining* with age or follows a "(non-inverted) u" life cycle. The "inverted-u" life cycle emerges only in the case of questions concerning willingness to pay for reduction in the risk of heart-attack.

Finally, Persson's estimates of willingness-to-pay based values of avoidance of non-fatal injuries involving (a) facial lacerations and (b) concussion were SEK 5,200,000 and SEK 4,200,000 respectively based on mean responses and SEK 1,000,000 and SEK 800,000 based on medians. In 1992 prices these convert to figures of £570,000, £460,000, £110,000 and £88,000.

(iv) Maier et al (1989)

The Maier et al Austrian study also employed a questionnaire based on that used by Jones-Lee et al. This study involved a small non-random sample drawn from Vienna and Neulengbach (a rural area close to Vienna). Direct interviews produced 98 completed questionnaires. Again, the results were very similar to those of the Jones-Lee et al study. For example, the value of "own" statistical life for road safety was about AS 40,000,000 (presumably in 1988 prices) which converts to about £2,300,000 in 1992 prices. there was also evidence of an "inverted-u" life cycle and of a positive income elasticity for the value of statistical life - though no numerical estimate can be obtained since the regression analysis employed a simple medium/high income dummy.

(v) Miller and Guria (1991)

Miller and Guria's New Zealand study was again in part based on the Jones-Lee et al survey instrument and Persson's refinements of it, though their questionnaire also contained a number of additional questions concerning, for example, willingness to pay to drive on a safer road (by paying a toll), willingness to pay to take a course on road safety and willingness to pay to live in a neighbourhood with improved road safety. The survey was carried out during 1989 and 1990 as part of the New Zealand Ministry of Transport's periodic travel survey and the stratified random sample of 655 produced 568 useable responses. Miller and Guria's estimates of the value of statistical life for New Zealand yield a mean of about NZ$1,900,000 and a median of NZ$1,400,000 which convert to about £780,000 and £590,000 respectively in 1992 prices. It should, however, be noted that these figures include a component reflecting willingness to pay for one other "average" family member's safety. Again, there was some evidence of an "inverted-u" life cycle and a positive income elasticity of the value of statistical life of broadly similar magnitude to that found in other studies.

The various empirical estimates of the value of statistical life for road risks are summarised in Table 1 :

Table 1
Values of statistical life for road risks

Study	Value of Statistical Life (£-sterling, 1992)	
	Based on mean marginal rate of substitution	Based on median marginal rate of substitution
Blomquist	£530,000	−
Jones-Lee et al	£2,600,000	£850,000
Persson	£1,800,000-£2,200,000	£500,000-£1,000,000
Maier et al	£2,300,000	−
Miller and Guria	£780,000	£590,000

Clearly, then, Blomquist's and Miller and Guria's results together with the "median" findings of the Jones-Lee et al and Persson studies fall in a range from about £500,000 to £1,000,000 in 1992 prices, while estimates of the value of statistical life for road risks based on mean responses from the various other questionnaire studies range from about £1,800,000 to £2,600,000. Thus, if credence is given primarily to the Blomquist, Miller and Guria and median estimates then a value of statistical life in the region of £750,000 would seem to be warranted, whereas reliance on the other mean figures would produce a value nearer to £2,200,000. But which sort of figure should actually be used? Clearly, if one sticks rigidly with the kind of argument developed above in Section 2, then it is the value based on *mean* marginal rates of substitution that is appropriate. However, there are two quite independent arguments in favour of picking a value closer to the figure implied by the *median* responses to the questionnaire studies. First, a standard "median voter" type of argument indicates that in situations in which the distribution of individual valuations is single-peaked and highly skewed (as it clearly is in the case of the Jones-Lee et al and Persson studies) then the median valuation will lie closer to the "majority wish" than will the mean.[10] Furthermore, even if one wished in principle to retain the definition of a value of statistical life based on the mean marginal rate of substitution of wealth for risk, there is still an argument in favour of paying careful attention to the median responses to questionnaire studies. The rationale for doing so is quite simply that extreme observations in the top and bottom tails of the distribution of responses to a contingent valuation question are inherently more suspect than the others, so that one might wish to discount, or even discard, responses in these tails. If one actually "trims out" such responses this will leave a distribution whose mean approaches the median of the original distribution as the proportion of "extreme" responses trimmed out is increased. In fact, in the Jones-Lee et al study it was found that trimming out the top 10% and bottom 10% of responses produced a distribution whose mean was substantially closer to the original median than to the original mean. This sort of trimming would therefore favour a value of statistical life for road risks that was closer to £850,000 than to the figure of some £2,600,000 implied by the mean of the untrimmed distribution.

It has to be conceded that the range of possible values of statistical life for road risks implied by the results of the various studies is uncomfortably wide. To a substantial degree, this is a result of the considerable difficulty associated with empirical work in this area. However, given such a range, three observations would seem to be pertinent. First, the imprecision associated with the estimated value of statistical life is no greater – and, in some cases, significantly *less*-than the imprecision associated with engineering and scientific estimates of risk effects themselves. Second, even the lowest estimates of willingness-to-pay based values of statistical life are more than *double* their gross output counterparts. Third, if one accepts the arguments in favour of the willingness-to-pay approach in principle, then there is simply no alternative but to exercise critical, informed judgement in selecting a particular value from the range concerned. This is precisely the approach that has been adopted by the U.K. Department of Transport (DoT) which in 1988, following a comprehensive review of the literature and careful consideration of the various issues involved[11] decided to adopt a willingness-to-pay based value of statistical life for road risks of £500,000 in 1987 prices in place of its former gross output based figure. Since then this value has been increased in line with inflation and the growth of real income per capita and currently stands at £715,330 in 1992 prices.[12]

However, it is important to appreciate that the DoT's decision to adopt a willingness-to-pay based value of statistical life that lay towards the bottom end of the range of estimates available in 1988 almost certainly reflected a desire to temper a radical change of methodology with an element of caution in the selection of a specific numerical value. With the passage of time, the willingness-to-pay methodology has gained substantially increased acceptance and is indeed the approach to the valuation of safety recommended in the Treasury "Green Book".[13] In the light of this it may be that in due course the DoT will become amenable to the suggestion that it might set its value for the prevention of a road fatality closer to the sort of figure implied by the mean estimates from the U.K. and European contingent valuation studies summarised above in Table 1. Were it to do so then a value well in excess of £1,000,000 and probably nearer to £2,000,000 in 1992 prices would seem to be well-warranted.

(v) Other Studies

Having considered the results of those empirical studies that were explicitly concerned with the valuation of road safety, it would seem appropriate to provide a brief summary of empirical work that has dealt with the valuation of safety in other contexts. In fact, the bulk of this work has been concerned to elicit values of statistical life from observed wage premia for risk in labour markets. As such, these so-called "compensating wage differential" studies are clearly revealed preference exercises and typically involve the use of multiple regression analysis to "control" for the variety of other factors, besides job risk, that can be expected to influence equilibrium wage rates in labour markets.

As might be expected, like their transport counterparts, these and other studies have produced a wide range of estimates of the value of statistical life, again reflecting the inherent difficulty of empirical work in this area, but in broad terms these estimates are largely consistent with the findings of the studies discussed above. Thus for example, Marin and Psacharopoulos (1982), in the most extensive compensating wage differential study so far conducted for the U.K. estimated the value of statistical life for occupational risks to be £1,250,000 in 1981 prices (i.e. about £2,300,000 in 1992 prices) which is remarkably similar to the Jones-Lee et al questionnaire based estimate reported above.

Furthermore, in summarizing the results of 21 revealed preference studies carried out prior to 1986 (7 for the U.K., 13 for the U.S.A. and one for Austria) - 14 of which were compensating wage differential studies - Jones-Lee (1989) found the estimates of the value of statistical life from the various studies to have an overall mean of £1,720,000 and a median of £600,000 in 1987 prices, which convert to about £2,400,000 and £830,000 respectively in 1992 prices. However, as Miller (1986) points out, most of the compensating wage differential estimates have been derived from data concerning pre-tax wages, whereas what is actually required under the willingness-to-pay approach is willingness to trade *net-of-tax* income for risk. The compensating wage differential estimates, which are essentially of marginal rates of substitution of gross-of-tax for risk, should therefore be scaled down by the relevant marginal income tax rates. Nonetheless, the findings of these studies are clearly broadly consonant with those of the predominantly questionnaire-based studies used in the road context. This is encouraging because *a priori* one would expect that individual valuation of safety on the roads and in the workplace would not be grossly dissimilar.

Finally, considering all of the estimates that have been discussed in this section, there do not appear to be marked international differences, at least as far as the U.K., U.S.A., Austria, Sweden and New Zealand are concerned.

4.   The Value of Preventing Non-Fatal Road Injuries

Partly because fatal injuries constitute a "worst case" and partly because death is a clear-cut, if tragic event, research on the economics of safety has tended to concentrate upon the value of preventing fatal, as opposed to non-fatal injuries. However, given the growing tendency for government and related agencies in various countries to adopt the willingness-to-pay approach to the valuation of safety, there has emerged a growing impetus to redress the balance and direct research efforts explicitly at the empirical estimation of willingness-to-pay based values of preventing *non-fatal* injuries of varying severity, in order to ensure consistency of approach in the treatment of fatal and non-fatal casualties.

The first and most obvious difficulty in obtaining empirical estimates of willingness-to-pay based values of preventing non-fatal injuries is that the latter cover a spectrum ranging from minor cuts and bruises with no admission to hospital, through to injuries resulting in severe permanent disability. In view of this, there would seem to be little chance that the revealed preference approach could be successfully employed. While some revealed preference studies based on labour market data have attempted to elicit such values, their success has been very limited, and the resultant figures apply to an unsatisfactorily wide and heterogeneous class of injuries. Not surprisingly, the major difficulty in applying the revealed preference approach to non-fatal injuries is that the risks of such injuries tend to be highly correlated with each other and with the risks of fatality, so that disentangling the respective willingness to pay for risk reduction is extremely difficult, not to say impossible. Furthermore, individual willingness to pay for reduction in the risk of non-fatal injury is likely to be obfuscated by the existence of statutory and other forms of compensation, particularly in labour markets.

This leaves some variant of the contingent valuation approach as the only serious possibility. To the best of the author's knowledge, so far only two substantial programmes of research aimed at obtaining willingness-to-pay based values of preventing non-fatal injuries have been initiated. As it happens, both programmes have focused exclusively upon road accidents. The first programme, directed by Dr. Ted Miller of the Urban Institute in Washington, was carried out in the United States, while the second, which involved two sub-programmes, was carried out under the auspices of the Department of Transport and the Transport Research Laboratory in the U.K. The authors were Directors of the Management Team of one of these two sub-programmes.(14)

(i) Non-Fatal Road Injuries in the USA

The principal procedure employed by Miller et al [1988] to obtain estimates of willingness-to-pay based values of preventing non-fatal road injuries was as follows. First, a "best estimate" of the value of statistical life was derived from the empirical literature. The figure employed was, in fact, $1.95 million in 1986 prices. This figure was then used to obtain a "value per life year" of $120,000 using a discount rate of 6% per annum. Finally, the value per life year was applied to the number of lost "life years of functioning" that would be avoided by the prevention of one of each of the various categories of injury. Lost years of functioning were estimated essentially on the basis of expert medical opinion, though in Miller et al [1988] it is reported that the U.S. National Highway Traffic Safety Administration has "... research underway to confirm [the estimated lost years of functioning] based on a broader range of medical expertise". As a check on the results of this procedure Miller et al also used an approach based on the utility loss due to particular non-fatal injuries relative to the utility loss due to death. Estimates of utility losses were obtained from various studies in the health care literature.

Finally, Miller et al supplemented their estimates of the pure willingness to pay values for each category of injury by an allowance for avoided direct costs, such as output losses and medical costs. The resultant overall willingness-to-pay based values, converted to £-sterling in 1992 prices, are shown in Table 2 for injuries classified on the basis of the Maximum Abbreviated Injury Scale (MAIS).

Table 2 : Miller et al, (1988) estimates of values of preventing various severities of non-fatal road injury (£-sterling 1992)

	Injury Type	Percentage of all non-fatal road injuries	Value	Value ÷ value of statistical life*
MAIS 1	Minor injury (eg superficial laceration/fractured toe)	84.73	3,500	0.002
MAIS 2	Moderate injury (eg deep laceration/fractured foot)	10.83	27,400	0.016
MAIS 3	Serious injury (eg fractured femur/amputation below the knee)	3.72	101,000	0.058
MAIS 4	Severe injury (eg amputation above the knee/crushed pelvis)	0.46	331,000	0.188
MAIS 5	Critical injury (eg penetration of the brain/laceration of the spinal cord)	0.27	1,345,000	0.763

*   £1,760,000 in 1992 prices.

(ii) Non-Fatal Road Injuries in the U.K.[15]

The work carried out by the authors and others for the U.K. Department of Transport and Transport Research Laboratory was based on the results of a nationally representative sample survey of 891 respondents from England, Scotland and Wales. The survey, which was conducted by professional Transport Research Laboratory interviewers during July, August and early September 1991, employed two distinct questionnaire versions, each administered to approximately half of the sample. The first version employed contingent valuation questions of the type used in the studies summarised above in Section 3. The second version employed what are known as "standard gamble" questions. Essentially these ask respondents to suppose that they have incurred a particular non-fatal injury and then to assess the largest risk of failure that they would be prepared to accept in a treatment which, if successful, would completely cure the non-fatal injury but if unsuccessful would result in death. Denoting the sample mean response to this sort of question by θ, it can be shown that the value of preventing the injury is given by θ times the value of statistical life. Since the Department of Transport has already set the value of statistical life for use in road safety decision making at about £715,000 in 1992 prices, this clearly provides a basis for estimating the value of preventing non-fatal injuries of various severities.

In fact, the results of the survey suggest that the contingent valuation responses were subject to a number of substantial upward biases. By contrast, the standard gamble responses appear to have been largely free from bias. The willingness-to-pay based values of preventing various serious[16] non-fatal injuries implied by the mean standard gamble responses are shown in Table 3. Respondents were, in fact, asked to indicate the *largest* risk of treatment failure that they definitely would accept, the *smallest* risk they definitely would not accept and, within the resultant range, to identify a best estimate. The ratios reported in Table 3 are based upon the latter. As in the case of Miller et al's results, these figures include an allowance for avoided direct costs such as output losses and medical costs.

Given that injuries classified as serious in U.K. road accident statistics effectively exclude those classified as MAIS 1 in Miller et al's study and given that Miller's MAIS 5 injuries include some cases of paraplegia/quadriplegia/severe head injuries, the results reported in Table 3 are broadly consonant with those given in Table 2, at least viewed as ratios in relation to the value of statistical life.

On the basis of the results reported in Table 3, the Department of Transport has recently increased its value for the prevention of a serious non-fatal road injury from about £20,000 (based on the gross output approach)

32

to some £75,000 in 1992 prices, the latter being a weighted average of the values shown in Table 3.

Table 3 : Values of preventing various serious non-fatal road injuries estimated from standard gamble responses (£-sterling, 1992)

Injury Type	Percentage of all serious non-fatal road injuries	Value	Value + value of statistical life*
Full recovery in 3-4 months (outpatient)	19	17,900	0.025**
Full recovery in 3-4 months (inpatient)	15	17,900	0.025
Full recovery in 1-3 years	31.5	40,000	0.056
Mild permanent disabiity (outpatient)	6	40,000	0.056**
Mild permanent disability (inpatient)	13	102,000	0.143
Some permanent disability with possibility of some scarring	14	153,000	0.214
Paraplegia/Quadriplegia/ severe head injuries	1.5	715,000	1.000**

*   £715,000 in 1992 prices.

**   These estimates have been interpolated from respondents' rankings of the various injuries in terms of relative "badness"

5. Willingness-To-Pay Based Values of Safety for London Underground and British Rail[17]

Given that they are nationalised industries, it is tempting to suppose that London Underground Limited (LUL) and British Rail (BR) should both use the same monetary value of safety as does the Department of Transport for the roads. However, while it is clear that the case in favour of employing the willingness-to-pay approach applies with as much force to LUL and BR as it does for the roads, it is equally clear that the willingness-to-pay methodology leaves entirely open the possibility that the numerical magnitude of values of safety for the Underground and the railways may differ significantly from the value that is applicable to the roads. The reason for this is quite simply that willingness-to-pay based values of safety are explicitly intended to reflect the preferences and attitudes to risk of those members of the public who will be affected by the decisions in which the values are to be used, and there are no grounds for supposing that these preferences and attitudes need necessarily be the same for road users, Underground customers, and those who travel by rail. Indeed, this possibility is acknowledged in the Treasury "Green Book".[18] But why might the travelling public's preferences and attitudes towards safety display systematic differences between the roads on the one hand and the Underground and railways on the other? Elsewhere,[19] we have argued that there are two types of effect that might lead to a significant premium for a willingness-to-pay based value of Underground safety relative to its road counterpart, namely "scale" effects and "context" effects. Such effects will almost certainly also apply in the case of rail safety.

(i) Scale Effects

Scale effects derive from the fact that, whereas accidents involving ten or more fatalities account for only a very small proportion of all road fatalities, over an extended period such "large-scale" accidents are likely to be the cause of a much greater proportion of the total number of Underground and rail fatalities, as the accidents at Moorgate in 1975 (43 fatalities), Clapham Junction in 1988 (34 fatalities) and King's Cross in 1987 (31 fatalities) remind us. Now there are those who have argued that the loss of, say, 30 lives in a single large-scale accident is inherently worse than the loss of 30 lives in separate small-scale accidents and that for this reason a significant premium is called for in valuing each fatality prevented in a large-scale accident vis-a-vis the corresponding value for

small-scale accidents. This argument is almost
certainly misguided, in that there would seem to be no
satisfactory moral or ethical grounds for regarding the
simultaneous loss of 30 lives as being any worse - or,
for that matter, any better - than the separate loss of
30 lives. Indeed, this view appears to have been shared
by a substantial majority of respondents in a
questionnaire-based study commissioned by LUL and
described more fully below. In particular, a sample of
225 Underground customers were asked to indicate whether
they agreed or disagreed with the following statement :

> *"25-30 deaths in a single Underground accident is
> worse than 25-30 deaths in separate Underground
> accidents*

The frequency distribution of responses is shown in
figure 1.

A related, and equally dubious argument in favour of
setting the value of statistical life for large-scale
accidents at a premium with respect to its small-scale
accident counterpart appeals to the media attention and
political reaction that are typically generated by
accidents involving catastrophic loss of life. However,
such attention and reaction probably say more about
ghoulish curiosity, a desire to apportion blame and a
determination to prevent any recurrence of the
particular circumstances that gave rise to the accident,

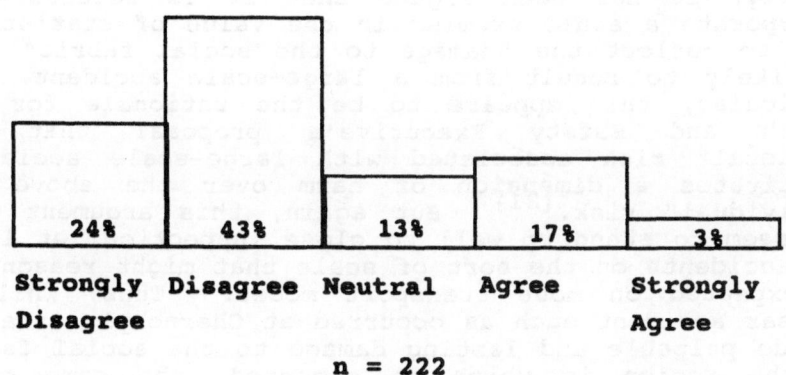

Figure 1 : Frequency Distribution of "Agree/Disagree"
Responses.

35

rather than a carefully considered concern to ensure an appropriate allocation of scarce resources to safety improvement. This conclusion gains some support from the widespread belief that the safety countermeasures recommended by recent transport disaster enquiry reports are, in some cases, extremely difficult if not impossible to justify in terms of cost-effectiveness and "value for money".[20]

A third, and again almost certainly flawed argument in favour of setting the value of statistical life at a significant premium in the case of large-scale accidents relates to the job security of senior management in public and indeed private sector transport modes. For example, following the King's Cross fire, the Chairmen of LUL and London Transport both lost their jobs. While it would be entirely understandable if this were to provide an incentive for senior management in organisations such as LUL or BR to set the value of statistical life used in appraising proposed Underground or rail projects at a substantial premium in relation to the figure used for the roads, the provision of such an incentive hardly seems to constitute the ideal way in which to determine levels of expenditure on rail and Underground safety. Indeed, it is arguable that if regulatory bodies, such as the U.K. Health and Safety Executive or the Department of Transport, are satisfied that decision makers in LUL or BR are taking their safety investment decisions in an informed and responsible manner, then senior management jobs should *not* be at risk if and when chance happens to produce an unhappy coincidence of human and system failure.

Finally, it has been argued that it is necessary to incorporate a scale premium in the value of statistical life to reflect the "damage to the social fabric" that is likely to result from a large-scale accident. In particular, this appears to be the rationale for the Health and Safety Executive's proposal that the "societal" risk associated with large-scale accidents constitutes a dimension of harm over the above the "individual" risk.[21] But again, this argument does not seem to stand up well to close inspection, at least for accidents on the sort of scale that might reasonably be expected on most transport modes. Thus, while a nuclear accident such as occurred at Chernobyl certainly did do palpable and lasting damage to the social fabric of the region in which it occurred, the same could hardly be said of the rail and Underground accidents at Clapham Junction and King's Cross. No doubt rail and Underground customers were somewhat unsettled in the weeks immediately following these accidents, but for the vast majority of people these effects probably wore off quite quickly. If this is so, then apart from the system damage costs associated with large-scale transport accidents, there would seem to be little if any basis for believing that such accidents give rise to extensive and lasting damage to the social fabric, over and above the physical and psychological injuries sustained by the accident victims themselves.

Having disposed of the spurious arguments in favour of a scale premium on the value of statistical life for large-scale accident risks, let us now turn to what is almost certainly a *good* argument. This runs as follows. Because large-scale accidents happen with such relative rarity and because any such accident is the result of a quite unique combination of human and system failure, there is inevitably a degree of imprecision or ambiguity associated with the predicted probability of occurrence of such accidents and uncertainty concerning the scale of loss of life and injury that will ensue, should such an accident happen. If members of the travelling public are averse to this ambiguity and uncertainty *per se*, then one might reasonably expect that their willingness to pay to reduce their exposure to large-scale accident risks would display a significant premium in relation to their willingness to pay to reduce other more accurately predictable risks, such as those on the roads.

(ii) Context Effects

As their name suggests, context effects are a result of people's perceptions of, and attitudes towards, the context or circumstances in which an accident might happen. Factors related to context that might be expected to influence individual willingness to pay for a risk reduction almost certainly include people's perceptions of the extent to which the risk concerned is (or is not) :

- under their own control;
- voluntarily assumed;
- their own responsibility;
- borne above ground (as opposed to "deep in the bowels of the earth").

Furthermore, since Underground risks are typically perceived to be outside passengers' control, involuntary, management's responsibility, and often borne deep below ground in a dark and confined environment – whereas road risks are usually seen to be much more under the road user's control, voluntary, largely his or her own responsibility, and normally borne above ground – it would not be at all surprising if context effects were substantially to reinforce scale effects. This would produce a significant premium on the willingness-to-pay based value of statistical life for Underground risks in relation to the corresponding figure for road risks. With the exception of the "above/below ground" distinction, much the same factors would seem to apply to rail risks.

(iii) Estimating the Magnitude of Scale and Context Effects

Since the 1987 King's Cross fire and publication of the Fennell Report,[22] LUL has given high priority to the development of its procedures for appraising proposed safety projects. As part of this process, LUL commissioned the authors to undertake a programme of research aimed at deriving an estimate of the willingness-to-pay based value of Underground safety. Following a pilot study conducted in the latter part of 1992, the main phase of this programme, undertaken in the early part of 1994, involved a questionnaire-based study designed principally to provide estimates of scale and context premia for the value of Underground safety relative to road safety.

The study was structured around thirty "focus group" meetings, each involving between six and eight members of the public, the total number of participants being 225. The focus group meetings were held during the evenings and took place in suitably sized rooms rented from hotels. Focus group participants were required to be both Underground users *and* car drivers or passengers and were selected in such a way as to reflect a reasonable spread of ages, gender and social class. In addition, the locations of the focus group meetings were spread around the area served by the Underground so as to obtain a reasonable coverage of the various lines.

Broadly speaking, the focus group meetings - each of which lasted for a little over an hour - proceeded along the following lines. Participants were first presented with various short statements concerning safety in general and invited to consider these statements in an open-ended discussion, guided by the group moderator, and to explore the extent to which they agreed or disagreed with the statements. In turn, prior to each of the questions intended to elicit quantitative estimates of scale and context premia, participants were presented with four statements concerning large-scale as opposed to small-scale Underground accidents and four statements concerning small-scale Underground accidents as opposed to small-scale road accidents. These statements were based on typical views expressed by participants in the pilot study focus group discussions and related to the main considerations that seemed to weigh with people in choosing between the prevention of large-scale and small-scale Underground accidents or between the prevention of Underground and road accidents. In each case, participants were asked to indicate the extent of their agreement or disagreement with the statement concerned - this was done on an individual, rather than group basis - and to discuss the various issues raised by the statements.

Following completion of the relevant block of "agree/disagree" questions, participants were then presented with quantitative questions - again, to be answered on an individual, rather than group basis - designed to elicit estimates of scale and context premia for the value of Underground safety in relation to its road counterpart. Essentially, the first of these questions sought to determine the number of single-fatality Underground accidents that respondents would require to be prevented by a safety programme if that programme were to be judged "equally as good as" a programme that would prevent one large-scale Underground accident involving 25-30 fatalities. This question was intended to form the basis for estimating the scale premium. In turn, in order to estimate the context premium, respondents were asked about the number of single-fatality *road* accidents that they would require a safety programme to prevent if that programme were to be judged "equally as good as" a programme that would prevent 25-30 single-fatality Underground accidents. In all cases respondents were asked to regard the safety programmes concerned as involving the same overall costs and as being effective over the same time period. The results of this exercise - which were broadly in line with the findings of the pilot study[23] - were, to say the least, somewhat surprising given the expectations that we had held prior to commencing empirical work in this area.

Thus, while the open-ended discussion and the questionnaire responses both confirmed our belief that Underground safety would command a clear context premium in relation to road safety (the questionnaire responses giving a mean context premium of 51% with a standard error of 7%), *there was no evidence in favour of a significant scale premium*. In particular, the "agree/disagree" responses and comments in the open-ended discussion suggested two quite distinct kinds of reaction to the risk of large-scale accidents. The first was very much along the lines of the "ambiguity/uncertainty" aversion discussed above. However, the other reaction, which worked in the *opposite* direction, related to doubts about the effectiveness of expenditure aimed at attenuating the risks of large-scale accidents, with a consequent preference for directing limited resources towards the more regular and predictable single-fatality incidents.[24] More specifically, those who showed this latter reaction seemed to feel that one would have no real way of knowing in advance that funds expended with the aim of reducing the risks of large-scale accidents would actually be effective in preventing the unique combination of human and system failure that typically gives rise to such accidents. Moreover, if such expenditure were undertaken and, in the event, no large-scale accident occurred, then, in retrospect too, one would have no real way of knowing whether such an accident had been avoided because of the expenditure or whether it would not have occurred anyway.

In the event, this second type of reaction appears to have effectively offset the "ambiguity/uncertainty" aversion phenomenon, so that on balance responses to the questionnaire produced a mean scale premium of *minus* 2% with a standard error of 6%. Taken together, therefore, the estimates of the scale and context premia that emerged from the study point towards a willingness-to-pay based value of statistical life for Underground risks that is some 50% larger than its roads counterpart, with *all* of this premium being attributable to context effects. Given that the DoT's current willingness-to-pay based value for the prevention of a road fatality is £715,330 in 1992 prices, this 50% premium would entail a willingness-to-pay based value for the prevention of an Underground fatality in the region of £1.1m. However, were the DoT to raise its value closer to the U.K. and European mean contingent valuation estimates summarised in Table 1, then an Underground figure well in excess of £1.5m and probably closer to £3.0m would be implied. Since it seems probable that scale and context premia for values of rail safety would be much the same as the figures for the Underground, one suspects that a similar value of statistical life would be appropriate in the case of BR's safety project appraisal.

Pending the outcome of further research, it is understood that LUL and BR are both working with values of statistical life in a range for which the lower end is £2m. However, given that both bodies are still at a relatively early stage in this whole exercise, such values are apparently being applied with a measure of flexibility and circumspection. Nevertheless, it would appear that both LUL and BR are moving towards a position in which values of statistical life of at least £2m will be employed in rail and Underground safety project appraisal. Given that this sort of value is more than twice the figure currently used by the DoT in its roads project appraisal, some are inevitably asking the question : is the balance right? In the light of the results reported above, the short answer would seem to be that it may well be. For various reasons discussed above, on average, members of the travelling public do seem to be more averse to Underground (and hence, by implication, rail) risks than to road risks of comparable magnitude, and are hence willing to pay substantially more for their attenuation. This on its own would be ample justification for a significant premium on the value of rail and Underground safety relative to the roads.

But could the use of such values result in fare increases that would drive many rail and Underground passengers onto the (more dangerous) roads, thereby on balance causing more deaths and injuries and, in addition, exacerbating the already serious problems of congestion and pollution? Again, the short answer is that as long as the LUL and BR values of statistical life are broadly accurate reflections of Underground and rail customers' willingness to pay for safety, then this sort of switch should not occur to any significant extent. After all, if Mr. A would genuinely be willing to pay up to an additional £X per annum in order to enjoy a particular improvement in the safety (or comfort, or convenience, etc) of his rail travel, then, provided that BR actually delivers such an improvement in the standard of its service, it would be somewhat perverse - not to say irrational of Mr. A to point to an annual fare increase of up to £X as his reason for choosing not to travel by rail and taking to the roads instead.

6.   Summary and Concluding Comments

The main conclusions reached in this paper are as follows :

●   As far as public sector transport modes are concerned, safety should be valued on the basis of the *willingness-to-pay* approach.

●   Empirical estimates of the value of statistical life for road safety for the U.S.A., U.K., Sweden, Australia and New Zealand partition, broadly speaking, into those that point towards a figure in the region of £750,000 in 1992 prices and those that put the value at nearer to £2,200,000. In part, this divergence reflects the difference between mean and median estimates implied by the highly right-skewed nature of the frequency distributions of individual marginal rates of substitution.

●   In the light of the wide range of empirical estimates of the value of statistical life for road safety, it is clear that an element of critical judgement is required in the selection of a single figure for any particular mode. In exercising such judgement in relation to road safety, the U.K. Department of Transport has opted for a value towards the bottom end of the range of estimates, reflecting a desire to temper a radical change of methodology with an element of caution.

- Empirical work in the U.K. and U.S.A. on the *relative* values of preventing serious non-fatal and fatal road injuries suggests that the willingness-to-pay based value of preventing a serious non-fatal road injury should be in the region of 10% of the value of preventing a road fatality, given the breakdown of serious non-fatal road injuries into differing severities.

- There is no a priori reason why willingness-to-pay based values of statistical life for public transport modes such as London Underground or British Rail should have the same numerical magnitude as their roads counterpart. Indeed, recent work by the authors suggests that the value of statistical life for Underground risks should be set at a premium of about 50% relative to the figure for the roads. Somewhat surprisingly, and contrary to popular wisdom, the whole of this premium is attributable to the public's concern about responsibility, control, voluntariness and other matters related to context, rather than concern about the possibility of large-scale Underground accidents.

To the extent that the value of preventing non-fatal road injuries and the value of preventing an Underground fatality have both been estimated *relative* to the Department of Transport's willingness-to-pay based value of preventing a road fatality, it is clear that the latter constitutes a vitally important benchmark in the formation of transport safety policy. It is therefore a matter of some concern that this figure has been deliberately set at the most conservative end of the range implied by existing empirical evidence. While it is quite understandable that the Department of Transport should have exercised a degree of caution in selecting a specific monetary value when it decided to effect a radical change of methodology by adopting the willingness-to-pay approach in the late eighties, it is arguable that in the mid-nineties, such caution is no longer necessary or appropriate. Indeed, given the widespread acceptance of the willingness-to-pay approach both in this country and abroad, there would now seem to be a persuasive case for setting the value of preventing a road fatality at a figure well in excess of £1m and probably nearer to £2m.

FOOTNOTES

1. For a comprehensive account of output-based traffic-safety costs and values that have been adopted in other countries, see Silcock (1982).

42

2. These weights, if inversely related to income or wealth, would reflect the judgement that a marginal £-worth of benefit to the poor would be socially more valuable than a marginal £-worth of benefit to the rich. The objection to the use of such weights is that if the distribution of income or wealth is held to be suboptimal, then it would be more appropriately and efficiently adjusted by redistributive taxes and transfers, rather than by "tampering" with the inputs to a cost-benefit analysis.

3. For a more comprehensive discussion of this point, see Jones-Lee (1989), Ch. 1.

4. The discounted present value of the excess of an individual's expected future output over his future consumption is defined as his 'net output' and would constitute a direct loss to the rest of society should he die prematurely.

5. For a more detailed account of empirical work in this area, see Jones-Lee (1989), Ch. 2.

6. In fact, an earlier study by Ghosh *et al* (1975) for the UK also produced an estimate of what is effectively a value of statistical life for road safety. Essentially, this was obtained by examining the time/fuel consumption/safety trade-off implicit in determining an optimal motorway speed. On the assumption that the then average motorway speed of 58.8 mph was the result of private optimization decisions by motorists. Ghosh *et al* obtained a value of statistical life of £94,000 in 1973 prices which updates to about £550,000 in 1992 prices. However, it is important to bear in mind that during the period from which the data for this study were obtained a speed limit of 70 mph was in force on motorways in the U.K. It may well be, therefore, that the observed average motorway speed owed as much (or more) to this limit as to private optimization decisions by motorists.

7. Interestingly, Blomquist's figure is very similar to estimates of the income elasticity of the demand for medical services in the U.S.A.

8. For a summary of the findings of the feasibility study, see Hammerton *et al* (1982).

9. For a discussion of the grounds on which one might wish to base values of statistical life on median – rather than mean-marginal rates of substitution of wealth for risk.

10. In particular, more people would vote in favour of the median valuation than the mean in a pairwise contest between the two.

11. See Dalvi (1988).

12. See Department of Transport (1993).

13. See H.M. Treasury (1991), Annex B.

14. The other member of the Management Team for this subprogramme was Dr. Peter Philips of the Centre for Health Services Research, University of Newcastle upon Tyne. The second UK sub-programme was carried out by a team at the University of East Anglia and was principally aimed at evaluating the feasibility of employing the type of approach used by Miller and his colleagues in the US.

15. The material in this section is based on Jones-Lee et al (1993).

16. Non-fatal injuries are classified as "serious" in the UK only if they involve admission to hospital a an in-patient or fall within a specified list, including fractures, internal injuries, severe cuts and lacerations and so on. Because whiplash neck injuries are not on this list and typically do not involve admission to hospital as an in-patient, they are somewhat anomalously classified as "minor" in the majority of cases recorded in UK accident statistics.

17. The material in this section is based on Jones-Lee and Loomes (1994a, 1994b). Opinions expressed do not necessarily reflect the views of London Underground Limited.

18. In fact, Savage (1993) provides evidence of substantial differences between willingness to pay to reduce the risks from various different kinds of hazard - specifically, road and aviation accidents, domestic fires and stomach cancer - and shows that willingness to pay is significantly affected by various psychological factors, including perception of the "dread" and "unknown" attributes of the hazard concerned.

19. Jones-Lee, M.W. and Loomes, G. (1993, 1994a).

20. See, for example, Hope, R. (1992).

21. See, for example, Health and Safety Executive (1989), p.14.

22. See Fennell (1988).

23. The findings of the pilot study are reported in Jones-Lee and Loomes (1993 and 1994a).

24. Savage (1993) suggests that much the same kind of reaction may account for his finding that in some cases willingness to pay to reduce risk tends to be negatively related to the extent to which the hazard concerned is perceived to be "unknown".

REFERENCES

A Panel Study of Income Dynamics 1968-1974.

Blomquist, G.   "Value of Life Saving : Implications of Consumption Activity",   Journal of Political Economy, 87, pp.540-58   (1979)

Dalvi, M.Q.,   The Value of Life and Safety : A Search for a Consensus Estimate, London,   Department of Transport (1988)

Department of Transport   Highways Economics Note No. 1, London, Department of Transport   (1993)

Fennell, D.   Investigation into the King's Cross Underground Fire, London, HMSO (1988).

Ghosh, D., Lees, D. and Seal, W.   "Optimal Motorway Speed and Some Valuations of Time and Life", Manchester School, 43, pp.134-43 (1975).

H.M. Treasury   Economic Appraisal In Central Government.   A Technical Guide for Government Departments 'The Green Book', London : HMSO. (1991).

Hammerton, M., Jones-Lee, M.W. and Abbott, V.   "The Consistency and Coherence of Attitudes to Physical Risk", Journal of Transport Economics and Policy, 16, pp.181-99 (1982)

Health and Safety Executive   Quantified Risk Assessment : Its Input to Decision Making, London, HMSO (1989).

Hope, R. "'Rational Spending on Safety Brings Results" Railway Gazette International, May, pp.345-349 (1992).

Jones-Lee, M.W.   The Value of Life : An Economic Analysis,   London, Martin Robertson,.   Chicago, University of Chicago Press (1976)

Jones-Lee, M.W., Hammerton, M. and Philips, P.R. "The Value of Transport Safety : Results of a National Sample Survey", Economic Journal, 95, pp.49-72 (1985).

Jones-Lee, M.W. The Economics of Safety and Physical Risk, Oxford, Basil Blackwell (1989).

Jones-Lee, M.W. "The Value of Transport Safety", Oxford Review of Economic Policy, 6, pp.39-60 (1990).

Jones-Lee, M.W. and Loomes, G. The Monetary Value of Underground Safety : Results of a Pilot Study ('The Phase 1 Report') Report to London Underground Limited. (1993)

Jones-Lee, M.W., Loomes, G., O'Reilly, D. and Philips, P. The Value of Preventing Non-Fatal Road Injuries : Findings of a Willingness-to-Pay National Sample Survey, Contractor Report 330, Crowthorne, Transport Research Laboratory (1993).

Jones-Lee, M.W. and Loomes, G. "Towards a Willingness-to-Pay Based Value of Underground Safety", Journal of Transport Economics and Policy, 28, pp.83-98 (1994a).

Jones-Lee, M.W. and Loomes, G. The Monetary Value of Underground Safety : Results of the Main Study (The "Phase 2 Report") Report to London Underground Limited (1994b).

Maier, G., Gerking, S. and Weiss, P. "The Economics of Traffic Accidents on Austrian Roads : Risk Lovers or Policy Deficit?" mimeo, Wirtschaftuniversitat, Vienna (1989).

Marin, A. and Psacharopolous, G. "The Reward for Risk in the Labor Market : Evidence from the United Kingdom and a Reconciliation with Other Studies", Journal of Political Economy, 90, pp.827-53 (1982).

Miller, T.R., "Benefit-Cost Analyses of Health and Safety : Conceptual and Empirical Issues"', Working Paper, Washington, The Urban Institute (1986).

Miller. T.R., Luchter, S. and Brinkman, C.P. "Crash Costs and Safety Investment", Proceedings of the 32nd Annual Conference of the Association for the Advancement of Automotive Medicine, September 12-14, Seattle, Washington (1988).

Miller, T.R. & Guria, J. The Value of Statistical Life in New Zealand, Wellington, N.Z., Land Transport Division, New Zealand Ministry of Transport (1991)

Mishan, E.J., "Evaluation of Life and Limb : A Theoretical Approach", Journal of Political Economy, 79, pp.687-705 (1971).

Needleman, L., "Valuing Other People's Lives", Manchester School, 44, pp.309-42 (1976).

Persson, U. "The Value of Risk Reduction : Results of a Swedish Sample Survey"', mimeo. The Swedish Institute of Health Economics (1989).

Savage, I. "An Empirical Investigation into the Effect of Psychological Perceptions on the Willingness-to-Pay to Reduce Risk", Journal of Risk and Uncertainty, 6, pp.75-90 (1993).

Silcock, D.T. "Traffic Accidents : Procedures Adopted in Various Countries for Estimating their Costs or Valuing their Prevention", Transport Reviews, 2, pp.79-106 (1982).

Viscusi, W.K., Magat, W.A., and Forrest, A. "Altruistic and Private Valuations of Risk Reduction", Journal of Policy Analysis and Management, 7, pp.227-45 (1988).

"Programming Languages and Safety-Related Systems"

Les Hatton

Programming Research Ltd.

Abstract

This paper considers the necessary attributes of a safety-related programming language against a backdrop of numerous recent reliability measurements made on real systems. These measurements indicate that no particular language is strongly favoured for safety-related work and the paper concludes that the number of lines implemented and the way they are implemented are far more strongly correlated with the ultimate reliability. Until more measurement-based evidence is in place, appropriate advice is very difficult to give to the safety-related programmer as existing measurements and intuition frequently conflict. Effective standardisation is consequently sparse.

1. Introduction

The choice of programming language in any system, let alone one which is safety-related, is unfortunately highly emotive. Each language supporter will argue fervently in favour of their own choice on grounds which frequently, have more to do with familiarity than intrinsic safety. Historically, most safety-related systems have been written in assembly language, known in some quarters as "the ultimate dead language". The primary reason for this is that assembly language is sufficiently close to the machine for the engineer to believe that they have direct control. As an example, to implement the high-level statement:

```
      x = a + b;
```
might require the following hypothetical assembly instructions:

```
    LR      2, a   ; Put contents of a in register 2
    LR      3, b   ; Put contents of b in register 3
    AR      2, 3   ; Add register 3 to register 2
    SR      2, x   ; Store register 2 in address x.
```

In other words, there is only the assembler in the way, rather than something more complicated like a high-level language compiler. In a real-time system, this of course is an important issue, as each assembly language instruction is normally accompanied by a time of execution. Unfortunately, as can be imagined from the above example, direct programming in assembly language leads to many problems due to its low-level nature. Conventional wisdom used to hold that programmers made about the same number of errors per 1000 lines whatever the language, e.g. (Moller and Paulish 1993). Hence, given that many common high-level languages have a compression ratio of around 5, i.e. they tend to generate about 5 assembly lines for each of their own lines, (the example above generates 4), a considerably higher degree of unreliability for a given functionality would be expected. The work of (Moller and Paulish 1993) supports this view when comparing a highly structured assembly language with a rather less structured version. In that case the compression ratio was 2.

As a result, in recent times, there has been a rapid growth of safety-related systems developed in high-level languages such as Pascal, Ada, C, C++, Fortran and the Modula languages. The arguments are very attractive: the systems are more compact (in source form), more easily maintainable, more portable, and last but not least, more reliable for the reason stated in the previous paragraph. There are of course disadvantages also. The primary problems are: timing issues, in that high-level instructions are not usually accompanied by an estimate of execution time; space issues, in that programs written in assembler are usually smaller and more efficient than ones written in high-level language, (although frequently not much!); and also numerous potential problems associated with use of the high-level language itself. For example, compiler errors may be encountered (i.e. the output of the compiler is not logically equivalent to the input) and also the ease in which the high-level language can be abused, for example if there is critical dependence on known weak areas of the language[1].

In this paper, timing issues will not be discussed further, beyond noting that the arguments presented in favour of using assembly language usually

---

[1] To a lesser extent, these also occur with assembly languages as some commonly used microprocessors are known to contain horrendous errors. In addition, assemblers themselves may contain errors.

revolve around time and space considerations in an age when microprocessor performance and memory storage capacity have made dramatic advances. Such arguments should be scrutinised very carefully as they may simply not hold water. If using a more powerful microprocessor with more memory allows the use of a high-level language rather than an assembly language, existing experience strongly suggests that this is a more appropriate route and there are regrettably numerous examples where the practice of spending huge amounts of software resource on assembly language development to "shoe-horn" an application into a computer of rather limited size and performance led to complete project failure and cancellation, for example, the RAF's planned Nimrod based successor to AWACS reported by the BBC in the 1980's.

Attention will therefore be focussed on those problems arising from the direct use of the high-level language itself either through compiler error or through inadvertent abuse of the language. Note that this paper does not recommend any particular language for use in safety-related work and will not degenerate into detailed comparison of language features. It simply discusses important principles of safety and reliability, and relevant properties of certain languages widely-used in this context, and therefore does not require a detailed knowledge of these languages to follow the exposition.

2. Language features, safety and reliability

2.1 Intrinsic safety and reliability

It is easy to get lost in a welter of detail when discussing the intrinsic safety of a language, i.e the degree to which it is unambiguously defined and correctly implemented. An early example to compare different languages in a safety context was given by (Cullyer, Goodenough et al. 1991), which compares Ada, C and Modula-2. Although an important discussion, this paper compares these languages only for their intrinsic properties of safety and reliability, using a somewhat arbitrary set of criteria, without regard for either typical compiler quality or tool support. These criteria comprised the following:

- Potential for wild jumps
- Potential for overwrites
- Semantics
- Model of maths
- Operational arithmetic
- Data typing

50

- Exception handling
- Safe subsets
- Exhaustion of memory
- Separate compilation
- Well understood

They came to the conclusion summarised in table 1 below:

Item	Ada	Ideal Ada subset	Modula-2	Modula-2 subset	C
Wild jumps	*	*	?	*	?
Overwrites	?	*	?	*	X
Semantics	?	?	*	*	X
Model of maths	?	*	*	*	X
Operational arithmetic	?	*	?	?	X
Data typing	*	*	?	*	X
Exception handling	*	*	?	?	?
Safe subset	X	?	?	*	X
Exhaustion of memory	X	?	?	?	?
Separate Compilation	*	*	*	*	X
Well understood	?	*	*	*	?

Table 1: Comparison of Ada, Modula-2 and C reported by (Cullyer, Goodenough et al. 1991). X means no protection provided, ? means some protection provided and * means sound protection provided.

In addition, some inaccuracies in the description of C, (for example, that it is not well understood compared with Ada or Modula-2, and that there is no safe subset), could lead to the perhaps tongue-in-cheek conclusion that there is only a remote chance of writing anything reliably in C. This generally fits the first reaction that many people have about C but simply does not fit with observed facts about the safety and reliability of C systems in general compared with those written in other languages[2].

---

[2]For example, the AT & T portable C compiler, itself written in C has a reputation for being one of the most reliable pieces of software ever written. Bug reports appear to be rare to the point of non-existence.

From the point of view of both safety and reliability however, (the two are not synonymous), all that really matters is the time-honoured engineering dictum:

"The theoretical safety and reliability of a system is irrelevant; it is how safe and reliable that system can be made in practice".

So it is also for programming languages. The question as to whether, from its definition, Ada, Modula-2 or indeed any other language is safer than another, is irrelevant. What is important is how safe and reliable the use of a particular language can be made in practice. *In other words, a programming language, its tool support and its typical compiler quality must be considered together, along with the effectiveness of enforcement of reliability inducing practices, for example, the use of safe well-defined language subsets.*

After incorporating this notion and correcting the inaccuracies with respect to C, (Hatton 1994c) arrived at the following re-marked table:

Item	Ada	Ideal Ada subset	C + tool support	Modula-2 subset	C
Wild jumps	*	*	*	*	?
Overwrites	?	*	*	*	X
Semantics	?	?	*	*	X
Model of maths	?	*	?	*	X
Operational arithmetic	?	*	?	?	X
Data typing	*	*	*	*	X
Exception handling	*	*	?	?	?
Safe subset	X	?	*	*	X
Exhaustion of memory	X	?	?	?	?
Separate Compilation	*	*	*	*	X
Well understood	?	*	*	*	?

Table 2: Modified comparison of Ada, Modula-2 and C reported by (Hatton 1994c).

2.2 Standardisation

Standardisation is a very important and necessary property of a candidate language for safety-related development, a point made by both (IEC 1991) and Def Stan 00-55, for example. When a language is standardised, this implies that each part of a language has been formally reviewed by (usually) many people. They may not be able to define what happens very well, but at least it has been reviewed. There also will exist *validation suites*, which must be passed to prove that a language compiler implements the language properly[3]. *In a safety-related environment, the use of a non-standardised language (or compiler) should be reviewed most carefully and allowed only if there is no reasonable alternative, with an appropriate assessment of risk.* This would normally preclude the use of either assembler or C++ for example. This latter language is unlikely to be standardised before 1998 or so and assemblers cannot easily be standardised by their very nature.

The assiduous user can go further than mere validation by running such programs as *paranoia* to assess the behaviour of floating point arithmetic on the target machine, if this is relevant to the safety-related application. This was originally developed in BASIC by W.M. Kahan of the University of California at Berkeley in Pascal and has been translated into C by David Gay of AT & T Bell Labs and Thomas Sumner of the University of California at San Francisco from the Pascal version written by Brian Wichmann of the U.K. National Physical Laboratory. It's C source is available by sending the message "send paranoia.c from paranoia" to the Internet address, netlib@research.att.com. The resulting source file can be then compiled for the local machine and run as appropriate.

2.3 Problematic linguistic features in safety-related systems

One of the central problems facing programmers working on safety-related systems is the lack of international standardisation regarding such developments. Although, there are a number of draft, military or industry-specific standards, for example, (IEC 1986), (IEC 1991), (IEC 1992), and Def Stan 00-55/56, addressing this issue, there is no formally accepted international standard as yet. It should be noted that (IEC 1991) is specifically marked as being non-referenceable owing to its draft nature, even though it is widely-referenced by programmers working with safety-related systems, in the author's experience. Of these, both Def-Stan 00-55 and (IEC 1991) discuss features of programming languages which are likely to be problematic in a safety-related system. These and other features will be discussed now in a rather general context.

---

[3]Unfortunately for some languages, different validating authorities use different validation suites. For example, the Plum-Hall validation suite is used to validate C in Europe whereas the Perennial validation suite is used in the United States. As with everything else, commercial interest inevitably intervenes.

2.3.1 Static v. Dynamic detectability

It is easy to understand why the average software engineering group's reaction to poor reliability is to live test more thoroughly. Unfortunately, it is far less efficient than static inspection, (i.e. inspection of source code) in removing fault, c.f. (Grady and Caswell 1987), (Beizer 1990) amongst many others. The principle reason for this is that software systems are typically very complex and live testing can only achieve a very small coverage of all possible paths. In contrast, inspection sees all code by definition. Hence, any language which is designed to favour static detection of error rather than dynamic detection of error has a significant advantage. In general, language designers are not very good at this although Ada designers took considerable pains to maximise such detection and, perhaps by chance, most of the poorly defined parts of C are statically avoidable, (Hatton 1994c).

2.3.2 The influence of language complexity

One of the most important attributes of any programming language is that it be well-defined, and its effects easily understood. Remarkably few languages achieve either of these ideals, and arguably no recent ones. It may come as a surprise to the average reader that programming languages are so poorly defined in general. However, the reader should realise that fashion, politics and proprietary interest play a not insignificant role in the standards-making process. For example, most commonly used languages are currently undergoing revision to incorporate what has come to be known as *object-orientated* features. The idea here is to promote re-use, information hiding and lots of other things believed to improve the reliability of programs. The essence is to break a design down into small, loosely-coupled and therefore easily-managed components. Interestingly, there is an almost breathtaking lack of evidence anywhere in the world to suggest that this makes things better. It is simply a belief. The computer industry does this frequently, with many other tarnished silver bullets like CASE in the 1980's and quality systems in the 1990's, neither of which have delivered the promised improvements. Now it is the turn of object-orientation to be promoted vigorously without evidence. In fact, very strong empirical evidence has recently emerged to suggest that large modules are proportionately *more* reliable than small components across a wide range of languages, (Basili and Perricone 1984), (Hatton and Hopkins 1989), (Davey, Huxford et al. 1993), (Moller and Paulish 1993). According to (Hatton 1994b) the relation is logarithmic over a wide range of module sizes.

When considering a programming language for use in a safety-related system, several things should be considered:

2.3.2.1 The difficulty in defining unambiguous subsets

Defining a well-defined subset of a language for use in safety-related systems as recommended by documents such as (IEC 1991) and Def Stan 00-55 is surprisingly difficult. Languages may be poorly defined for many reasons. As an example, the standardised language commonly known as ANSI C[4], a candidate for use in safety-related systems owing to the huge amount of practical experience with it in different environments has the following areas of poor definition:

- *Unspecified* behaviour. This covers 22 legal constructs which the ANSI committee decided not to specify. A relatively commonly occurring example in real C code is *reliance on evaluation order*. For example, in the expression:

    ```
    x = f(a) + g(b);
    ```

    The order in which the two functions f() and g() are called before their sum is taken, is not defined. If either affects identifiers used elsewhere in the expression, different implementations might arrive at a different result. This problem is not limited to C. In fact a form of it appears in most commonly used languages including Ada and Fortran, for example.

- *Undefined* behaviour. This covers 97 illegal constructs for which the ANSI committee were unable to define what should happen.

- *Implementation-defined* behaviour. This covers 76 items which must be defined, i.e. can be subject to formal argument, but which may differ between implementations still conformant to ANSI C. In other words, these areas are not addressed by validation suites.

- *Locale-specific* behaviour. These 6 items are associated with internationalisation issues such as character sets.

Each of these four categories is specifically defined by ANSI C, for example, to the extent that a program which contains none of them is known as a *strictly-conforming program*, in the sense that its behaviour is meant to be unambiguously defined and therefore consistent across *conforming*, i.e. validated implementations, (see later). All languages in common use contain items which could in principle, be categorised as one of these four, but the attractive thing about ANSI C is that it explicitly defines them, allowing them to be avoided, usually by means of tool support. Some languages such as Ada, go to considerable pains to avoid some of them. In ANSI C, their explicit description has allowed the development of numerous supporting tools which can enforce their absence, (Hatton 1994c).

---

[4]American National Standards Institute. The language is now an ISO language and should be correctly known as ISO C. Its official reference is ISO/IEC 9899:1990

To these categories can be added another, again occurring in all modern programming languages:

- *Accidentally-undefined* behaviour. Committees simply forget things or don't understand the full implications of some features at the time of standardisation. When discovered, these lead to the concept of *interpretation requests*, whereby a compiler writer or programmer will formally ask the committee to decide what a construct, not apparently covered by the standard, actually means. The number of these forms an excellent metric for the simplicity of a language. The more complex the language, the more examples there are of issues which are not addressed by the standard. The following table, Table 3 shows the number outstanding for certain popular standard languages:

	Pascal	Extended Pascal	C	Ada
Number of interpretation requests	1	~80	50	~2000

Table 3. This table shows the approximate number of interpretation requests outstanding for several modern standard languages. There are no prizes for guessing which is the most complex language here.

Suitable safe(r) subsets of C are described in (Hatton 1994c) and for Ada are described in (Carré 1990), (Hutcheon, Jepson et al. 1993) and (Forsyth, Jordan et al. 1993).

2.3.2.2 The ease of use of the language

This of course is a key issue. Programmers will find it difficult to express ideas safely in a complex language. Other programmers will find it difficult to inspect them for correctness and finally of course, there can be expected to be a higher risk of compiler error due to the difficulties of implementing the compiler. This is a particularly important issue in Ada, as discussed for example, by (Forsyth, Jordan et al. 1993). By contrast, implementing a high-quality C compiler is relatively simple, and consequently C compilers have an excellent reputation for consistency and reliability, however, this is unlikely to be the case for its errant offspring, C++, owing to the extraordinary complexity of this language compared with its parent C.

2.3.2.3 The influence of grammar

The simplicity and therefore ease of use of a language is inversely related to its expressiveness and its grammar. The more ways there are of saying the same thing, the richer and more interesting the language is and *the more potential there is for misunderstanding and ambiguity*. This is actively sought after in a spoken language of course for the sake of interest. Consider for example the syntactically legal English sentence "Fruit flies like a banana.". This has numerous different semantic interpretations and is highly ambiguous. Irregularity is also encouraged in spoken languages for interest. This accounts for the fact that Esperanto, in spite of its considerable claims to be a highly regular and therefore easy to learn language, is hardly ever spoken. Unfortunately, this same paradigm has been taken into programming languages with simple regular languages such as Pascal being largely ignored in commercial systems development in favour of massively complex ones such as PL/1, Ada and more recently C++. (C is simple but not very regular).

To see how easy a casual growth in grammar of a language can lead to a combinatorial explosion of complexity and irregularity, consider the following simple grammar.

Possible nouns:	MAN DOG
Possible verbs:	BARKED PANTED
Grammar:	noun verb

This is very similar to the way that programming language syntax is defined. The above leads to the following four possible sentences

MAN BARKED

MAN PANTED

DOG BARKED

DOG PANTED

Each sentence is syntactically correct and furthermore has sensible semantics. Supposing we now tire of such a simple and regular language in favour of something more, shall one say, "expressive", by adding the noun CAT.

This then leads to the following six possible sentences:

MAN BARKED

MAN PANTED

DOG BARKED

DOG PANTED

CAT BARKED

CAT PANTED

So the cost of one extra noun is a 50% increase in the number of syntactically legal sentences, half of which in the unfortunate absence of any known species of barking cat, is absurd. Hence in our semantics we now have an exception. In a further search for more "expression" to cover an internationalisation issue, that of a yodelling man, the verb YODELLED gets added by the now growing standardisation committee. This then gives a total of nine syntactically legal sentences:

MAN BARKED

MAN PANTED

MAN YODELLED

DOG BARKED

DOG PANTED

DOG YODELLED

CAT BARKED

CAT PANTED

CAT YODELLED

So the original language has more than doubled in terms of the sentences which can be built using it, at the expense of introducing three absurd ones, giving a notional $3/9 = 1/3$ irregularity ratio. So it goes on, culminating in something like PL/1, "the language to end all languages", which it nearly did because nobody could understand it, or Ada which invited such eloquent and informed condemnation in (Hoare 1981), or C++, a language which daily becomes more incomprehensible in its pursuit of orient objection[5], or Fortran 90 which took so long to produce that it spanned two decades and many ambitious names such as Fortran 82, 84, 86, 88, 8X, 9X [6], and finally 90, when everybody got sufficiently fed up as to release it, or early versions of what came to be known as Algol 68. Such languages contain countless ways of constructing absurd "sentences", some of which find their way into safety-related systems. If the reader would like to peruse a beautifully written, very humorous but ultimately gloomy picture of the development of modern, "expressive" programming languages, they should consult the Turing Award paper of Tony Hoare, referenced above, (Hoare 1981).

To summarise, in software "engineering", simplicity has not yet found its way to the heart of things as it has done in more mature engineering disciplines, and, in general, like governments, people currently get the programming languages they deserve. Hopefully, in the case of programmable safety-related systems, humanity will get the opportunity to learn from its mistakes.

---

[5]or something like that. This will nicely cover the rather arbitrary treatment of wide characters and so on, commonly used in oriental countries, however.
[6]The author, in some despair, suggested XX at one stage but was not taken seriously.

2.3.3 *Some empirical data on reliability*

Most advice given in computing is anecdotal or intuitive in nature rather than based on any experimental observations. As such, it is roughly in the position of science prior to the Renaissance. Such recent experimental evidence as there is appears to support the following:

- As was mentioned earlier, there is now strong evidence to support the view that *small components of a software system are proportionately more unreliable than large components*, (Basili and Perricone 1984), (Hatton and Hopkins 1989), (Davey, Huxford et al. 1993), (Moller and Paulish 1993). This directly contradicts advice normally given to improve the reliability of a software system.

- In a famous study of IBM operating system errors, (Adams 1984) reported amongst other things that 15% of all software "fixes" created a problem at least as large; a significant percentage of errors had very long mean times to failure; and that most software failures of a large system are caused by a minority of faults.

- If a language allows irregular and inconsistent sentences to be constructed, they will occur. In his study of C and Fortran, (Hatton 1993) found that statically detectable faults (i.e. potential errors) occur at about the rate of 12 per 1000 source lines in C and 8 per 1000 source lines in Fortran. Note that this data showed that neither safety-related applications nor applications developed using a formal certified quality system performed any better than average.

- If a programming standard exists, it will be transgressed at rates typically around 50 per 1000 source lines, (Hatton 1994c).

- Error rates seem to be relatively independent of programming language and more highly correlated to the number of lines actually written, e.g. (Moller and Paulish 1993). Even in Ada, a language widely touted as one suitable for safety-related systems, error rates reported in a communications system by (Compton and Withrow 1994) averaged around 7 per 1000 source lines across a wide range of component sizes. *Put simply, error rates in real systems strongly favour using the smallest number of lines possible to implement any functionality but, given this constraint, the language used is more or less irrelevant.*

- If N versions of the same algorithms are developed independently in the same programming language, and given the same input data with the same parameters, in at least one major study, disagreement grew approximately 100,000 times more quickly than differences in floating point implementations would suggest and about 1000 tines quicker than those resulting from the same software ported to different machine types, (Hatton and Roberts 1994).

- Even in NASA, one of the most necessarily quality-orientated development environments known, error rates across all languages show around 6 errors per 1000 lines on average as is shown in Figure 1 below, produced by the University of Maryland's Software Engineering Lab when analysing NASA-supplied data.

Errors per 1000 lines at NASA Goddard
1976-1990

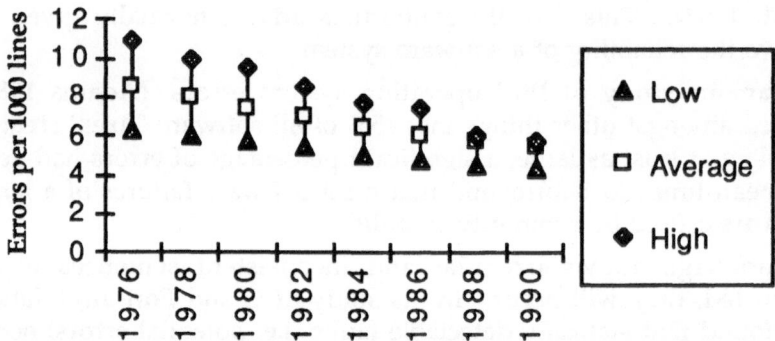

Figure 1. Graph published in the December 1991 special issue of Business Week showing the drop in software errors at NASA Goddard.

Note that in spite of the Trojan efforts, great concern over quality and generally excellent resources, *most of the improvement has been achieved by improving the bad ones, not the good ones.*

- Formal methods appear to be one way of improving the error rate somewhat[7] as described by (Ostrolenk, Southworth et al. 1994), who developed a COBOL parser written in C from a formal specification in Z. They also used robust parser generating tools such as lex and yacc. They achieved an estimated error rate of 1.3 per 1000 lines as compared for example to IBM "cleanroom" techniques which achieved around 3.4 errors per 1000 lines, (Dyer 1992), (Hausler, Linger et al. 1994). This can be compared with the automatically enforced safe-subset environment for C and C++ described by (Hatton 1994a) which is currently running at an estimated error rate of 0.8 per 1000 lines.

---

[7]But not as much as was hoped. Two formally developed systems which the author has seen recently both exhibit around 1 error per 1000 lines, or about 6 times better than the average. This is relatively small comfort when faced with a 200,000 line safety-related system however.

With this measurement backdrop, various linguistic features considered poor practice in safety-related systems will now be discussed.

2.3.4 *Apparently dangerous features*

The following are frequently cited as being unsafe in safety-related systems.

2.3.4.1 Dynamic objects and recursion

The argument against dynamic objects and recursion is based essentially on the problems of running out of dynamic memory or pointer abuse associated with dynamic memory. Whilst the warning is reasonable in one sense, forbidding any such occurrence would automatically require the foundations of modern data structuring in terms of trees, lists and so on to be completely re-written, with all the attendant risk of throwing the baby out with the bathwater, (Forsyth, Jordan et al. 1993), (Hatton 1994c). There seems to be no empirical evidence to support the ban. More relevant advice would seem to be to either ensure that there is enough dynamic space for any recursion or that pointers do not point anywhere naughty using the extensive tool support available in some languages such as C.

2.3.4.2 Multiple entrance / multiple exit

The argument for restricting all components to be single entrance / single exit is based on the foundations of structured programming. There appears to be no empirical evidence comparing the reliability of systems built with and without this restriction.

2.3.4.3 Overloading

Overloading in programming language terms means using the same notation to mean different things at different times. This occurs in most programming languages. Consider, for example,

    x = a + b;
    n = i + j;

where x, a and b are floating point numbers and n, i, and j are integers. In this case, in common with normal mathematical notation, the + means different things. In the former it adds floating point numbers and in the latter it adds integers together - two distinct operations. This particular example would on the face of it cause no confusion; not so. Consider the following example,

    x = y + d;

where y and d are double precision floating point variables and x is a single precision floating point variable. A staggeringly expensive occurrence of this was reported as being responsible for the Shuttle Endeavour failing to rendezvous with the Hughes/Intelsat 6/F3 spacecraft in 1992. The problem was traced to a loss of precision when a double precision floating point value was inadvertently stored in a single-precision variable within an expression. Although this could be equally well viewed as a type mismatch, this is statically detectable and quite rightly appears as a forbidden construct in numerous safety-related programming standards which the author has seen.

In more modern languages such as C++, overloading has reached the point where it may present a serious threat to the comprehensibility and inspectability of C++ programs in that nearly all the operators of the language can be overloaded. Overloading certainly has the potential to cause problems but its supporters would argue that it also has potential to remove problems. Again, there are no experimental studies to guide advice here, and until there are, it is difficult to say anything beyond the above.

2.3.4.4 Inheritance

This relates to the ability of a programming object to inherit the properties of another object explicitly and perhaps an arbitrary number of others implicitly. It is at the heart of object-orientated techniques which, as was mentioned earlier, appear to be completely unsubstantiated with reference to the improvement of reliability. Experiments are certainly needed here also.

3. Conclusions

Many conclusions can be drawn from the above discussion, the most important of which are:

- Error rates typically seem to be around 6 per 1000 lines across a wide range of languages.

- No modern programming language in wide use is a natural candidate for use in a safety-related system without severe and automatically enforced restriction to safe well-defined subsets. Defining such subsets requires great care and considerable experience gleaned from many disparate programming application environments.

- Existing fault studies do not favour any particular programming language over another. They are much more strongly correlated to the number of lines which the programmer actually writes and the way they are produced.

- Getting better than around 1 error per 1000 source lines appears to be very difficult by *any* known technique, in *any* language. This strongly supports the conclusions of (Littlewood and Strigini 1992), (Littlewood 1993) who argue that software by itself is unlikely to achieve anywhere near acceptable failure rates for safety-related systems.
- There is little if any measurement-based evidence as yet in favour of supporting any partial or complete ban on algorithmic techniques such as the use of dynamic objects, recursion, multiple entrance / multiple exit and others cited as having potentially dangerous aspects. This may simply because the disadvantages are more or less equally balanced by the advantages.

4. Acknowledgements

The author would like to acknowledge his colleagues at Programming Research and also the many stimulating conversations on these topics which he has had with delegates and friends at meetings of the Centre for Software Reliability at City and Newcastle Universities in the U.K. Norman Fenton, Bev Littlewood and Peter Mellor at City University brought some of the above data to the author's attention.

5. References

Adams, N. E. (1984). "Optimizing preventive service of software products." IBM Journal Research and Development 28(1): 2-14.

Basili, V. R. and B. T. Perricone (1984). "Software Errors and Complexity: An Empirical Investigation." Comm. A.C.M. : 42-52.

Beizer, B. (1990). Software Testing Techniques. Van Nostrand Reinhold.

Carré, B. A. e. a. (1990). SPARK - The SPADE Ada Kernel. Program Validation Ltd.

Compton, B. T. and C. Withrow (1994). Improving Productivity: Using Metrics to Predict and Control Defects in Ada Software. Second Annual Oregon Workshop on Software Metrics, Oregon,

Cullyer, W. J., S. J. Goodenough, et al. (1991). "The choice of computer languages for use in safety-critical systems." Software Engineering Journal (March, 1991): 51-58.

Davey, S., D. Huxford, et al. (1993). Metrics Collection in Code and Unit Test as part of Continuous Quality Improvement. EuroStar'93, London.

Dyer, M. (1992). The Cleanroom Approach to Software Quality Development. John Wiley and Sons.

Forsyth, C., D. Jordan, et al. (1993). A Study of High Integrity Ada: Trusted Ada Compilation. York Software Engineering / British Aerospace.

Grady, R. B. and D. L. Caswell (1987). Software Metrics: Establishing a Company-Wide Program. Englewood Cliffs, N.J., Prentice-Hall.

Hatton, L. (1993). The quality and reliability of scientific software. Software Quality Management, Southampton, Computation Mechanics Publications, Elsevier.

Hatton, L. (1994a). A case history of automated improvement of software product quality. Software Quality Assurance and Measurement Ed. N. Fenton. London, Chapman-Hall.

Hatton, L. (1994b). Is modularisation always a good idea ? CSR' 94, Dublin, Ireland, Chapman-Hall.

Hatton, L. (1994c). Safer C: Developing for High-Integrity and Safety-Critical Systems. McGraw-Hill, Dec. 1994.

Hatton, L. and T. R. Hopkins (1989). Experiences with Flint, a software metrication tool for Fortran 77. Symposium on Software Tools, Napier Polytechnic, Edinburgh,

Hatton, L. and A. Roberts (1994). "How accurate is scientific software ?" To appear in IEEE Transactions on Software Engineering, late 1994.

Hausler, P. A., R. C. Linger, et al. (1994). "Adopting Cleanroom software engineering with a phased approach." IBM Systems Journal 33(1): 89-109.

Hoare, C. A. R. (1981). "The Emperor's Old Clothes: 1980 Association of Computing Machinery Turing Award lecture." Comm ACM 24(2):

Hutcheon, A., B. Jepson, et al. (1993). A Study of High Integrity Ada: Tool Support. York Software Engineering and British Aerospace.

IEC (1986). Software for Computers in the Safety Systems of Nuclear Power Stations.

IEC (1991). Software for computers in the application of industrial safety-related systems. International Electrotechnical Commission: Drafts only - cannot yet be referenced.

IEC (1992). Functional Safety of Electrical / Electronic / Programmable Electronic Systems: Generic Aspects. International Electrotechnical Commission: Drafts only - cannot yet be referenced.

Littlewood, B. (1993). The Need for Evidence from Disparate Sources to Evaluate Software Safety. Directions in Safety-critical Systems Eds. F. Redmill and T. Anderson. London, Springer-Verlag. 285.

Littlewood, B. and L. Strigini (1992). "Validation of Ultra-High Dependability for Software-based Systems." Comm ACM to be published:

Moller, K.-H. and D. J. Paulish (1993). An empirical investigation of software fault distribution. CSR'93, Amsterdam, Chapman-Hall.

Ostrolenk, G., M. Southworth, et al. (1994). Cost-effective evaluation of a COBOL Parser using an operational Profile. CSR'94, Dublin, Ireland.

ON THE QUALIFICATION OF SAFETY CRITICAL STRUCTURES - THE SAFESA APPROACH

N.C. Knowles
WS Atkins Science & Technology, Epsom, UK

J.R. Maguire
Lloyds Register Industrial Division, Croydon, UK

Abstract

A formal, disciplined approach to the qualification of safety critical structures is described which includes an analysis of error sources. The central role of idealisation error is explored and the relationship between uncertainty and error is also discussed.

1. Introduction

Over the last two decades the engineering design environment has been changed radically by the emergence of increasingly sophisticated computer-based simulation, modelling and analysis tools. For the design and analysis of engineered structures the pre-eminent tool in this category is the Finite Element Method. Its adoption within the design process is giving rise to situations in which increasing reliance is being placed on the results of uncorroborated analysis. This brings with it the question of how to qualify structures by analysis alone and the extent to which supporting tests are required. (The term qualification is used herein to describe the process of demonstrating a structure's fitness for purpose in respect of its ability to carry loads [SAFESA 94a]).

Many practitioners argue that finite element analysis methods are mature enough to be used routinely and confidently. The mathematical basis is well established and great progress is evident in the control of discretisation and procedural error [Szabo, 91, Zienkiewicz, 87]. At the same time standards covering the management and use of finite element methods in industrial practice are emerging [NAFEMS, 93]. However, notwithstanding these developments there can still be very real difficulty in achieving a structural model which is an adequate representation of the real world. The difficulty is compounded by the power and utility of modern graphics-based pre-processors, which, although facilitating the structural modelling process, tend to mask its underlying details and make it more difficult to provide a satisfactory audit trail. Moreover, the widespread availability and apparent ease of use of such preprocessing packages tends to encourage people with inadequate training to undertake finite element analysis. Accordingly, there is still considerable anecdotal evidence of significant differences between measured behaviour of structures and analytical predictions.

2. Uncertainty & Errors in Structural Qualification

Inevitably many of the differences between the actual behaviour of a structure and that predicted by analysis is due to uncertainty in the physical description of the structure used in the analysis. Uncertainty exists because of the natural variability in the real world and our lack of understanding of it. Thus in the present state of knowledge it is not possible to define precisely quantities such as material strength, or actual seismic loading. Differences due to uncertainty in the physical description should be distinguished from those due to errors. The latter are associated with the way the description of the real world is manipulated into a computational model and the accuracy with which the subsequent analysis is carried out.

In carrying out the analysis one moves from the real structure (with its uncertainties) to a description of it and eventually to a structural model (Figure 1). This process we term idealisation; it involves a series of approximations and assumptions about the structural behaviour which are consistent with the use to which the result of the analyses are put (i.e. the qualification). The description of the real structure necessarily involves identifying and in some cases, removing sources of uncertainty in the real world structure. The description is influenced directly by the qualification criteria against which the structure is assessed, and a certainty is often explicitly accounted for in the qualification codes. For example, most building codes of practice address the uncertainty in wind loading by

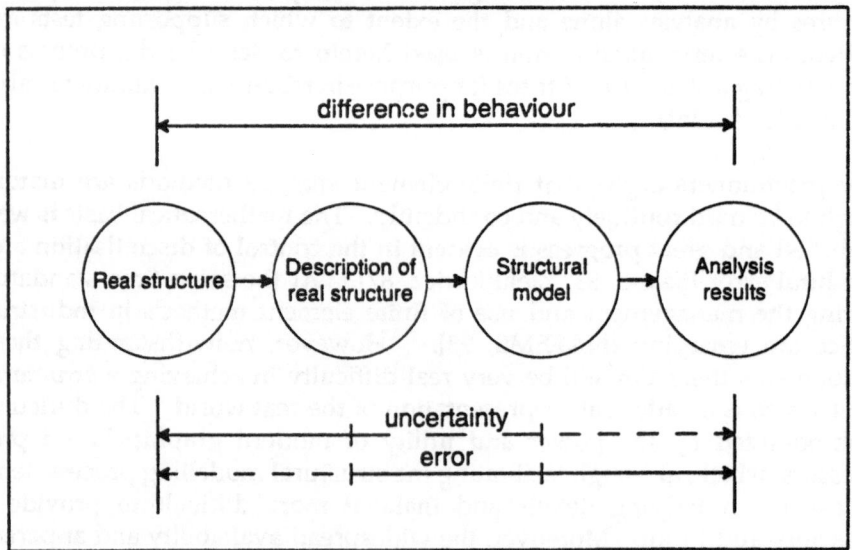

Figure 1 Uncertainty and Error

prescribing a design wind speed based on conservative probabilistic considerations. Similarly in some codes material strength variability may be accounted for by stipulating 'lower shelf' values. However, not all uncertainties are so treated and others have to be addressed directly in the qualification. In this context it may be convenient to recognise four basic types of uncertainty, namely, physical (from natural sources), measurement (identify and to measure accurately the right quantity), epistemic (e.g. dearth of information about extreme events/future events) and model.

Various sources of uncertainty will be identified in the course of a structural qualification. With these identified and treated, all other differences between the results of an analysis and the real behaviour are attributed to idealisation 'error'. Sources of error can be identified concurrently with uncertainties and formal error treatment (e.g. the SAFESA approach [SAFESA 94b]) applied to limit the error to acceptable bounds.

3. Approach to Structural Qualification

The SAFESA approach to the structural qualification process initially involves defining the scope of the problem, then, performing a detailed assessment from which qualification conclusions are drawn, as shown in Figure 2.

An examination of this figure reveals a series of distinct sub-processes. These begin with an examination of the real structure to generate an idealised model, continues with a process of discretisation and meshing and finally leads to the generation of a solution model. This sequence of sub-processes continues until we eventually finish with a set of stresses, displacements, etc. for a de-idealised structure. At each stage in this complex process the errors relevant to the assessment process are identified, quantified and controlled.

A hierarchical classification system has been created as an aid to the identification of errors. This categorises errors into those involved with the modelling process (modelling errors) and those associated with the total process of solving the finite element model (formulation and solution errors). Within this broad classification individual errors can be named, located within a group and identified.

This broad classification is useful to the analyst in identifying the type of errors which can occur in each stage of the decomposition process where the model is decomposed according to load paths, structural behaviour and boundary conditions. It does not, however, locate the error source in a very exact way within the FE analysis or allow its effects to be assessed. In order to do this the concepts of features and primitives have been introduced into the analysis process. A feature is a component of a design exhibiting a specific characteristic whilst a primitive is the smallest entity in a feature which is irreducible.

These concepts are familiar to all analysts who traditionally decompose a structural model according to its characteristic behaviour patterns. However, in the present application the role of the primitive is to identify error sources. For example, in a specific analysis a structural joint might be regarded as a primitive if it is analysed as a beam and the associated error is the difference between the assumed beam behaviour and the actual behaviour of the joint. It should be noted that the detailed assessment requirements have a critical bearing on the selection of the primitive: If the joint is rivetted and the qualification requires assessing the growth of a crack under the rivet head, then the primitive could be the head region. The latter would obviously involve a more detailed finite element model than that obtained by beam theory. Thus, the decisions relating to the error control process are the key issues in deciding on the level of idealisation.

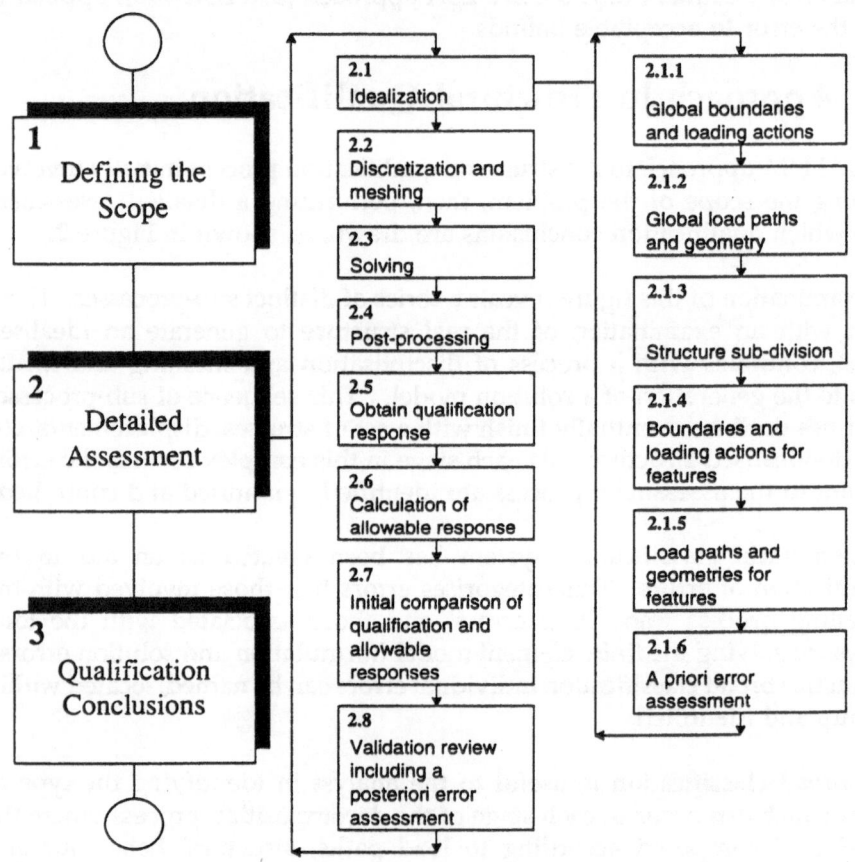

Figure 2 The SAFESA approach to structural qualification

Whilst the actual errors associated with a given analysis can be located in specific parts of the model using the process described above, they could be due to a number of causes. The main identified errors can be attributed to:

68

the mathematical model, the domain, the boundary conditions, the material properties and any dimensional reduction performed. The procedure being advanced in SAFESA envisages the user taking responsibility for identifying such errors, as described above, then treating them. Error treatment is affected by a variety of techniques including application of experience, simple calculations, comparison with tests, hierarchical modelling and sensitivity analyses.

4. Conclusions & Future Work

The ideas expressed in this paper have evolved in the course of a major UK based collaborative research project which has created the SAFESA Quality Standard [SAFESA 94a]. This has led to the following results:

   i.   A formal and disciplined approach to the finite element analysis process has been established.

   ii. A broad link between the finite element analysis of a structure and the requirements of the qualification process has been outlined.

   iii. The errors associated with the finite element analysis have been classified.

   iv. A method, based on the use of primitives and features, has been set up to facilitate the location of errors as a vehicle for idealisation error control.

   v.   Error sources have been defined and procedures established to control error.

Further work is continuing in a number of areas including:

   i.   objective (cf. subjective) error measures.

   ii. formal methods for incorporating experience into the qualification process.

   iii. human factors and subjective error control.

   iv. development of a knowledge based system to incorporate these developments.

Acknowledgements

The authors are indebted to their co-workers in the SAFESA project. Dr. D.C. Mackay of Lloyds Register, Dr. M.J.H. Fox of Nuclear Electric, Dr. A.A.

Rahman of Assessment Services Limited, Professor A.J. Morris and Dr. R. Vignjevic of Cranfield University and Mr. C. Kaucky-Lawrence of WS Atkins have all contributed substantially. The financial support of the DTI under the framework of the Safety Critical Software Programme and parent organisations is gratefully acknowledged.

References

[NAFEMS 93] NAFEMS QSS "Quality Systems Supplement to ISO 9001 relating to Finite Element Analysis in the Design and Validation of Engineering Products" - NAFEMS, East Kilbride 1993.

[SAFESA 94a] "The SAFESA Quality Standard" - published by Assessment Services Limited on behalf of the SAFESA Consortium 1994.

[SAFESA 94b] "The SAFESA Approach to the Qualification of Safety Critical Structures" - published by Assessment Services Limited on behalf of the SAFESA Consortium 1994.

[Szabo 91] Szabo B. & Babuska I., "Finite Element Analysis" - J. Wiley 1991.

[Zienkiewicz 87] Zienkiewicz O.C. & Zhu J., "A simple error estimator and adaptive procedure for practical engineering analysis". International Journal for Numerical Methods inEngineering, Vol. 24 1987.

FRESCO - An Investigation into a Framework for the Assessment of Safety-Critical Systems

Hunt, J.R., Lucas, P.R., Wingate, G.A.S.
Eutech Engineering Solutions Ltd, Chilton House, Billingham, TS23 1LD, UK

Abstract

There is a clear need for guidance on cost-effective assessment and certification of safety-related and safety-critical systems within the process industries. This paper introduces FRESCO, a collaborative project sponsored by the DTI's Safety Critical Systems Programme, which is investigating suitable assessment frameworks. The project intends to use the chemical industry to verify its recommendations. Interim results are presented and initial feedback from the project's industry interest group, drawn from the chemical and pharmaceutical industries, are reported.

1 Introduction

Industry needs dependable systems to be competitive. Dependability may be defined as a combination of four attributes; availability (accessible service), reliability (continuous service), security (preservation of integrity) and safety (no catastrophes). [Wingate 94] Shortfall in any of these attributes reflects ultimately on the bottom line figures. Particularly in the case of safety, this can be seen as a licence to operate issue, and in this sense environmental performance is every bit as important as avoiding spectacular accidents. It may be tempting to duck the issue by not using programmable systems, but this would be to miss out on the flexibility, and hence competitive advantage, that their use can bring to a manufacturing process. What is needed is a framework within which safety-critical and safety-related software systems can be assessed and certified.

There is a parallel with the situation that pertained in the last century regarding boilers. To counter an unacceptable number of explosions, insurance companies set up an assessment and certification scheme. A manufacturer was then able to buy a certified boiler with confidence in its safety, and incidentally enjoy reduced insurance premiums.

2 An Industry Need

A comparison between today's factory and its predecessor of say ten years ago shows how pervasive programmable electronic systems have become. Much equipment is now controlled by programmable systems, where once a box of relays or a human operator would have been in control. Often this PES is supplied as part of a package deal, with the ultimate user having little or no idea of the standards that have been applied in its design, or perhaps even of the potential hazards should it malfunction. In plant monitoring, the PC has become a common sight in the control room.

There is currently a lack of clarity as to how industry can demonstrate not only to its regulators, but also to the public, shareholders and employees, that it is operating safely where programmable systems are concerned. Even the meaning of the terms themselves, safety-critical and safety-related, continues to be debated.

In the chemical industry, the drive is coming from the need to comply with IEC standards for Safety, Health and Environment (SHE)-related programmable systems. These are IEC65A parts 1, 2 and 3, which are now available in draft form. In the UK, HSE guidelines for Programmable Electronic Systems (PES) have been available for some years. [HSE 87]

The pharmaceutical industry has the same HSE PES drives, but in addition it must satisfy the Good Manufacturing Practice regulatory bodies[1] requirements for assuring product quality through process validation.

In recent years, manufacturing industry has concentrated more on its core capabilities, and has shed many of its own experts who in the past used their skill and experience to assure the safety of systems they installed and operated. Instead, many companies rely on external suppliers to provide software based systems. By the same token, these system suppliers now lack skills in process technology. The availability of an assessment and certification framework would therefore provide a means for manufacturing industry to buy programmable systems with confidence. The distinction here between safety-critical and safety-related systems is one of degree, not kind.

The confidence imparted by independent assessment of safety-critical programmable systems has been recognised in several recent standards, but there is lack of guidance on how to carry this out, and on the level of detail required for the assessment. Although there are some schemes that cover the assessment of software (eg TickIT, ITSEC[2]), these are not specific to safety-related systems.

It is difficult for an industrial user or supplier of programmable systems to select an assessor, because of widely differing approaches by different assessment bodies. On the one hand, too superficial an assessment will fail to provide the assurance required, while on the other hand too detailed an assessment, or the imposition of

---

[1]   eg Food and Drugs Administration (FDA) in the USA, Medicines Control Agency (MCA) in the UK.

[2]   TickIT is BS5750 interpreted for software development.
      ITSEC = IT Security Criteria.

unnecessarily stringent standards, will add costs without bringing additional benefits. The availability of a widely accepted assessment framework is seen as an important step in enabling UK industry to be competitive in European and world markets.

3 The FRESCO Project

The name FRESCO is an acronym for FRamework for the Evaluation of Safety-Critical Objects. The project is sponsored by the DTI, and is a collaboration between Eutech Engineering Solutions, Admiral Management Services, Lloyd's Register, Eversheds Hepworth and Chadwick, and Railtrack. The project duration is 30 months, ending in June 1996. [Admiral 93]

FRESCO plans to establish UK industry requirements for the assessment and certification of safety-related programmable electronic systems, and to propose a framework for meeting these requirements. This will include codes and procedures for assessment bodies, assessment criteria and procedures, and assessor skills.

Safety-critical systems may be defined as systems which provide the final assurance for the safety and health of people, and protection of the environment. This is a special case of safety-related systems, whose function or non-function may affect safety, health or the environment. In practice, FRESCO has widened its role to cover safety-related as well as safety-critical systems.

3.1 Organisation

The project organisation is shown in diagram 1.

Eutech was formed in 1993 out of ICI's internal engineering functions, to provide high quality engineering solutions to both ICI and external customers. It brings to the project extensive experience in applying safety assessment techniques to manufacturing plant, complying with the HSE PES guidelines, and quality and validation programmes for process computing installations. In addition, its wide network of contacts gives it an industry-wide perspective. Eutech chairs the FRESCO Steering Committee and the Interest Group.

Admiral has extensive experience in both the safety assessment and security evaluation fields. It has been instrumental in developing standards for security evaluation practices, working with UK and European standards bodies. It has carried out assessments of safety-related PESs in the nuclear, railway and avionics industries. Admiral holds the project management and technical chairs.

Lloyd's Register has a world-wide reputation for its work in certifying shipping and maritime systems against national and international standards. More recently, it has expanded into the assessment of industrial plant, and the assessment of

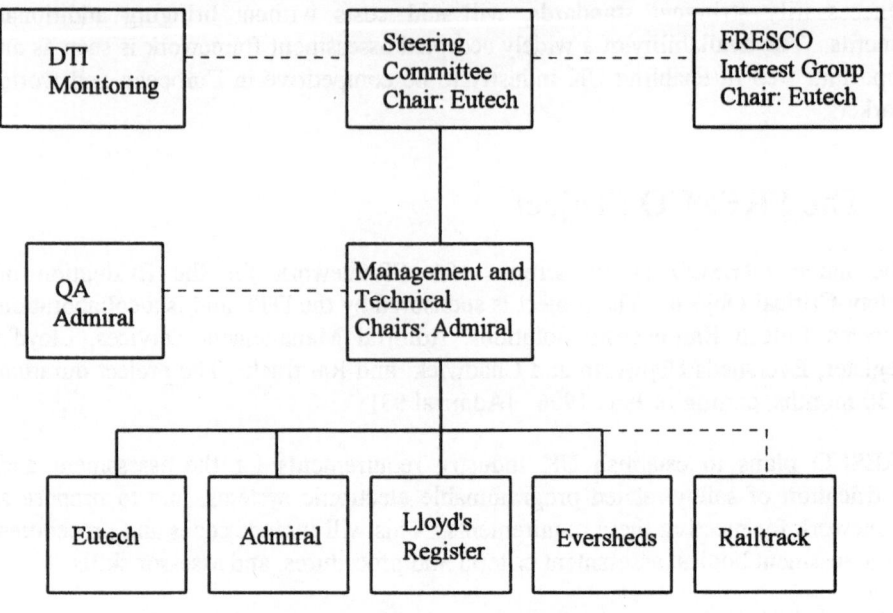

Diagram 1 - Project Organisation

software for marine, industrial and automotive use. It has also developed a set of guidelines for the development of dependable systems, and an assessment methodology for software based systems. Lloyd's Register is providing technical input to the project.

Eversheds, Hepworth and Chadwick is a leading firm of solicitors with a reputation for its work in the area of product liability and safety legislation, and its understanding of the legal aspects of safety-related systems and software.

Railtrack has a long history of operating assessment schemes for safety-critical systems, and in recent years has done much work on the methods required to establish the dependability of software based safety systems. Railtrack's role is as independent reviewer to the project; it does not receive DTI funding for this.

3.2 Work Packages

The main technical activities of the project have been divided into ten work packages, supported by three management work packages.

Work Package 1 (WP1) is project start-up, whose activities are carried out primarily by Admiral and Lloyds Register. Deliverables are management, quality and technical plans, and a report on current practices, which is known as the safety assessment baseline.

Work Package 2 (WP2) identifies the parallels between safety assessments and security evaluations. This is done mainly by Admiral, and delivers a report following interviews with practitioners in both the safety and security fields.

Work Package 3a (WP3a) determines the process control requirements for assessment and certification, and Work Package 3b (WP3b) studies generic requirements for assessment and certification.

This work is concerned with the overall assessment and certification process, and the objectives to be achieved by its application. The emphasis is on those organisations which would benefit from such a scheme, rather than those who would provide it. This work is led by Lloyd's Register, with input from other project partners, in particular Eutech and Admiral. The deliverable is a report describing the marketplace for an assessment and certification scheme, and the preferred scope for such a scheme.

Work Package 4a (WP4a) develops a process control organisational framework, and Work Package 4b (WP4b) studies the feasibility of developing a generic organisational framework.

Given the establishment of a marketplace for a safety assessment and certification scheme, this work package addresses possible approaches to the overall operation of the scheme. It is performed primarily by Admiral, with input from Lloyd's Register and other partners. The deliverable is a report proposing an organisational structure for an assessment and certification scheme, and mechanisms for its operation.

Work Package 5 (WP5) develops a technical approach to assessment.

This work package addresses the technical aspects of performing the assessment, as opposed to the organisational framework within which it will operate. It is carried out by Admiral in conjunction with all the partners. The deliverable is a report describing the findings.

Work Package 6 (WP6) investigates legal aspects.

This is carried out by Eversheds Hepworth and Chadwick. The deliverables are reports on existing law on liability for defective safety-critical systems, commercial implications of this law, analysis of the legal effects of proposed frameworks, and final recommendations.

Work Package 7 (WP7) liaises with other bodies, to provide data for other work packages, to keep the project focused on industry needs, and to inform interested parties such as professional bodies, regulators and insurers. One or more formal presentations will be given at the end of the project to disseminate project findings and proposals.

Work Package 9 (WP9) prepares the final proposals. This comprises preparing and providing presentations on the project's results, both at a point one year into the project, and at the end of the project, and issuing a final report. The work is led by Admiral, with contributions from Eutech, Eversheds and Lloyd's Register.

Work Package 11 (WP11) is the establishment of a process sector interest group, further described below. It is chaired by Eutech. The group is producing an end of year report, and an end of project report.

Work Package 8 (WP8) is project management, and Work Package 10 (WP10) is quality assurance. Both are provided by Admiral.

3.3 Project Timescale

The project timescale is shown in diagram 2.

The principal deliverables within the first project-year are the baseline report (WP1), the parallels report (WP2), the report on relevant existing law (WP6), the requirements report (WP3a), the framework report (WP4a), the interest group end of year report (WP11), and a presentation of interim results at the end of the project-year (WP9). At the time of writing, only the first three have been completed, in line with project plans. The results of the others will be presented at the conference.

4 Industry Collaboration

A FRESCO interest group (WP11) has been established under the chairmanship of Eutech, and has proven successful in eliciting industry participation in the project. The companies involved are BASF, Boots, BP, Courtaulds, DuPont, Fisons, Hickson, Sandoz and Wellcome. It can be seen that this represents a wide coverage of process industry, with emphasis on chemicals and pharmaceuticals.

The group is producing a report at the end of the first project-year, commenting on whether the requirements captured meet industry's needs and whether the proposed framework for safety-critical software assessment and certification is practicable and appropriate. It also comments on the possibility of project success. There will be another report at the end of the project. These reports are being produced independently of the project management and steering committees, to provide an external assessment of the project.

5 Results to Date

The technical baseline (WP1) has been produced. This presents the current situation in a number of industries, including automobile, marine, medical,

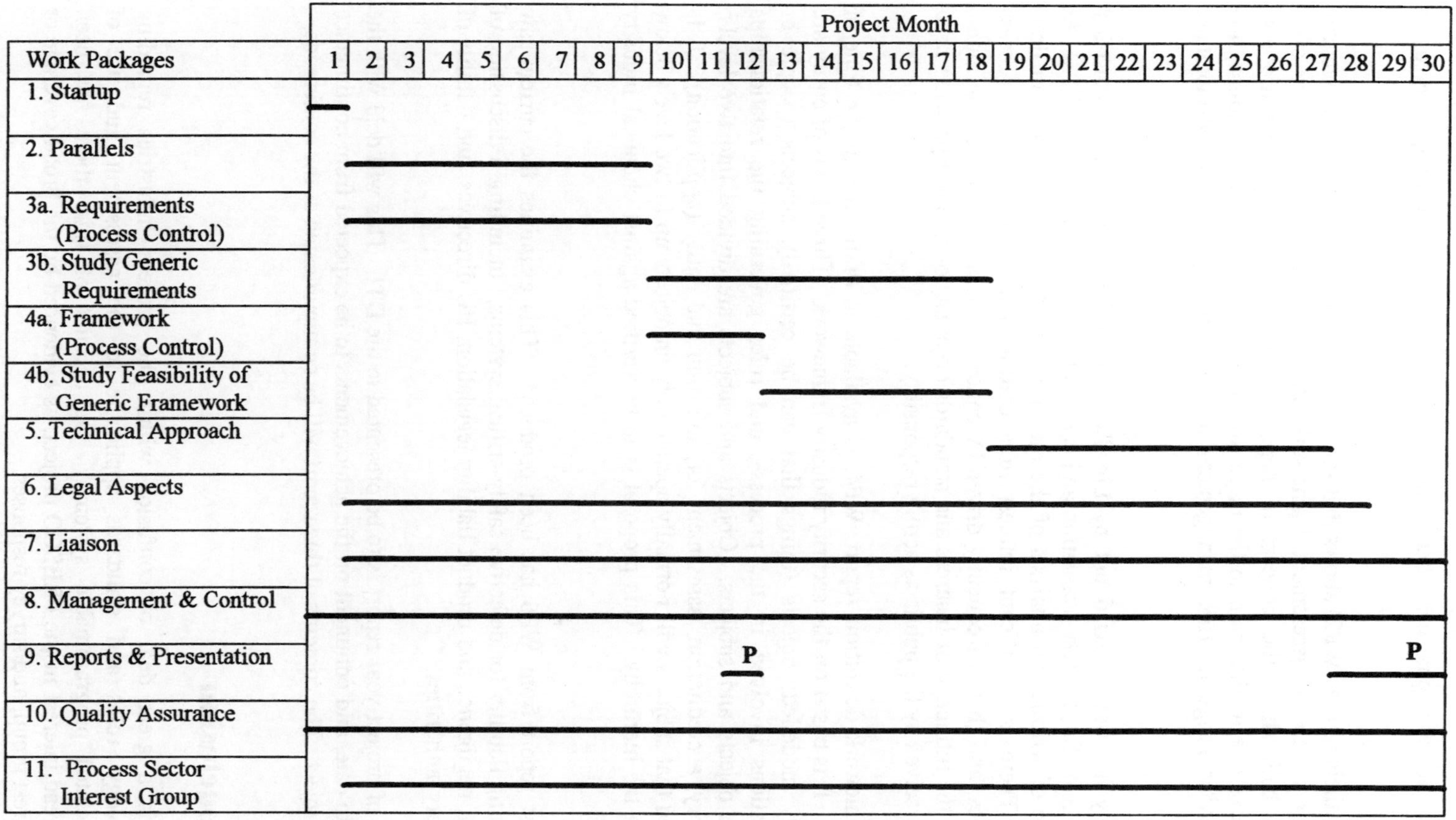

Project Month

Work Packages	1	2	3	4	5	6	7	8	9	10	11	12	13	14	15	16	17	18	19	20	21	22	23	24	25	26	27	28	29	30
1. Startup																														
2. Parallels																														
3a. Requirements (Process Control)																														
3b. Study Generic Requirements																														
4a. Framework (Process Control)																														
4b. Study Feasibility of Generic Framework																														
5. Technical Approach																														
6. Legal Aspects																														
7. Liaison																														
8. Management & Control																														
9. Reports & Presentation												P															P			
10. Quality Assurance																														
11. Process Sector Interest Group																														

P=Presentation to DTI

Diagram 2 - Project Timescale

77

offshore, nuclear, pharmaceutical, chemical and railway, in varying degrees of detail.

The parallels report (WP2) shows that many parallels do exist between the security area, for which an assessment framework already exists, and the safety area, because both share the concept of dependability. However, there are also differences due in the main to the more diverse nature of safety applications, and the broader range of risk management activities throughout the development process.

A survey has been carried out by Lloyd's Register into the process industries' requirements for a safety evaluation and certification scheme, as part of WP3. At the time of writing, an analysis of the questionnaire replies is available in draft form. There is significant interest in a certification scheme for the process industries, and this is frequently driven by concerns wider than legislation alone, such as the influence of insurers and standards organisations, and a growing need to be perceived by the public as acting responsibly.

The framework definition report, WP4, is available in draft form at the time of writing. It is based on the security industry framework. Three kinds of entity are currently modelled; objects (things that can be certified), subjects (people or organisations involved in the process), and rules governing the relationships between objects and subjects. Objects and subjects are divided into five levels; regulatory, accreditation, assessment, organisation and entity (application). It is proposed that subjects will normally operate only on objects up to two levels below them in the hierarchy. This proposal is to be verified against chemical industry practice.

The first report from WP6 has been produced. This examines the current law concerning liability for defective safety-critical systems, including a discussion of contract, negligence and product liability legislation, EU directives, and liability of regulatory authorities.

An end of project-year report is to be presented to the DTI. This will deal with the results to date, and comment on the achievements to be expected from continuance of the project. The contents of this report will be communicated at the conference.

6 Conclusion

There is a great deal of confusion within the process industries regarding terminology, roles and standards applicable to the assessment and use of safety-related programmable systems. On account of this, there has been widespread interest in the FRESCO project, as evidenced by the broad coverage of the interest group, and survey responses.

An assessment structure would be seen as a valuable deliverable, but there remains uncertainty whether this should be generic in nature, or industry specific. Whatever the outcome, the proposals will be verified against the chemical industry before they are finalised.

The FRESCO project is succeeding in building up UK expertise in cost effective assessment structures. It is building on parallels with security, is raising awareness in industry of the potentially beneficial role of assessment, and is publicising its existence and activities through means such as this paper.

At the time of writing, the project has another 20 months to run. During that time, the deliverables of the remaining work packages will be completed, culminating it is anticipated in final recommendations for an assessment and certification framework, in June 1996.

7 References

[Admiral 93]   Framework for the Evaluation of Safety Critical Objects (FRESCO), Admiral Management Services Ltd, Document Ref 9280/1, June 1993.

[HSE 87]   Programmable Electronic Systems in Safety Related Applications (Parts 1 and 2), Health and Safety Executive, 1987.

[Wingate 94]   Wingate, G.A.S. & Preece, C., "The Importance of Dependability", Mechatronics: The Basis for New Industrial Development, (Ed. Acar, Makra & Penney), Computational Mechanics Publications, 1994.

Independent Safety Assessment of Rail Systems in their Operational Environment

Morris Chudleigh
Cambridge Consultants Limited
Cambridge, England

James Catmur
Arthur D. Little Limited
Cambridge, England

Abstract

There is an increasing use of Programmable Electronic Systems (PES) in safety related applications: some of these, including many rail systems, are large and complex. Emerging standards, and some of the regulatory authorities, require that the safety of such systems is checked independently: this is the role of the Independent Safety Assessor (ISA) which we are carrying out on behalf of two railways in Hong Kong. We believe it important that the ISA should examine the technical changes, human operation changes and technical/human changes required as a consequence of the use of the new system in the operating environment of the railway because these changes may impact safety. Our work as ISA covers audits of the development process, auditing of plans and of the adaptation of the system to the railway and assisting with preparation of the safety case. Our auditing approach is an open, exploring one using experienced staff which we believe is far more effective than a closed approach with extensive use of checklists. This conclusion is confirmed both by our clients and by those companies we have audited.

1 Introduction

There is an increasing use of Programmable Electronic Systems (PES) in safety related applications and ensuring that such systems are conceived, designed and produced with appropriate attention to safety is not easy. Some of these systems are large and/or complex including many of the railway systems in the world.

Some rail systems (for example those in the U.K. and France) require some form of regulatory approval before they are put into service, so the operators need to assure themselves (and the regulatory authorities) that the safety needs have been met.

Standards are emerging for the development of PES based systems for safety-critical applications [IEC 91 and MOD 91] and they require that the safety of the system is checked independently. This is the role of the Independent Safety Assessor (ISA).

Rail signalling systems normally involve a number of sub-systems, often supplied by different companies. The sub-systems are often existing products, perhaps with some modification (we are not aware of any two installations on different railways that are the same). This can lead to a problem because in assuring the safety of the total system, it may not be sufficient to rely on earlier checks of the original sub-systems. A design that might be safe in one environment may be unsafe in another.

We believe that the work of an ISA (and certainly for rail systems) should cover the operational system as well as the basic design of the system.

The remainder of this paper will give some of the reasons why we believe this to be true and describe our approach to such ISA work. The paper is based on our experience working as ISA for two different railways in Hong Kong (both of which are subject to regulatory approval very similar to that in the U.K.). Clearly, however, we will not be able to disclose any results of our work to date as that remains confidential.

2 Why the ISA must examine the environment

A PES system may cause a hazard either through its own technical failure, or through human failure in using the system correctly. We believe it essential that any safety assurance work for an operational railway system should consider the changes in operational environment as well as the changes in the basic design of the PES system for the following technical and human reliability reasons.

Technical Reasons:

- The interfaces to other systems/sub-systems change
- The operational requirement will change from implementation to implementation
- The interactions with other equipments (generation of harmonics etc) change
- The environmental conditions of temperature, EMC etc change

Human Reasons

- There may need to be changes to the railway existing rules and procedures
- The operational philosophy is different from railway to railway
- The fall-back requirements in the event of failure change from railway to railway

Technical/Human Reasons

- The data requirements change
- The installation and change-over process may introduce new hazards

Each of the above issues will now be addressed.

2.1 Technical Reasons

2.1.1 The interfaces to other systems change

The equipment the system interfaces to will change in each application. This will have impacts on the actual design of the interface as well as on the safety. In some cases new interfaces will be required which must be safety critical, or the criticality of existing interfaces will change. Examples of these interfaces include such issues as:

- The requirements for brake functions, like automatic train protection (ATP) brake control, can vary substantially from railway to railway.
- The interlocking systems will have differing physical as well as safety interfaces.
- Simple issues such as the need to have an unmotored, unbraked axle may cause problems.
- In some cases of existing rolling stock basic features such as the number of spare train lines may affect the new application.

2.1.2 Operational requirements will change

The operational requirements that change will have an impact on the safety of the system and need to be taken into account when assessing its use. Typical items that change include:

- The number of trains per hour
- The headway between trains
- The run time, or journey time
- All trains running on the railway may not be equipped with ATP
- The conventions for temporary speed restrictions vary

All of these have some bearing on the way the safety of the system is reviewed and evaluated.

2.1.3 Interactions with other equipment

There will be much equipment on the railway (or adjacent to the railway) that the system being procured does not need to interface with but with which it must be compatible. Thus, possible effects of harmonics and interference need to be checked, both the effect of those generated by the system and those generated by the adjacent equipment.

2.1.4 The environment changes

Environment issues such as temperature, temperature range, humidity, and EMC are different on almost every railway.

2.2 Human Reasons

2.2.1 The existing rules and procedures may change

When the new system is fitted rules and procedures will need to change to cover operation and maintenance as well as working after failure. In some cases these changes can have major safety implications, especially if significant re-training of staff in new working practices is required.

2.2.2 Operational philosophy changes

Although there are broad similarities in the way that railway signalling is carried out around the world, there are many differences from country to country. Thus, the signalling principles used on French railways are different from those used on British railways and from those used on German or American railways. The systems we are working with need to operate in Hong Kong, where the signalling principles are similar to the British but with many detailed differences. One of the ATP systems has been developed in France, the other in Belgium. It is important to check that the system:
* can interface safely with other equipment
* staff can understand it
* it can be maintained
* it does not invalidate other fundamental safety principles

2.2.3 Fall-back requirements change

When the system fails to a 'safe state' the railway must continue to operate. This may be achieved through a back-up system or the rules and procedures. In Hong Kong an operational break of 10 minutes would be considered to have severe safety implications (one of the reasons being the sheer number of people crowding onto the platforms). Again, the impact of how a system fails, what back-up exists and how it is used are critical in defining the overall safety.

2.3 Technical/Human Reasons

2.3.1 The data requirements change

These PES based signalling systems are generally data driven systems

which need to react to the different line layouts, gradients etc:
- there is a need to prepare the data, input it in a system compatible form and check that it is correct
- the required data may not exist currently, or may be inaccurate
- the procedures for updating of data are critical and should be taken into account

2.3.2 Installation and change-over

Safety during installation and change-over is important and must be carefully checked. If safe practices cannot be generated then the system cannot be installed.
- the system must work in parallel to the existing system and be compatible with it throughout the installation process
- the process of switching between the old and the new system can be critical and should be checked in detail

3 How we carry out the role of ISA

In fulfilling the role of ISA we find that we have to carry out three different but linked activities:
- Assess that the development is carried out to an adequate level to ensure safety - this is the normal role of an ISA as envisaged in the emerging standards [MOD 91 and IEC 91].
- Review and assess the plans for all phases of the work from development to final operation of the railway. The plans are not just those from the system supplier: the plans of the railway project team and those of the operating railway must also be assessed. This concentrates on the adaptation of the system to the railway.
- Assist the operator in preparation of safety case documents to present to the regulatory authorities.

Each of the above areas is described in more detail below.

3.1 Audits of the development process

Normally, the system to be supplied will not have been developed specifically to meet the needs of the railway, but is an existing system which will be modified to meet those needs. In order to ensure that the Contractor delivers a system which is safe we carry out what can be called classical ISA audits of their development process. These audits are carried out by a team of two or three and aim to review for all the safety-critical elements of the system:
- How any changes to the basic hardware and software are carried out. This will include the design, coding, testing, verification, validation

and manufacture of the changes.
- What techniques are being used and how they compare to the techniques discussed during any pre-tender audit and with emerging standards. This will include evaluation of the adequacy of the techniques being used.
- Compliance of the contractor with their proposed techniques.
- Documentation of the results of their work, in order to ensure a full record is maintained.
- Compliance of the manufacture with the safety requirements of the railway.

We develop protocols for each audit before carrying it out and base the approach on sampling the technical detail of the most critical areas. This does not guarantee that all problems will be found, but we believe that the dedication and time allowed is sufficient to gain confidence in the contractor's work.

At the end of each audit an Audit Report (AR) describes the work carried out, the understanding gained and any findings. The ARs concentrate on the findings noted and the method for resolving each finding. Our current view is that audits in four sections are needed:
- Audits once software and hardware design work is well under way.
- Audits when the code has been written/hardware developed and module testing is complete.
- Audits once final development is complete. These will review the modified ATP and any other supplied systems and begin to review the manufacturing processes.
- Audits of the manufacturing process.

A final report describes the work carried by the ISA, the results of each audit and how findings were resolved, and the overall conclusions on the design, development, testing and manufacturing processes. This report is one of the critical inputs to the final pre-operation safety reports.

In our experience, the ISA work is not mechanical auditing but requires experienced, intelligent auditors who explore the operational environment. Merely auditing against checklists will not, in our opinion, give a full assessment of the safety of the system. This is key to our approach and is described in more detail in Section 4 below.

3.2 Auditing of plans and adaptation

During the process of designing the new system for the railway and installing it and commissioning it the contractor will:
- Submit a series of design submissions and other documents to the railway
- Prepare plans and procedures for installation, testing, commissioning, training and start-up of the new system
- Install the new equipment

- Test the system
- Train railway staff
- Commission the system

During this work both the railway project team and the operating railway are involved in reviewing the submissions from the contractor and in reviewing and witnessing the work carried out.

In our opinion, the ISA's role is to review the actual documentation involved in this process in order to provide confidence that sufficient measures are taken to protect the railway during installation, testing, commissioning and operation, and to audit the actual railway review and contractor installation and commissioning processes as they occur. This work can be broken down into four principle areas:

- Attending the contractor's initial and detailed safety analyses for the line on which the system is to be installed. The ISA should be present during the studies but, in order to maintain independence, we only review and comment on the results and report our conclusions to the railway, after discussing them with the contractor.
- Reviewing design submissions and plans. After each review we identify any areas of concern and report to the railway on the results of the reviews.
- Audits of the review, installation, testing and commissioning activities of both the contractor and the railway. These audits involve the ISA in reviewing the work being undertaken and at the end of each audit producing an AR with the results of the audit.
- Carrying out informal hazard analyses of all the issues to do with the above issues relevant to adaptation of the system to the railway. This area is described in more detail in Section 5 below.

The ARs throughout the life of the project are summarised in a final report which reviews all the audits carried out, the results of the audits, and how the findings were resolved. This final report is a critical document in completing the overall safety report.

3.3 Assisting with the safety case

It is usual for the operating railway to consult with the regulatory authority when the project is in the earliest stages as to what safety requirements would be imposed on their project. A common response may be that the new system (containing PES) should not result in a lower level of safety than the existing system. In other cases more stringent requirements are placed on the system. To demonstrate that an adequate level of safety has been achieved, and in order to receive permission from the regulatory authority to operate the new system, the railway will almost certainly be required to submit a detailed report outlining how safety standards will be maintained.

It is our practice to submit a number of Project Safety Reports (PSR).

For our work in Hong Kong we supply three reports: an Installation PSR which demonstrates how installation and testing is carried out safely; a Commissioning PSR which shows how the system can be commissioned safely and an Operational PSR which provides the demonstration of safety levels for the final, fully commissioned system. In all the PSRs the documents produced will be limited to the equipment and management systems which have changed as part of the Project.

4 Approach to development and manufacturing audits

There are two contrasting styles of auditing in common use in the rail industry, one based on a closed, checklist approach and the other based on an open, exploring approach. The open approach is much used in France and the closed approach is much used in the U.K. We believe that a combination of the open approach with the checklists used only as a self-audit of the auditor is the most effective way to assess systems. A brief description of our approach is given in [Chud 92]. Our experience, confirmed both by our clients and by those we have audited, is that the open approach gives a far more intensive and searching audit of a system.

The characteristics of the two approaches are now outlined, giving what we feel are the advantages and disadvantages of each.

4.1 The Open Approach

4.1.1 Advantages of the Open Approach

The open approach uses experienced engineers who work with an open, short protocol to guide extensive interviews which explore the way work was done and gives the freedom to investigate issues as they see fit. Checklists are used but only as a self-auditing tool by the auditors. The investigations thus go well beyond the checklist, while ensuring that all the normal checklist issues are covered. This approach thus gives a better guarantee that the important issues affecting safety actually get addressed.

The freedom to explore allows interesting interactions to be investigated, allows the auditors to follow instinct and to probe in detail. When done well the approach is risk based and there is a homing in on detail concerning the most critical parts of the system.

4.1.2 Disadvantages of the Open Approach

The open approach, while having the above advantages, does have a drawback.

The approach requires experienced staff who fully understand the problems, these staff tend to be more costly than those who are used to audit against a checklist.

4.2 The Checklist Approach

4.1.1 Advantage of the Checklist Approach

The advantage of the checklist approach matches the disadvantage of the open approach. Thus, audits tend to be carried out by auditors with less experience and so work out cheaper for an audit of the same number of man days.

4.2.2 Disadvantages of the Checklist Approach

A disadvantage of the checklist approach is that it only covers the material given in the checklist and so tends not to probe in detail.

The approach concentrates on checking compliance with standards or specifications included in the checklist but does not usually check the depth of the compliance. We often find issues where there is a superficial compliance with best practice but an in depth exploration finds that the reality of the detail is rather different. Similarly, we find situations where there may not be strict compliance but the route taken by the supplier gives an equivalent level of rigour.

An area that is difficult to address with checklists is the homing in on the major areas of risk. If the checklist is extremely well designed it may be possible to build in this focus on areas of risk but this is normally not the case.

5 Adaptation audits

Adaptation audits are used to see how the new system interacts with the proposed operational environments. For each of the points given in Section 2 there is a need to review systematically the impact of the new operational environment. The audit aims to identify safety problems as well as reliability/maintainability as these latter issues will affect the human interaction with the new system and so it is vital to identify them as early as possible.

We use a special study, done first prior to the contract being let and checked later. It is a study of the whole system, the new system together with all interfaces, internal data and the study addresses compatibility with existing systems.

We use a modified Hazard and Operability (HAZOP) methodology: a combination of HAZOP and brainstorming. HAZOP analysis [CIA 87, Chud 93] is based on systematic analysis of the deviations from normal design or operating intent at the interfaces between systems, sub-systems or equipment items. The initial study uses high level system block diagrams and also uses a general equipment check-list to ensure that the main features of the existing and final system configurations are covered. Later studies check the areas of concern found in earlier studies and check that no new concerns have arisen.

The team comprises the ISA as leader, railway personnel directly involved in procuring the new system as well as railway operators and maintainers and experts from the contractor organisation.

The team searches for concerns of the safety, service and reliability of the new system in its operational environment. It is normally assumed that existing trains and any signalling equipment that is being maintained are safe in their own right. The study focuses on the potential hazards and operational/maintenance/reliability problems that arise from changes to both the existing equipment and signalling system and the new system both during the change-over period and the final configuration. A need for major changes to rules and procedures is likely to have a significant impact on overall safety and must be identified early.

One approach used is to consider particular scenarios envisaged during change-over such as modification of track circuits, mixed fleets of upgraded and existing rolling stock etc.

The concerns are ranked to help determine the urgency with which they need to be addressed.

6 References

[MOD 91]   Ministry of Defence: "Interim Defence Standard 00-56/Issue 1 - Hazard Analysis and Safety Classification of the Computer and Programmable Electronic System Elements of Defence Equipment." MOD, April 1991.

[IEC 91]   Functional Safety of Electrical /Electronic /Programmable Systems. Generic Aspects. IEC 65A (Secretariat) 123. 1991

[CIA 87]   A Guide to Hazard and Operability Studies. Chemical Industries Association Limited, 1987.

[Chud 93]   Chudleigh M: "Hazard Analysis using HAZOP: A Case Study" In: Górski (ed) Proceedings of the 12th International Conference on Computer Safety, reliability and Security, October 1993 (Safecomp '93) pp 99-108.

[Chud 92]   Chudleigh M, Catmur J: "Safety Assessment of Computer Systems using HAZOP and Audit Techniques" In: Frey (ed) Safety of Computer Control Systems 1992 (Safecomp '92) pp 285-292.

Enhancing Safety Assurance Using Security Concepts

John Elliott, Andy Lovering
SRC (Consultants) Limited

Chris Gerrard
Gerrard Software Limited

Abstract

Despite previous evidence of similarities, the safety and security related IT application domains have developed separate cultures and standards in their approach to achieving an acceptable assurance that their high integrity requirements are satisfied. This paper presents a review of safety engineering prior to a perspective on how advances in computer security concepts and techniques may assist in enhancing the current best practice approach to safety assurance. These early findings are from current challenging research involving a comparison between the two domains of safety and security in developing and applying the concept of Safety Policy and deriving suitable Safety Models. The ultimate goal is to develop a Safety Policy Method for application to all types of safety sectors to be utilised throughout any safety systems life cycle.

1 Introduction

1.1 Context

This paper seeks to outline some aspects of the UK DTI/ESPRC-sponsored Safety Policy and Models (SPAM) project. The paper is concentrating on presenting the current concepts and direction for how system safety assurance can be advanced using concepts and techniques from the computer security domain, one of the project's major themes that makes this research both challenging and unique.

Whilst the engineering discipline of computing is still in its infancy, all sectors (industry, commerce, military, government) are applying computer-based systems in an ever-increasing range of high integrity applications, in fact faster than the creation of new techniques to meet increasing high integrity demands. In particular, computer controlled safety and security related systems are now in widespread use, with their user communities demanding ever more varied and complex computer-based applications. In these, as in other areas, society has become very dependent on technologies that are not yet sufficiently dependable. With this increasing demand for high assurance systems, it is clear that the safety engineering techniques used for software-controlled systems must be founded on better approaches, where the safety integrity requirements are of the highest priority.

1.2   SPAM Project Overview

Within this context, the aim of the SPAM project is to contribute to meeting the need to improve safety assurance practices for computer-based safety systems in the following key areas:

- Eliciting, defining, modelling and validating safety requirements through the development of a System Safety Policy framework.

- Utilising policy-based methods within safety management activities to provide verification support to demonstrate safety assurance throughout the systems life cycle.

Having accepted the common view that safety is primarily a system issue, initial system studies are directed at exploring the state of the art and associated concepts within the safety and security domains. The aim is to fully understand the fundamentals of both domains, and to understand the relative strengths and weakness of their approaches, and then apply useful concepts and techniques from the world of security to enhance safety practices. Over the last decade, various analogies between the safety and security domains have been postulated [CSR 86, Brewer 93] and experience has emphasised that there are likely to be significant benefits in bringing the two communities together. The SPAM project is a serious attempt at exploring these parallels initially for the benefit of the safety community.

Following the system studies, this project will develop methods that build upon current safety system practices to be subsequently refined and applied to the specific technical areas such as software to improve its contribution to the system's safety integrity. An additional aim is to provide support in the safety management area by developing a more objective approach to demonstrating safety and to defining safety integrity for systems and their software components. Also, the investigations so far have emphasised the fact that many safety systems also have attendant security requirements, so a subsidiary aim is to provide some groundwork to define and adopt a common approach for such dual safety and security systems.

It needs to be stressed that IT security concepts and practices may not be the only source for enhancing safety techniques. Much has been done and learnt over the past two decades or so as to how to go about building sound, assured safety related computer systems. Hence, the approach is to develop methods for the achievement of safety, based upon these mature foundations allowing for the increasing demands and complexities of future innovative and beneficial computer based safety related systems.

This paper, therefore, presents some of the early findings from the SPAM project. Section 2 provides a review of the state of art for safety engineering drawing on fundamentals and on identifying and evaluating the current best safety practice. After setting the scene for where safety is today, the overall aim is to strengthen any identified weakness using other approaches, including those used in computer security. Hence, Section 3 presents an equivalent high level review of security system practices with the aim of highlighting the major concepts and techniques that seem to be worth exploring in order to strengthen any current weaknesses in the safety domain. Section 4 discusses the potential for harnessing security techniques, combining them with the existing strengths in current safety techniques, to enhance the system safety life cycle. Section 5 concludes with a discussion on the direction for this dual safety and security research.

2 Safety Systems Engineering - Current Issues

2.1 General

This section discusses the safety system context and the issues involved when understanding, defining and implementing safety. After these reflections, a general examination of the state of the art in safety systems development is presented. This state of the art review will identify best practice and key safety concepts with the emphasis on the strengths and weaknesses in the current approaches involved within a safety life cycle. This review will be subject to continuing appraisal based on different perspectives across the safety system community. There now follows a general discussion on safety expanding on the topics mentioned above[1].

2.2 Socio-Technical Context for System Safety

Before developing approaches to improving system safety, it is important to recognise wider socio-technical issues. Safety systems are multi-layered and overall system safety cannot be divorced from the system's

---

[1] Note, the reader may wish to move on to Section 2.7, a summary of current safety practice, if they feel the contents of the next few sections are familiar.

local environment including organisational and physical attributes (plus human and non-IT engineering components) all of which interact with the IT-system. This does not mean that all safety concerns are being monitored, or controlled by IT systems as some safety enforcing mechanisms may not be feasibly implementable by conventional engineering, or by an IT system, and so in these cases we must rely upon human intervention.

The "socio" element of safety concerns the influence that humans and their different organisations place on the perception of what technical systems are beneficial for, whilst providing an acceptable level of safety. Fundamentally, humans belong to system procurer and supplier organisations as well being system users. All of their safety objectives need to be taken into account when creating a "technical" system such as a computer based safety related system.

Human organisations, e.g. a safety system supplier, show some commitment to safety as is evident in their statements about safety objectives, policies and procedures (where they exist). Furthermore, within any such supplier organisation, project teams will need to define their equivalent safety statements. Derived from their customers and system users, will be a further set of safety statements. The main point is that there may be at least three safety statements involved when developing a safety system and it is of great importance that all such safety statements are accounted for, as well as being consistent, explicit and demonstrable.

2.3   What is System Safety?

Within a socio-technical context, understanding the fundamentals of system safety is extremely important when attempting to build on current established practice. Whilst the intuitive understanding of system safety is associated with causing harm to its environment, the choice of definitions which best characterises the meaning of safety are often tailored to suit a context or interest. This is particularly apparent in safety standards. Other contributions to the more philosophical debate can be helpful in understanding what safety really means [Dobson 92, Leveson 86]. However, there is still no universal definition of safety even though all those associated with safety systems would benefit from starting with a common template of understanding.

Alongside understanding system safety is the need to establish the dependability relationships between safety and the many other system attributes such as reliability, availability, security, functionality, human factors etc. Achieving the desired level of safety often requires resolving conflicts with regard to other system attributes. Overall, this safety resolution is a question of cost benefit accounting for attendant risks and, as

93

with other attributes, partly depends on the use of effective assurance techniques.

Fundamentally, achieving safety is uncertain. Typically, safety requirements concern the avoidance of causing harm within an acceptable and tolerable level. This is often expressed in terms such as "the risk of an accident needs to be less than 10^{-9}". This uncertainty is primarily the result of a lack of human understanding of, and control over, safety. For example, understanding system safety requires determining all system-environment interactions through knowing all conditions in the environment. If engineers could always predict the behaviour of their systems in response to all known stimuli then safety risks would be reduced. However, this is not possible when the human is included as a part of a system safety problem.

These uncertainties about safety give rise to a number of important properties which distinguish it from other system attributes, namely:

- Safety has no absolutes; nothing is 'safe'. Thus safety is relative and non-deterministic as assuring safety particularly depends on the completeness of human understanding and associated assumptions concerning the system environment.

- Safety cannot be measured directly. Quantification of incident rates, i.e. probabilities of events which can cause accidents, need to be inferred statistically from practical observations of trials or in-service data. As a countermeasure, qualitative approaches are used to classify systems using notions of risk and integrity.

The combination of these two safety properties identify key difficulties: safety is relative and uncertain, and it cannot be measured directly. This provides project managers with one of the most demanding technical risks when demonstrating compliance with any safety requirement. In particular, showing that explicit deterministic safety objectives have been met and this involves ensuring the effective management of the uncertainties. The latter has attracted most of the system safety world's attention with techniques focusing on hazard and risk notions. Meeting safety objectives is an area where the engineering community appears to have paid less attention to, and this is where the SPAM research has been largely targeted.

The emphasis on the uncertainty problem has led to a generic culture amongst safety engineers, worrying about what may go wrong rather to initially define what explicit safety objectives need to be met. Both cultures are necessary in that whilst aiming to achieve a defined objective, engineers must still question and demonstrate the achievement of safety through the use of analysis and testing.

2.4   Defining and Implementing Safety

As indicated, there have been many attempts at defining safety of which two interesting examples are :

"*Safety* is the freedom from unacceptable risk of harm."   IEC Draft Standards [IEC 92].

"*Safety* is a measure of the degree of freedom from risk or conditions that can cause death, physical harm, or equipment/property damage." [Leveson 86]

These words introduce the concepts of harm and risk. In addition, there are the related ideas of safety levels and safety integrity that have been defined by the draft IEC standard [IEC 92]:

 "The *level of safety* is how far safety is to be pursued in a given context assessed by reference to an acceptable risk, based on the current values to society. "

"*Safety integrity* is the probability of a safety related system satisfactorily performing the required safety functions under all the stated conditions within a stated period of time (related to the performance of the safety related system in carrying out the safety functions[2])."

Safety levels are derived from a perceived need whereas the use of safety integrity is more linked towards the implementation of satisfying the achievement of the safety levels. The degree of tolerable risk that safety functions do not operate as intended leads to defining the level of safety integrity to be the basis of a target within implementation (hence the term *safety target*).

Assessing the safety risk requires an assessment of the likelihood that any cause (e.g. hazards, accident sequences) arising together with evaluating the severity associated with any consequence (e.g. harm caused by accident). Hence, safety integrity can be used to define and form a safety target, i.e. what measurable level of safety is to included in the implemented system.

---

[2] A safety function is one that is controlling a system that can harm, or avoid the impact of another system causing harm, through some protection or monitoring capacity.

Deriving safety integrity has two key elements in that account has to be taken of the relevant systematic and random nature of system failure. Random means that some inherent uncertain effect may arise, hardware failure for example. Systematic means that the effect is always present and is repeatable although it may not be evident, e.g. design fault. Hence, harm can be cause by two sources, a system failing due to wear out and built-in design faults. Therefore, the measurement of safety is often through the use of integrity levels, [IEC 92, MoD 91], that is a classification of a system, or associated component, in terms of the overall tolerable safety risk and whether the system can directly cause harm, or whether it can contribute to causing harm.

The purpose of this reflection on definitions is to emphasise key safety concepts and to progress the theme that socio-technical driven safety statements, i.e. strategy and objectives, need to be in explicit terms. The notions to identify are: potential harm, risk, safety levels, safety targets and safety integrity. The safety target notion is useful as it provides detailed implementable safety integrity objectives, applicable to each level of design.

2.5   Overview of Safety Life Cycle - Process and Concepts

This section provides an overview of agreed current practice within system safety development. One major area is the recognition that there exists a separate and generic safety life cycle that identifies a number of safety processes within a system development life cycle. Example safety and development life cycles can be found in safety standards [IEC 92, MoD 91].

In general, the safety life cycle comprises of the following phases:

- Safety Planning.

- Preliminary Hazard Identification.

- Safety Requirements.

- Hazard Analysis and Risk Classification.

- Safety Compliance Verification.

- Safety Case / Safety Certification.

The above phases reflect current traditional practices and procedures used to ensure that safety is achieved through a rigorous safety assurance process where the potential for a system to enter service with unacceptable safety characteristics is minimised.

The primary purpose of a safety assurance process is to maximise the visibility of equipment safety characteristics at all stages of a project life cycle. This is achieved through selected analytical safety assessments and verification activities, carefully integrated with the overall project programme in accordance with a Safety Programme Plan.

The results and record of the on-going analytical and verification activities provide increasing levels of confidence that the desired safety characteristics have been designed into the system. The combined results of a systematic approach to these activities provide a logical argument with supportive evidence, sometimes referred to as a safety case, for the deployment of the system into service.

It should be noted that a system procurer or user has an incomplete, and not fully defined, perception of any safety requirements in terms that may be demonstrable. A detailed safety requirements specification usually relies on the existence of the detailed system information; hence only high level expressions of safety are expressed in terms of undesirable happenings and the associated maximum acceptable risk of them arising, as discussed in Section 2.3.

However, in order to determine whether the levels of confidence have been achieved, safety requirements must be defined to determine what is acceptably safe for any system, i.e. what risk is the procurer prepared to take that the system is indeed unsafe.

Traditionally, safety requirements can be specified in various ways and usually involve determining the work programme activities (the chosen safety process) and the quantitative probabilities for undesirable or hazardous events, e.g. accidents resulting in death, serious injury etc. Such requirements do not specify the system's safe behaviour as other statements are needed to guide development. Often designer's define the safe behaviour using architectural or equivalent design techniques, involving redundancy, protection, diagnosis and monitoring elements.

2.6   Evaluation of Safety - Process and Concept Issues

2.6.1   General

Evaluating the strengths and weakness of the current safety life cycle approaches raises interesting issues. This section will make various observations identifying key areas where attention may be advantageous.

A general observation about safety systems development is that safety appears not to be planned as a detailed requirement to be met and fully demonstrated. The current, and largely successful, common safety culture is to rigorously manage safety, adding safety design features later in the life cycle. This is after analysing early designs in terms of what may go wrong and conducting extensive system validation tests.

The approach appears to apply management and analysis process at solving the safety problem, leaving designers to ensure what they create behaves safely. A contrast may be to find ways of improving the safety requirements process by being more explicit when demonstrating that the safety requirements have been met. Also, it is acknowledged that due to the relative and uncertain aspects of safety, the quantitative (probability of causing harm) and qualitative (levels of safety integrity and risk) aspects are necessary and based around reasonably well development concepts and techniques. However, it seems that the current approach to defining a safety life cycle process is too biased towards showing conformance with any quantitative and qualitative requirements. These observations cover the major life cycles areas:

- Safety Requirements.

- Safety Compliance.

- Safety Management.

- Safety Integrity Levels and Targets.

- Safety Standards, Life Cycles and Techniques.

- Other Safety Issues.

2.6.3   *Safety Requirements*

Achieving a validated safety requirements definition is a major element towards gaining any necessary safety assurance. The nature of safety requirements and the state of the art in how such requirements are defined are briefly highlighted as a pre-requisite to the current evaluation conclusions. Here are some salient features concerning safety requirements capture and definition:

- The relative nature of safety leads us not to try to define it in any absolute terms. However, there is a deterministic component to a safety requirement, namely a specification of safety behaviour or property, ultimately to be implemented by safety functions.

- Initial safety requirements are usually only expressed in quantitative and qualitative terms and not in behavioural aspects.

- Safety requirements appear to be deduced from the various levels of analysis of specification or design details to determine whether hazards arising from any system failures or equivalent system mis-behaviour may lead to a safety incident. This may lead to defining the safety integrity level requirement.

- Assuring the complete capture of a safety requirements is difficult.

- Generalising about safety requirements characteristics across all types of safety system is a difficult problem. There is no agreed form of categorising such requirements.

- Safety requirements are not defined in sufficiently explicit terms at the beginning of a project life cycle, even within a Safety Plan. Safety practitioners claim that safety cannot be defined fully at the beginning of a project owing to the nature of safety.

- Insufficient use is made of the safety assumption concept. These assumptions are usually external to the system and are statements that must be argued to be true to demonstrate that a correct implementation of a safety system possesses safe behaviour.

- There is no standard template for safety specifications. Hence safety requirements may be currently specified in a number of different ways, some of which are difficult to interpret.

- Safety requirements may be expressed in a form consistent with the level of detail available within the development stage being considered.

- Safety specifications are not always separated from that for other system attributes.

- There are no systematic or formal techniques specifically tailored for defining and validating safety requirements.

These observations highlight some of the aspects that justify the close scrutiny of how safety requirements are created and managed, most of which imply inherent weaknesses that are worth improving. This a major aspect which is being investigated and compared with the approach taken for managing security requirements.

2.6.4   Safety Compliance

Once the safety requirements are defined, the next major aspect is to verify that these requirements are being satisfied at all levels of design and development. Safety compliance encompasses the broader aspects of safety verification, taking in account the inherent uncertain nature of safety, in order to gain the necessary safety assurance.

The current approach to safety compliance involves two aspects: quantitative and qualitative compliance, and behavioural compliance. Here are some observations:

- Safety compliance involves monitoring against the less intangible quantitative and qualitative measures and is judged by the provision of a variety of process and management evidence.

- Safety compliance involves a rigorous verification that the system behaves safely, assuming that explicit safety function specifications exist. Such verification is achieved through the use of extensive system validation testing (for behavioural verification) and is generally not achieved or attempted using rigorous analysis techniques for validation and verification including proofs and testing.

This latter lack of rigor in behavioural verification may be enhanced using:

- *Objective measures,* requiring adherence to defined safe states given related (normal and abnormal operating) conditions. In fact, a safe state not only represents a behavioural measure, but also attempts at defining an absolute component of the safety concept.

- *Safety assumptions,* (statements to be true to provide additional confidence about safety) which reference the system's environment. This may actually reduce uncertainty by supporting the development of an absolute safety component.

- *Logical safety rules,* which must hold true for safety to be preserved. These rules may be a more explicit, narrower form of a safety case or argument. Establishing these rules contributes to the capture of safety requirements as early on as possible in the life cycle.

The use of objective measures, safe states, logical safety arguments or rules, and safety assumptions may all help develop a comprehensive safety case principle to apply throughout the safety life cycle.

2.6.5   Safety Management

Whilst the approach to safety requirements and compliance significantly contributes to assuring safety, the cornerstone of achieving safety is in the use of rigorous safety management practices. In fact, safety management practices are well established through the use of safety plans and subsequent rigorous safety control activities leading to certification activities. The major observation is that although safety management is perceived to be a strength, the introduction of any new methods will have to take account of and be integrated with existing safety management principles.

2.6.6   Integrity Levels and Safety Targets

The concept of safety integrity levels and safety targets are a necessary measurement tool but they can suffer from being too subjective or hard to demonstrate that such requirements have been satisfied.

Also, integrity levels have inherent problems:

• Assigning integrity levels can be subjective and inconsistent, as the application of the available rules may vary between different engineers.

• Integrity level achievement usually requires following defined process requirements only rather than using standard acceptable safety system architecture's.

• The required processes for any integrity level are not agreed, and is subjectively implemented. Often only highest safety critical levels explicitly define practices, such as using formal methods [MoD 87].

• Where defined, the required design architectures for any integrity level are not agreed for all levels nor do all integrity levels require specific architectures to be used.

• There is no standard scale of integrity levels.

Of course, complying with an integrity level requirement does not imply a safe system although extra assurance will be gained through measuring against such safety requirements. However, there should not be an over-reliance on achieving integrity levels. One of the key issues in this research is to consider these inherent weaknesses and attempt to propose improvements in the measurement of safety as a result of seeking improvements in the definition of system safety. The major aspect is to attempt to define and apply integrity levels on a more objective basis, enabling improvements in how integrity apportionment rules may be applied and validated.

Closely related to qualitative integrity levels are safety targets, whose current composition is limited to: quantitative and qualitative elements, and the recognition of the separate random and systematic elements of safety targets.

An interesting approach taken by the DRIVE [Jesty 93] project relates the uncontrollability features of a system's environment to its integrity requirements; this is an area to be examined further.

2.6.7   Safety Standards, Life Cycles and Techniques

Safety standards [IEC 92, MoD 91] are geared to supporting the identification of safety processes. However the emphasis is on traditional safety related system development whilst providing some guidance for assessment activities, irrespective of who conducts them or how such assessments are undertaken. Current standards embrace the full safety life cycles and current best practice, using: safety management, safety integrity and targets, safety compliance including verification, assessment and certification.

One of the many other strengths of safety practices is the existence of safety life cycles and associated generic safety processes that provide a strong basis for safety standardisation. Also many safety techniques used in the life cycle, such as Fault Tree Analysis, are well-understood, mature and in day-to-day practical usage.

Although existing safety techniques may be proven, they have been oriented towards systems of low complexity and a relatively narrow field of application. Furthermore, safety techniques originate from the field of reliability engineering and thus have to be adapted for use in safety studies.

Applying safety techniques, such as hazard identification and safety analysis, can be very subjective and are heavily dependent on those performing the analysis and their knowledge and understanding of the system and its environment. In addition, there are difficulties in assessing the risk associated with hazards early in the project life cycle when a quantitative approach is used for the probabilities of hazards.

2.6.8   Other Observations

Ultimately, it would appear that the safety of a final system cannot be fully demonstrated. Even if safety behavioural compliance was demonstrable with a high confidence, there is the uncertainty surrounding safety as discussed earlier.

Although safety standards are developing in a positive direction, they emphasise development, and although some clues of assessment criteria are provided, they do not provide standards for safety assessment practice, e.g. when compared with computer security. Although seen as an advantage, safety standards can sometimes be too generic about process and product issues. This is because different sectors and organisations have their own cultures and practices, and there may be considerable diversity in the detail of agreed procedures. In fact, the diversity of safety applications discourages the development of a common safety culture. This research will study this diversity problem.

The over-emphasis on costly testing and post-design/implementation analysis may reduce safety integrity achievements due to real world cost constraints. Cost savings may be possible by applying more analytical verification techniques a priori, before committing to more expensive development phases.

A final observation is that the compilation of a safety case is normally only conducted at the beginning and end of a project. This is often because of the formal arrangements between Procurer, Design and Certification Authorities. However, there is some justification for developing "mini" safety cases containing logical safety arguments at various points along the development evolutionary path.

2.7 Safety Processes - State of the Art Summary

This paper is boldly challenging the effectiveness of the established and cultural aspects of safety life cycle, by highlighting its perceived strengths and its weaknesses. The impact of safety process improvements is likely to more significant if a genuine weak area is strengthened. Such a key area is that of safety requirements, closely following by safety compliance activities.

Major concepts such as safety integrity levels, safety targets, safety cases, safety management and life cycles are some of key areas for development. One of the general themes to be followed, throughout many of these concepts, is to improve the objectivity involved in building up a system safety case evidence, as well as using rigorous techniques that reduce the inherent uncertainty about safety.

Also, an improved approach that is more cost efficient, whilst increasing safety integrity, is likely to be an attractive way forward. This is a key criterion to judge whether this SPAM research meets its aims.

Despite many of its strengths, the safety processes needs improvement, if only to be more able to cope with the future demands on the diverse and advanced use of Programmable Electronic Systems (PES) technology. Such technology is generally used to enhance safety through providing innovative safety solutions even though there is the added risk that there are more functions to test and demonstrate.

A stronger approach to safety requirements and compliance will yield process improvements to provide perceived benefits that can be applied to areas, perceived as high risk, such as software. Such an achievement would be a tremendous step forward. Hence, the software safety aspect is a major goal for this research, even though, initially the system issues need resolving.

Having discussed the safety issues, it is appropriate that the next section discusses some key perspectives from the computer security world, where this research is intended to investigate where the security strengths can be used to develop the safety domain's weaknesses, whilst building on their current strengths.

3   Security Systems - Supplementary Approaches

3.1   Introduction

Section 2 emphasises that safety engineers have an abundance of techniques to help minimise the risk of failure in a safety critical system. The reason for this interest in security is because of the general acknowledgement that some areas of security are more advanced than their safety counterparts [CSR 86, Brewer 93], especially evident in standards. This section of the paper will endeavour to examine this more closely, whilst introducing the current best practices in computer security.

Section 3.2 will briefly introduce an overview of security to include some definitions of security to contrast with those of safety. Section 3.3 will look at the current security standards dealing with security techniques and their approach to security evaluation. The concepts described in these standards will be subsequently examined and a brief discussion given on how these might be applied to the safety domain.

3.2   Security Overview

The security evaluation document, ITSEC [ITSEC 91], defines security as the combination of confidentiality, integrity, and availability. Confidentiality is the prevention of the unauthorised disclosure of information. Integrity is the prevention of unauthorised modification to information, and availability is the prevention of the unauthorised withholding of information or resources. It should be noted from this, that security has been able to classify the types of requirement, applicable to all security problems.

Another useful definition for security considers the consequences of a security breach [Dobson 92]. This is:

"A security critical system is one whose failure could enable, or increase the ability of, others to harm us."

This definition is in contrast to those given in Section 2 for safety, where the concept of harm is generalised. Here the significance is that a security problem is only indirect and requires another action to cause direct harm, e.g. use of secret information, use of information whose integrity is damaged, access of information when security functions are not operating. One may claim there is a notion of indirect harm in safety, such as information systems concerning air traffic control. However, information security problems in a communication system, where safety is at risk if something interferes with such information, may present another perspective on any specific system security requirements.

One key aspect of safety is that the safety requirements are hampered due to uncertainty in the system environment. In security environments, one may say that they less diverse and more similar across different applications, as well as more structured and understandable than for safety environments. The implication here is that the uncertainty and risk considerations in security, as well as the possible consequences, is less directly severe.

This observation may be a result of culture or application differences between the two domains. The security domain appears to have less variation or complexity, this may have allowed the security community to identify its own commonalties to agree on the same concepts and techniques. Also the public profile of major accidents tends to be greater than that of security breaches, and hence it may be harder to gain consensus in safety. Whatever the motivation, the fact is clear that security standards are rich in concepts which are highly relevant to the safety domain. The next section summaries the major security standard initiatives that are indicative of the state of the art in security.

3.3   Security Standards

Although there is no formal security development life cycle or associated standards as there is in safety, there are many sources of reference for evaluation and miscellaneous security techniques. This section examines some of these established security standards.

Information Technology Security Evaluation Criteria (ITSEC). ITSEC draws on the experience of security specialists from throughout Europe and is now established as a provisional working standard. It aims to provide a set of common criteria for the evaluation of security systems and products. The basis of the document is to establish how much confidence one has in the way in which the security features of a system or product have been implemented, i.e. to try and evaluate how much *assurance* one has in the *security target* of the system or product. Assurance is gained by evaluating the system's functionality in terms of its *effectiveness* and *correctness* (these terms are described in more detail further on). As a result of an evaluation, the system or product is awarded a classification between E0 (no confidence) to E6 (high confidence).

The Rainbow Series. These books are published by the National Computer Security Centre, USA. They provide guidance on the evaluation of security controls built into computer system products, the *Orange Book,* and also interpretations of the requirements of any criteria (e.g., Green Book - password systems, Tan book - auditing, Red book - networks etc.).

The Orange book or Trusted Computer System Evaluation Criteria (TCSEC) [TCSEC 85] aims to establish the amount of trust that one can have in a secure system. It does this by using the security enforcing functions or *Trusted Computing Base* as a basis for evaluation to one of four categories (A - verified protection, B - mandatory protection, C - discretionary protection, and D - minimal security). *Security Policy, Security Models,* and *kernels* are all key concepts of the Orange Book, and these are briefly introduced later on in this paper.

The Common Criteria. Both ITSEC and the Orange Book have noticeable disadvantages which are being addressed by this document. The aim is to produce evaluation criteria that is widely adopted throughout the world, based on evaluation schemes of different nations.

Other standards bodies are looking into a variety of security techniques, such as cryptographic techniques, entity authentication and key management etc. The standards bodies working on these are: *JTC1 Information Technology SC27 Security Techniques* (this body has close links with IST33, a group dealing with banking and financial services), and *JTC1 Information Technology, SC21 Open Systems Interconnect, Data Management and Open Distributed Processing.* This body deals with OSI matters, however, some of their activities involve OSI security under the Security Frameworks in Open Systems banner.

3.4   Security Concepts and Techniques

After some initial study, security's perceived advancement over safety appears to be attributed two factors. Firstly, security evaluation is well established with a full set of criteria and advice on its application available. Secondly, key concepts and techniques are in place to ensure a system is inherently secure as opposed to just checked for security (via risk analysis) after the design has started. These concepts and techniques are now introduced:

Security Target. A security target specifies the security enforcing functions and the evaluation level that they expect to reach. The security target also states the security objectives of the system (the security that the intended system is to achieve), plus the threats to those objectives. The key components of the security target are:

• System Security Policy (SSP) or Product Rationale.

• Specification of the security enforcing functionality.

• Definition of the required security mechanisms.

• A claimed evaluation rating for the strength of these mechanisms.

• A target evaluation level for the security related components.

The Security Target document is very importance as it states early on in the life cycle precisely what the system must do to operate securely. This concept is of great interest to this project.

Security Policy. A security policy is a formal document that outlines the laws, rules and practices governing the use of a computer system. The contents of this document include a statement of the system's environment, a description of its security objectives, and a threats and countermeasures list. As part of the Security Target, the Security Policy states in formal terms, requirements that the system must adhere to remain secure. In the UK evaluation scheme (of which ITSEC is a part), the Security Policy is created by initially defining a Corporate Security Policy (CSP) for the organisation. This policy document states the rules, laws and practices of the organisation. During this process, the CSP will identify the corporate assets and general threats which help identify the security objectives of the proposed system. The System Security Policy (SSP) aims to satisfy these system security objectives and itself is partitioned into requirements for the security enforcing functions and procedural/personnel/physical security requirements. If required, the technical content of this policy may be partitioned into the Technical Security Policy (TSP). For high integrity systems, these requirements would have to be stated, as a formal model of Security Policy.

Security Models. The purpose of a Security Model is to present the Security Policy of a system concisely and unambiguously using mathematical notation to prove its authenticity. The benefit this technique has is that it proves that an implementation of the model will be secure providing the axioms of the model are upheld in it. This is very important when trying to produce an inherently secure design.

Security Kernel. The Orange Book is geared to systems based on the reference monitor concept, i.e. an abstract machine that mediates all access to objects by subjects. A Security Kernel is an implementation of this concept, and will enforce a set of rules (as described by the security policy) governing how the system data is modified and accessed by users of the system. The advantage to this approach is that regardless of the behaviour of the rest of the system, the Security Kernel (which has been verified for correctness) will ensure the system never enters an insecure state. This may be a valuable technique to adopt for safety related systems.

Trusted Computing Base (TCB - Orange Book) or Security Enforcing/Relevant Functions (SEFs - ITSEC). The TCB or set of security enforcing functions partition the security relevant functionality from the rest of the system components. It is this set that is then subject to evaluation. The evaluation then depends on the correct and effective implementation of each mechanism in the TCB.

Functionality Classes. As mentioned earlier, it is beneficial if each security enforcing function of the Target of Evaluation (i.e. the system or product that is undergoing evaluation) can be attributed to one of several types of functional class. For example, if it has been identified that there is a need for access control in a system, then one can select an appropriate mechanism from the access control functional class. This aids both the evaluator and system developer.

Assurance - Effectiveness and Correctness. Assurance is basically the trust that one can have in evaluated system that it meets its security objectives. It is subdivided into two components, namely effectiveness and correctness. Effectiveness deals with the proper construction and operation of the Security Target (i.e. is it secure?), whereas correctness deals with the processes used to implement the Target of Evaluation (i.e. does the design meet its specification?). This is an important concept that distinguishes the European Scheme (ITSEC) from the American TCSEC.

There are, of course, many other techniques involved in the development of secure systems, however, the above list does seem to represent approaches that would be both novel and innovative if applied to safety related systems. The key point is that these techniques bring out the fundamental security requirements at the initial stages of design, and state them in a manner that cannot be misconstrued. Providing they are implemented effectively, these properties will greatly enhance the chances of the system remaining secure in operation.

3.5   Applying Security Techniques to a Life Cycle

The above concepts and techniques may be applied to a typical life cycle [Lovering 94][3] which centres around a structured development life cycle (such as SSADM). This helps ensure the final system has been securely developed, whilst aiding the evaluation process.

Feasibility Study. The initial stages of product development concern itself with gathering information on the project from different sources. This may include relevant standards, the Corporate Security Policy, and the Project Initiation Document (PID). Also, a preliminary risk analysis will yield high level security requirements and the potential threats that the system may face. From this information, the fundamental security requirements can be partitioned into an initial System Security Policy and a formal policy model produced. The security objectives of the system can now be stated in the Security Target.

Requirements Analysis. This stage continues to build on the security requirements of the system implementing the countermeasures to threats identified in the preliminary risk analysis.

Requirements Specification. At this point, formal methods can be brought in to remove ambiguities and to verify the current design. System risk analysis should also provide a more stringent examination of the threats and vulnerabilities facing the current system.

System Specification. For highest evaluation levels there is a requirement to specify the security mechanisms and architecture formally. This results in a formal system model. Also the final System Security Policy should be produced which in itself, can be partitioned into the Technical Security Policy and a document detailing Personnel, Physical and procedural security.

---

[3] Although the life cycle described by this reference is relatively explicit, there does not appear to be an agreed standard security life cycle as there is in safety.

3.6   Initial Contrast with Safety

When this life cycle is compared to a typical safety related system life cycle [MoD 91], an important difference becomes evident. Fundamentally, when risk analysis techniques are removed from the safety life cycle, one is left with a standard development life cycle. However, when risk analysis is removed from the above security life cycle, there are still several other techniques and practices, primarily policy and modelling, that ensure security is always under consideration during design. This leads to the conclusion that safety related systems have to undergo continual verification in order to keep the design safe, whereas security critical systems utilise key techniques to positively drive the design in a secure manner.

In order to address the points established in the previous paragraph, the SPAM project hopes to bring the concepts of Corporate, System, and Technical Safety Policy to the safety domain along with Safety Policy Modelling in order to enhance the way safety critical systems are designed. This, of course, implies that the supporting techniques to policy development and policy modelling will have to be researched as well. These are typically: security targets, assurance levels, functionality classes etc.

4   Enhancing the Safety Life Cycle

4.1   Introduction

This section reflects on the topics covered in Section 2 and 3, in order to establish how the benefits identified in the security domain can address the weaknesses in current safety practices.

Initial comparisons between the two domains tend to suggest that safety related systems rely heavily on risk reducing techniques to verify that a developing system is safe. The concern with this approach is that one can never establish all the hazards facing the system (as the environment is uncertain and therefore unquantifiable), therefore, given the right set of mitigating circumstances, a fundamental design flaw may be exposed. This may be despite of rigorous checking and also the application of safety management principles.

In comparison with the security approach, one sees that the statement of fundamental security of the system (the Security Policy) and its formal policy model assures the inherent security of the system. This is theoretical, as the full extent of the environment (and hence threats) is never known, but secure systems are implemented in relatively known and restricted environments. This helps create a secure solution free from fundamental errors due to this policy modelling concept. The problem with applying policy and modelling to safety related systems, however, is that the environment is less easy to quantify with an almost unlimited set of threats to safety. This makes testing of any potential formal safety model a lot harder to progress than for its security counterpart.

The other area that could be of benefit to the safety domain is an established evaluation scheme. These ideas are now expanded upon.

4.2 Strengthening the Safety Approach with Security Techniques and Concepts

4.2.1 Categorising Security Requirements

Security requirements stem for using risk analysis and the initial Security Policy. It is of interest, that each requirement addresses threats to either confidentiality, integrity, or availability which may affect the type of policy model being developed. In order to find comparable safety properties, so that safety processes can benefit from this type of categorisation, the project is pursuing the idea of a Taxonomy of safety problems.

4.2.2 Applying Security Techniques

Section 2 listed several weaknesses in the current approach to safety engineering which will be addressed by the SPAM project. Initial security studies may have the solution to some of these problems, therefore this section will look at how these security techniques and concepts can be applied to the safety domain.

4.2.3 Applying Policy Models

Whilst it is appreciated that to construct a security policy model, one must first have an understanding of the intended environment (and hence the threats to the system), ideal behaviour is assumed and the impact of the environment is therefore restricted to key areas of concern. One may question the value of such a simplistic model, but a security model like Bell - La Padula [Bell 73] provides adequate protection against most threats to confidentiality. Threats left unaddressed can be considered later on in the design where cost benefit analysis may well eliminate them from the design. Hence, this approach focuses the mind on the most important threats early on in the design.

An analogous safety model similarly would hope to capture the fundamental safety requirement of a design in order to provide assurance

111

that a system is inherently safe. For example, a model based on the requirement that all civil aircraft should be able to glide, would ensure that serious failure in any part of the system would not result in the immediate destruction of the plane.

4.2.4   *Applying Policy Concepts.*

As mentioned in Section 3, the Security Policy contains the necessary requirements for the system to operate securely with adequate assurance. The Security Policy will contain: a statement of the environment, a description of the security objectives, and threats and countermeasures list. Applying this concept to the safety domain may result in the duplication of the hazard log, so the benefit of the Security Policy approach may just be limited to partitioning off the corporate, system, and technical safety requirements. It is expected to conduct further research on this matter.

4.2.5   *Applying the Security Target Concept*

The main benefit of the Security Target is that it contains all the security requirements of the system, plus detail on how these requirements were met. This notion is obviously well established in the security domain, however, in safety related systems there is no comparable document which is a drawback to their evaluation.

4.2.6   *Applying the Kernel Concept*

The two properties that make Security Kernels of interest may be applicable to the safety domain. These are: the Kernel itself which is small enough to be verified as correct, and its operation which assures the security of the system despite the behaviour of other system components. This project aims to understand what a Safety Kernel might mean, and how it might be applied.

4.2.7   *Partitioning the Security Relevant Functionality*

The Orange Book recommends that all functionality affecting the security of the design is partitioned into the Trusted Computing Base which is then subject to the evaluation.

ITSEC also follows a similar process with its Target of Evaluation. Likewise, safety systems already do this by partitioning their safety related components, but the evaluation process to provide the assurance that these components are safe could be improved by examining evaluation criteria in the security domain. Further work will be conducted in this area.

112

4.2.8   Functional Classes

This notion may be of interest to the safety community as different types of security mechanism can be placed under different generic headings. These headings address common problems associated with the development of a secure system, such as access control and system audit. How this might be of benefit to the safety domain is yet to be resolved, but it is possible that hazards common to different types of system can be addressed by safety mechanisms from the same functional class.

4.2.9   Assurance

As stated earlier, this project aims to compare the methods used to evaluate both safety and security critical systems. The security evaluation process gains confidence in the security of the design by examining both the *effectiveness* of the security functionality and the *correctness* of the development processes.

This approach leads to an overall evaluation level proportional to the trust in the system. It is possible that this approach to evaluation may enhance the safety evaluation process. Also, researching the contrast between security assurance and safety integrity is valuable in that they both aim to measure confidence in the system. This project will consider how assurance techniques can address the problems associated with using safety integrity levels.

5   The Way Ahead

5.1   General

This section seeks to emphasise the future direction and key project concepts to be researched in developing a safety policy oriented approach, to result in the SPAM method. Obviously from the contents of this paper, the role of security concepts and techniques will be a key focus.

5.2   Adopting Security Techniques and Concepts

Section 4 has given some indications of the possible commonality and cross-fertilisation between the security and safety IT communities. Detailed studies of security concepts and techniques with their applicability to safety will be examined. The prime candidates are: categorisation of security, security target, security policy, security models, security kernel, trusted computing base, functionality classes and assurance. Further detailed studies of the safety security parallels are continuing to ascertain further useful linkages.

5.3   Developing System Safety Policy and Models

As already established in security, operating at corporate, system and technical levels, the policy notion has yet to be exploited for safety systems. Defining safety, using the policy concept, early in the system development life cycle is seen to be a positive approach to safety definition and verification. This will encourage the argument of 'Is it safe?' as opposed to 'What are the consequences if things go wrong?' This is one of the main areas of work where the potential for this use has been indicated before [Brewer 93].

5.4   Develop Generic Approaches

Once safety applications are categorised and approaches defined, the policy concept can be applied to any sector's system development. Experience in the security world has shown that there are significant benefits, in terms of industry re-use, in producing generic models. The SPAM project will seek to provide guidance on how to take a particular Safety Model and produce a generic, easily re-usable counterpart.

5.5   Categorising Safety (via a Safety Taxonomy)

It is possible that safety applications may be categorised according to defined system attributes which exhibit similar safety characteristics. The SPAM project will study whether widely different problem domains (e.g. railway modelling, nuclear power generation) belong to the same class of safety problem and hence the same type of safety policy structure. The aim is to develop a SPAM Taxonomy. This will enable the categorisation of potentially diverse safety problems according to specific attributes. The SPAM method is expected to provide guidance upon safety problem categorisation, and then advise on appropriate policy development and modelling techniques for the allocated domain. It is possible that, in time, there may be pre-existing generic models which could be usefully applied for the domain.

5.6   Assigning Integrity Levels

The notion of safety integrity level [IEC 92], has gained significant ground in the last two years or so, and indeed does seem to offer benefits to developers and users of safety critical IT based systems. A current difficulty in applying integrity levels is to objectively assign of integrity level to components for a given design choice. The aim is to investigate this difficulty from the model-based approach with a view to developing rules and guidance as to how this process can be made more objective.

5.7   Application of SPAM Process to the Safety Life Cycle

Methods for introducing the policy concept and its attendant modelling framework are likely to change the safety prospective of a system life-cycle. The impact of any methods will be expressed as a process model linked to a generic safety life cycle. This will include guidance on safety requirements capture, safety compliance, and safety management. Also, the SPAM project will seek to define requirements for tool support that will assist in the application of the project-devised method.

5.8   Safety Standards

Following the provision of SPAM reference materials complete with new key concepts and an understanding of the life cycle impact, it is hoped to influence the standardisation of generic safety process and design approaches.

5.9   Software Safety

The SPAM project will implement the policy concept through refinements from the organisational, system and technical software levels. Major advantages to software safety verification are envisaged.

5.10   Safety and Security Harmonisation

There are a growing number of computer-reliant systems in use today which exhibit both safety and security critical features. For instance: Health Care Systems and Remote Nuclear Power or Process Plant Management These safety and security concerns need to be satisfied throughout the requirements capture and system verification stages. Given this current position, and the growing awareness within the two communities of IT safety and security that both the safety and security domains can learn and benefit from the other's experiences, then it would be advantageous to developers, users, and assessors if IT safety and security could be treated jointly. The advantages to developers (particularly for the growing number of mixed safety and security critical applications) would be seen in reduced development cost as well as improved integrity. During the SPAM project it is expected to provide further evidence to support this view, whilst helping the standards bodies in the process.

References

[Bell 73] Bell D E, La Padula L J: "Secure Computer Systems: A mathematical model. Volume II." Mitre Corp, Bedford Mass, November 1973.

[Brewer 93] Brewer D: "Applying Security Techniques to Achieving Safety." Directions in Safety Critical Systems, Ed. F Redmill, T Anderson, Springer Verlag, 1993.

[CSR 86] Centre for Software Reliability: "Safe & Secure Computing Systems." Ed. T Anderson, CSR Blackwell Scientific, 1986.

[Dobson 92] Dobson J, Burns A, & McDermid J: "On the Meaning of Safety and Security." Computer Journal, vol. 35, no. 1, February 1992.

[IEC 92] International Electrotechnical Commission: "Draft Functional Safety of Electrical/Electronic/ Programmable Safety Related Systems - Generic Aspects - Part 1- General Requirements." SG65A/WG10 Draft Document, Version 5, (IEC Reference: 65A Secretariat 123) January 1992.

[ITSEC 91] ITSEC: "Information Technology Security Evaluation Criteria, provisional harmonised criteria." Luxembourg: Office for Official Publications of the European Communities, June 1991.

[Jesty 93] Jesty P H, Buckley T F: "Towards Safe Road Transport Informatic Systems.", Safety Critical Systems - Current Issues, Techniques and Standards, Ed. F Redmill and T Anderson, Chapman & Hall, 1993.

[Leveson 86] Leveson N G: "Software Safety - Why, What and How?" ACM Computing Surveys 18, pp 125-163, 1986.

[Lovering 94] Lovering A S: "Formal methods, structured techniques and security - A unified methodology." National Physical Laboratory Report to be published, 1994.

[MoD 87] Ministry of Defence "Interim Def Stan 00-55: "Requirement for the Procurement of Safety Critical Software in Defence, Parts 1 and 2." MoD 1987.

[MoD 91] Ministry of Defence "Interim Defence Standard 00-56/Issue 1 - Hazard Analysis and Safety Classification for the Computer and Programmable Electronic System Elements of Defence Equipment." MoD 1991.

[TCSEC 85] TCSEC: "Trusted Computer System Evaluation Criteria." DOD 5200.28.STD, Department of Defense Computer Security Center, Fort George G. Meade, MD, December, 1985.

Extending a Security Evaluation Standard (the ITSEC) to Dependability

Alan Hawes
Admiral Management Services Limited
Camberley, England

1. Introduction

1.1 Purpose of this Paper

This paper reports on some work which was performed on the Eagle Project to look at the ways in which an existing security evaluation standard could be extended to address safety issues. The Eagle Project was project number S2114 in the European Commission's INFOSEC'93 programme of research.

The findings of the project are of interest to the safety community, because if these extensions are incorporated into security evaluation practice, a well established third party evaluation scheme, based on internationally agreed standards, will become available to developers of safety-critical systems.

Any discussion which bridges two fields - in this case safety and security - will inevitably hit difficulties of terminology, which will hinder understanding of the underlying concepts. While every attempt has been made to overcome this problem, the reader is asked to make allowances for any remaining ambiguities when assessing the proposals made in this paper.

1.2 Background to the Eagle Project

The Eagle Project was set up by the Commission of the European Community (CEC) during the two year period of trial use of the Information Technology Security Evaluation Criteria (ITSEC) Standard, as part of a number of initiatives to build on the experience of ITSEC in practical use. The standard was being applied in a number of community countries, and work was needed to coordinate the results of that experience in some recommendations for improvements to be made in the next version.

The ITSEC lays down an internationally agreed set of criteria to be used by evaluators when assessing the security characteristics of systems or products which include IT components. The standard formed the basis of security evaluation and certification schemes in Germany and the UK, and it was anticipated that its use would spread to other countries.

At this time, a companion standard the Information Technology Security Evaluation Manual (ITSEM) was being produced, and it had provided some clarification and interpretation of the ITSEC, which was being used differently by evaluators in different Evaluation Facilities (ITSEFs). However it appeared that some areas, such as Re-evaluation needed additional work beyond the ITSEM's clarifications. In particular, there were some errors and ambiguities in the ITSEC which had not been addressed by the ITSEM.

The CEC INFOSEC'92 programme was at that stage underway, and had already produced some results which made recommendations for the extension of the ITSEC, particularly in the field of the interface to Accreditation.

When the project started, some initial work had been done by the Common Criteria Editorial Board (CCEB) to establish common evaluation criteria for use in Europe and North America. It was felt that the work done by the Eagle Project could provide input to that initiative, if the CCEB wished to establish criteria with a wider scope of applicability than the ITSEC.

1.3 Objectives of the Project

The Eagle Project was set up to meet two major objectives:

a. "The primary objective of this project is to recommend the way ahead for enhancing the ITSEC to cover the re-rating and re-evaluation of products and systems."

b. "A secondary objective is to make recommendations for widening the scope of the ITSEC to cover additional topics..." where these topics were to be defined by the project.

1.4 Scope of the Project

It was recognised that the project would need to be wide-ranging in scope, to address all the issues related to the potential extensions of ITSEC. The project was set up to collect data on the practical use of ITSEC, by having representatives from ITSEFs in a number of countries where evaluations are being performed.

The project was designed initially to seek input from a wide range of experts in the field with knowledge of ITSEC, in order to ensure that a wide range of views were considered and that consideration was given to the most important areas for extension.

After this initial survey, the scope of the project was defined to address the four highest priority areas:

a. Re-evaluation

b. Clarification of Effectiveness Criteria

c. Alignment with Accreditation Requirements

d. Interfaces with Safety Criteria.

1.5 The Eagle Consortium

The project team consisted of representatives of five organisations, all of which were performing ITSEC evaluations. The partners were:

c. Admiral Management Services Limited (AMSL) UK

d. Industrieanlagen Betriebsgesellschaft mbH (IABG) D

e. Conception et Réalisation d'Applications Automatisées (CR2A) F

f. TNO Fysisch en Elektronisch Laboratorium (TNO-FEL) NL

g. Instituto Nacional de Tecnica Aeroespacial (INTA) ES

The staff employed on performing the technical work of the project were chosen to be those from each organisation with in depth knowledge of ITSEC and evaluation issues. The primary staff were:

a. Alan Hawes - AMSL

b. Helmut Kurth - IABG

c. Laurent Borowski - CR2A

d. Edward Hardam - TNO-FEL

e. Miguel Banon-Puente - INTA.

1.6 Method of Working

The method of working in each of the four areas was similar, and was based on the selection of experts, and interviews to collect their comments on initial project generated proposals and questions. The resulting proposals were reviewed regularly by the project team, and finally reviewed by selected experts who have made contributions to the work.

A large number of experts, twenty three in all, were interviewed on the project, bringing valuable input from Europe and North America to the project.

The major results of the project are documented in four reports, one for each of the areas. The work done in the safety area and the results achieved are summarised below.

2. Interfaces with Safety Criteria

The ITSEC has grown out of a need for governments to have a standardised approach to evaluating the security of IT products and systems, and a wish to harmonise the criteria that are used in order to facilitate trade. The primary issue addressed by these criteria is confidentiality - preventing unauthorised access to data - but the criteria also address some aspects of integrity and availability as understood in the safety field.

Safety Criteria have evolved in a completely different way, responding to the needs of specific application areas, such as Nuclear Power, Avionics, Rail, Air Traffic Control, Mining, Chemical Plants etc., and their legal and regulatory frameworks. These criteria are primarily concerned with Safety - loss of life or injury to humans or damage to the environment - and to a lesser extent with Integrity, Availability and Reliability through their contribution to Safety objectives.

Bringing these two concepts together produces the overall quality of Dependability. Many systems have requirements which bridge the two fields - for example preventing malicious interference with Air Traffic Control intended to cause air crashes. This points to the value of evaluating systems to criteria for both sets of characteristics. The concept of Dependability has existed for at least 6 years but in spite of this, development in each of the separate areas has progressed in isolation, and in some cases ignorance, of each other.

The major differences between the two field's approaches can be identified as "process" versus "product" evaluation. In the Safety field there is great emphasis on the integrity of the process of developing safe systems. High priority is given to planning for safety, together with continuous assessment for adherence to process standards. In the Security field the major concern is evaluation of the end product of development for conformance with criteria, its correctness and the effectiveness of its security functions. There is much commonality however, and a great deal of identical assessment work is performed.

In the EAGLE project a range of the recent developments in the Safety field have been examined in order to look for commonality or overlap with the ITSEC approach to evaluation. This investigation has identified changes to ITSEC, evaluation practice and the environment in which evaluation is performed which will enable evaluation under ITSEC to interface more effectively with safety assessment standards. It has also suggested changes in safety assessment practices, where lessons can be learnt from the security field, or other safety application areas.

The areas addressed by the study were:

a. dependability issues

b. lifecycle issues

c. risk analysis

d. functionality classes

e. effectiveness concepts.

3. Discussion

3.1 Dependability Concepts

IT Systems are increasingly being used in wider systems with inherent risks attached to them, and often the IT Systems are depended upon to operate correctly and effectively as a component of these systems. The risks associated with these systems can be concerned with death or injury to humans, damage to the environment, damage to property, loss of livelihood, or damage to society or national interests.

When building such systems, the designer and specifier need to consider how they are to respond to these risks, and consider how to perform compromises between the differing, and sometimes contradictory, requirements placed upon them. Currently, the tendency is to consider issues such as security, reliability, safety, etc., in isolation and perform trade-offs in an ad-hoc way.

As an example of this type of system, consider a military command and control system, which passes commands to operational units. It is potentially subject to hostile acts from an enemy, who wishes to disrupt its operation, but also has to prevent erroneous deployment of weapons or offensive action as a result of faults or misoperation. Access controls, put in place to enforce confidentiality may hinder or prevent recovery actions, contributing to reductions in safety levels.

To overcome this problem, designers need to use a common concept of Dependability to cover the full range of risks associated with a system, and need a common framework in which to assess security, safety, etc., as components of dependability. A useful definition of dependability is given in [Laprie 90] "Computer system dependability is the trustworthiness of the delivered service such that reliance can justifiably be placed on this service. The service delivered by a system is its behaviour as it is perceived by its users."

A recent paper [McDermid 94] has proposed a system for measurement of dependability in terms of its negative attribute, loss. It is proposed that all forms of loss can be given a quantitative expression, e.g. number of fatalities, loss of prestige, financial loss, loss of livelihood, which allows for the possibility of trade-offs being made in an objective way. This type of quantitative approach is common in the security field, and is becoming more widely applied in the safety field.

The process of Risk Analysis, which is described in more detail below, is used in both the safety and security fields to identify the measures that need to be taken to counter risks, and ensure that they are established at a tolerable level. Currently, however, two parallel analyses are performed, and they may result in incompatible recommendations, a unified risk analysis method based on dependability concepts would avoid this, by assigning combinations of countermeasures in a co-ordinated way.

Risk Analysis techniques used in the safety field are also concerned with establishing the level of assurance that is needed in the correct and effective operation of the countermeasures which have been defined. This linkage is largely missing in the security field, where assurance levels are derived using ad-hoc techniques based on confidentiality needs only. A common approach which links risk to assurance through the use of dependability concepts would be a great improvement.

There is scope for applying dependability concepts in the area of assessment of assurance levels, as the system of Safety Integrity Levels has clear parallels to the ITSEC's E levels and strength ratings. What is needed is a set of Evaluation Criteria which can be used to define assurance levels for a wide range of functions, appropriate to all aspects of dependability. The [CTCPEC 93] has made a move in this direction, by explicitly defining functions and assurance levels for Integrity and Availability countermeasures.

3.2 Lifecycle

The ITSEC is concerned with only a small part of the lifecycle of a system with security requirements, compared with the standards which are applied to safety critical systems. As a result it makes many assumptions about the lifecycle, many of them implicit.

In contrast, standards in the safety field are explicit about the type of lifecycle that is appropriate to achieving the required levels of assurance. In this respect, safety standards have a greater emphasis on the 'process' aspects of assurance, rather than the 'product' aspects. The infrastructure of control of the process of developing and operating dependable systems could with advantage be applied in the security field.

The draft IEC standards [IEC 65A 91] are intended to be generic across all safety application fields, and address the lifecycle issues of safety planning, the organisation of safety projects, expertise of staff used on safety projects, risk management and maintenance of safety through the operational life of a system.

These standards make explicit proposals for how the system lifecycle should be defined for a system, rather than requiring parties involved to follow a general quality standard, and by so doing, ensure that the specific actions appropriate to the management, assessment and certification of safety critical systems are performed.

It is recognised in the safety field that quality does not equal safety. If more prescriptive standards are put in place for a generic security lifecycle, (which includes all dependability aspects) it will become clearer how quality and security specific assessments are needed to assure security.

3.3 Risk Analysis

In the safety field risk analysis is performed to establish whether the level of risk inherent in a system is acceptable, and whether this can be reduced to an acceptable level by appropriate measures. The assets are human lives, human health or an undamaged environment. Present practice increasingly puts monetary values on these assets, in order to facilitate decisions about priorities.

The threats to safety critical systems are the causes of accidents or incidents which will result in losses with respect to the above assets, and tend to be well known in each of the application areas. The accidents or incidents themselves are equivalent to security breaches, in that they would not happen in an ideal world, but in reality have to be tolerated, at an acceptable level.

In both fields there is the concept of loss, which results from accidents or breaches, and the concept of risk, which combines the losses predicted for a system with the probabilities of those losses occurring through the system's use.

There is scope for quantitative assessments of loss and risk to be performed on the same basis in both safety and security contexts, so that a unified method of risk assessment, and consequent unified way of assessing the effectiveness of risk reduction measures can be applied.

Taking this concept further, the safety approach of linking risk levels to safety integrity levels could be used as a technique for establishing what evaluation level is required for security critical IT systems. This would fill a gap in the techniques needed by Accreditors of security and safety critical systems. See Eagle Accreditation Report [Eagle 94a].

This approach would provide a valuable, and currently missing, link between the range of risks to which a system is subject and the assurance levels that are needed for system components, as assurance is directly related to the confidence that is placed in them to reduce or control risk levels.

The process of initial risk assessment would consist of the following stages;

a. identification of potential accidents or breaches

b. assessment of the impact of each event in quantitative terms

c. determination of the probability of each event occurring

d. computation of a risk factor, based on impact and probability of all the events considered

e. assessment of whether the resulting level of risk is acceptable

f. selection of design techniques or countermeasures with sufficient strength to reduce the risk factor by an appropriate amount

g. determination of the level of assurance needed in the correct and effective operation of the countermeasures - ie its dependability.

Consider an example of how this scheme might work for just one incident:

a. loss of PC and work in progress through fire

b. financial value of loss = 8000 ECU

c. probability of fire = once in 20 years

d. computed risk factor = 400 ECU per year

e. tolerable risk = 100 ECU per year

f. smoke detector designed to be effective in preventing four out of five fires, reducing the risk factor to 80 ECU per year (ITSEC strength of mechanism = medium say)

g. assurance level required in correct and effective operation of smoke detector derived from required risk reduction, i.e. 400 ECU per year to 100 ECU per year (ITSEC assurance level E3 say).

If this type of analysis is repeated for all risks to which a system is subject, a definition of the types and strengths of all the risk reduction measures will be derived, together with required assurance levels. Including both safety and security related risks in the analysis will ensure that a balanced approach to risk reduction is taken.

3.4 Functionality Classes

The IEC Standards do not make recommendations for actual safety functions that should be used in particular risk scenarios, because of the very wide range of application areas to which they apply, but they do make proposals for the type of design approaches that should be used. A distinction is made between 'design approaches', which result in functions and mechanisms being implemented, and 'techniques', which are concerned with project and technical controls and practices. For example it is recommended that for high safety integrity levels an approach involving some form of diversity in the design is used (belt and braces).

These design approaches can be viewed as high level functions designed to achieve safety, when they are implemented in lower level functions and mechanisms. This is similar to the way that a high level function such as Access Control will be implemented in a variety of ways to achieve confidentiality or integrity objectives.

In this case, the function which votes on the outputs is an "integrity enforcing function", which is relied upon to reduce the risk of erroneous function outputs, and so increase their integrity.

The CTCPEC goes beyond other sets of evaluation criteria, by introducing high level security functions associated with integrity, such as Rollback and Self Testing, and with availability, such as Fault Tolerance and Robustness. There is clearly an overlap here with the functions which are used to achieve integrity, availability and reliability in a safety context.

The functions defined in the CTCPEC which are relevant to safety requirements are:

a. Physical Integrity

b. Rollback

c. Self Testing

d. Containment

e. Fault Tolerance

f. Robustness

g. Recovery.

It is proposed that the extension of the set of high level functions defined in the CTCPEC, to include additional functions associated with meeting safety requirements, would enable the designer of complex systems to choose from a wider palette of techniques when defining risk reduction measures.

Initial ideas for suitable high level functions are:

a. alarms

b. redundancy

c. diversity

d. fail-safe (or fail-predictable)

e. graceful degradation.

A wider set of functions, such as these, could be added into a risk analysis framework as described above, and result in the definition of a range of strengths for these functions, which are appropriate to countering a range of safety and security risks.

To take this concept further, it will be possible to define a set of functionality classes, or typical sets of functions, which will be frequently used, and can be pre-defined to be effective against commonly experienced risk scenarios.

An example of where this approach would be useful is the case where a database is used to hold information on hospital patients, which is used for administration of drugs. As well as classical safety risks, there is a chance that an attacker may use the system to harm another person. In this situation the designer may need a

126

high integrity database to counter this type of risk, and one employing dual redundancy and voting functions may meet his needs.

Market forces will decide whether this example, "high integrity dual redundant database" becomes a standard set of functions which is widely used. It should be possible to construct functionality classes using functions such as voting, and evaluate products which conform to them.

3.5 Effectiveness Concepts

A set of techniques which are used in the analysis of safety critical systems, collectively referred to as Hazard Analysis, have some relevance in the security field. The two considered here are Fault Tree Analysis (FTA) and Failure Modes and Effects Analysis (FMEA).

Fault Tree Analysis is a technique for connecting together in a logical and quantitative framework top level events, such as accidents and hazardous conditions, and the lower level causes, conditions and events which can contribute to them. It is a graphical technique, and is very good at showing both how failures, misoperation and coincidences can lead to accidents, and how safety functions can be used to reduce the probability of their occurrence.

When accreditors are assessing a system's ability to counter a particular set of attacks, this technique offers a way of connecting anticipated security breaches with all the events and conditions which could contribute to those breaches. It is a form of attack path analysis, showing all possible paths from initial security relevant events to the eventual security breach, and illustrating how security functions are deployed to reduce the probability associated with each path. See Eagle Accreditation Report [Eagle 94a].

The FTA technique is well supported with tools, to aid the analyst to handle the long complex trees which can arise, provide graphical representations of the trees, and do probability calculations across the trees. There is scope for using this technique as a component of Risk Analysis, during the system accreditation activity, and possibly as a part of evaluation, covering Suitability, Binding, Ease of Use and some part of Vulnerability Analysis within Effectiveness Analysis. See Eagle Effectiveness Report [Eagle 94b].

Failure Modes and Effects Analysis is a technique for identifying all the failure modes and sources of error that can occur in a system, and assessing how their effects can contribute to hazards at a higher level. It is essentially a bottom-up technique, requiring an understanding of the basic technology used, and the ways in which errors can occur and propagate through a system.

Recent work [Landwehr 93] has proposed a taxonomy for security flaws, which provides an introduction to the ways in which secure systems can fail, related to their cause, which could be used as a basis for FMEA. This is very closely related to the techniques used by evaluators to perform a vulnerability analysis.

These two techniques are examples of typical, commonly used, analysis methods, both of which are well supported by tools, which could be extended in their field of application, to include aspects of systems with security requirements. Their adoption would make the processes of accreditation and effectiveness evaluation clear to all involved, and provide a straightforward, tool-supported technique for performing them.

One of the most important benefits, however, of adopting these techniques in the security field would be the potential for uniform processes for accreditation and evaluation of systems with safety and security functions, in such a way that the effectiveness of combinations of these functions can be assessed.

4. Conclusions and Recommendations

This work on the possible extension of ITSEC to address both security and safety assessments has identified that many experts who are able to understand both fields consider that ITSEC has the potential for wider application if it is changed in the ways outlined in this paper.

The ITSEC has the potential for this extension as a result of the work done by its authors to generalise the process of evaluation away from specific functions or correctness issues, by introducing the concepts of assurance and effectiveness.

However, the ITSEC has weaknesses in its development of some of its underlying concepts, such as interfaces to accreditation and development of the Effectiveness concept, and these have been addressed by other work on the EAGLE project.

If some of these deficiencies are corrected, as proposed by that work, and as proposed in this report, then the ITSEC will be capable of extension to the wider field of evaluation of dependable systems.

Some aspects, such as the generalisation of Risk Analysis, are not easily changed, and require work to develop the ideas, and their practical application, further. There is a need for research to be undertaken to:

a. define a generic dependable system lifecycle

b. develop a common Risk Analysis and Accreditation framework

c. derive and define additional functions and functionality classes

d. demonstrate the usefulness of hazard analysis techniques in the security field.

It is recommended that this work is undertaken, and at the same time an editorial review of ITSEC is performed, to compile a set of changes to be made, to move the terminology used away from security to dependability.

The next generation of criteria [CCEB 94] are being developed currently without any intention to broaden their applicability beyond security, or to look beyond Evaluation to the wider lifecycle issues. The lack of clear links into Risk Analysis and Accreditation has also lead to the abandonment of the concept linking Effectiveness to Security Objectives in these criteria. This is a backward step, as it will reduce the scope for wider application of the criteria, and miss the opportunity for this provided by the ITSEC. It is recommended that the Common Criteria initiative re-examines these aspects, and broadens its scope to look at the wider field of Dependability.

It is recommended that work in the safety field to develop standards for safety-critical systems, such as the IEC initiative, take account of the common ground between safety and security, and look at the evaluation schemes that operate in the security field as a model of how third party assessment schemes can be operated.

A project in the DTI's current Safety Critical-Systems research programme - FRESCO - is looking at establishing a framework for third party evaluation schemes in the safety field, based on the ITSEC approach. It should provide further insights into the feasibility of some of the concepts proposed in this paper.

5. References

The following documents are referenced in the text of this report, and are defined more fully in this Annex.

[CCEB 94] Common Criteria Editorial Board - Common Criteria for Information Technology Security Evaluation, Version 0.6, Reference CCEB-94/039 and enclosures, April 1994.

[CTCPEC 93] - Canadian Trusted Computer Product Evaluation Criteria, Version 3.0e published by Canadian System Security Centre, January 1993.

[Eagle 94a] - Accreditation Issues, Eagle (S2114) Project Report No. 5269A/T/6/3, Infosec'93 programme, July 1994.

[Eagle 94b] - Clarification of Effectiveness Criteria, Eagle (S2114) Project Report No. 5269A/T/6/2, Infosec'93 programme, July 1994.

[IEC 65A 91] - two documents "Functional Safety of Electrical/Electronic/

Programmable Electronic Systems", draft version dated April 1991. "Software for Computers in the Application of Industrial Safety-related Systems", draft version dated November 1991.

ITSEC - Information Technology Security Evaluation Criteria - published by the European Commission, June 1991.

ITSEM - Information Technology Security Evaluation Manual - published by the European Commission, September 1993.

[Landwehr 93] - Carl Landwehr, Alan Bull, John McDermott and William Choi, A Taxonomy of Computer Program Security Flaws, with Examples, published by Naval Research Laboratory, Washington, Reference NRL/FR/5542-93-9591, dated November 1993.

[Laprie 90] - J C Laprie, Dependability: Basic Concepts and Associated Terminology, ESPRIT Project 3092, May 1990.

[McDermid 94] - John A McDermid, On Dependability, its Measurement and its Management published in High Integrity Systems/Oxford University Press. Volume 1, No. 1 1994.

A Framework for Enhancing the Safety Process for Advanced Robotic Applications

John Elliott, Steve Brooks and Peter Hughes
SRC (Consultants) Limited

Nik Kanuritch
Intelligent Systems Solutions Limited

Abstract

The growing importance of safety to the future applications of advanced robots has led to an intensive series of research activities. This paper presents one such activity aimed at developing a rigorous safety methodology tailored towards such advanced robot applications.

1   Introduction

This paper presents some of the research developments in producing a safety methodology to be applied to the development of advanced robots. This research is being carried out by the ROBUST project supported by the DTI and EPSRC's Safety Critical Systems Advanced Technology Programme.

The safety methodology will be based on enhancing current safety techniques to be more rigorous as well as being tailored to meet the special requirements for robot applications. Robot applications are becoming diverse and demand high performance and integrity, especially where human interaction is involved.

The scope of this paper is to present a generic "framework approach" being used to enhance the safety process and how the enhanced safety methodology will be relevant to robot developers.

Section 2 introduces the concept of advanced robots and Section 3 relates safety to advanced robots. Section 4 introduces the ROBUST project and the rationale and aims of the ROBUST safety framework. Section 5 discusses the framework in detail. Section 6 presents the enhancements that the ROBUST project is making to the safety assurance process and Section 7 presents the conclusions and further work.

2   Advanced Robots

2.1   What are Advanced Robots ?

Robots generally perform simple repetitive tasks especially in manufacturing environments. In such cases, the environment in which the robot operates is highly stable and structured, therefore safety precautions are simple and interaction with humans is kept to a minimum. These robots are pre-programmed to perform actions unless interrupted by human intervention.

Robotic applications are becoming increasingly more advanced and robots have to be more sensitive to changes in their environment. This provides a tremendous technical challenge due to the need to manage the application complexity as well as find technological solutions which provide effective control over the robot behaviour in all environmental conditions.

Whilst there is currently no agreed definition of what constitutes an advanced robot, the general consensus is that an advanced robot is one which has manipulative and/or mobile capabilities. The main functional difference between these robots and current industrial robots is that advanced robots must be able to operate in uncertain or unstructured environments. This lack of structure is not only in the physical environment but can also be a lack of a predefined task structure when a desired outcome is specified but without a precise set of actions for achieving that outcome.

As the technology and the requirements for advanced robots develops then greater autonomy of robotic systems is anticipated i.e. robots without human control. This is where the true advanced nature of robotics systems tends to take off. Advanced robots will possess some degree of autonomy and tasks will not be so well specified, the robots will need to be able to operate effectively and safely in a much more unstructured environment and will need to respond intelligently to uncertain or incomplete data. However, the use of artificial intelligence (AI) techniques and advanced sensors to interpret object images and associated dynamic changes in the environment is still in its early stages of development.

2.2   What is so Special About Advanced Robots ?

Having identified some of the main characteristics of an advanced robot, we need to determine what makes advanced robots different from other systems. Advanced robots can be classified as programmable electronic systems (PES), about which a considerable amount of research and other work has been, and is currently being, performed. [Pegman 94] indicates that the main characteristics which, when taken together, make advanced robotic systems different from other programmable electronic systems are:

- Advanced robots are generally complex, real time systems and the greater the complexity the more difficult it is to predict the behaviour of the robot under all circumstances.

- Advanced robots are designed to interact with their environment and as such are not systems with readily definable boundaries.

- Advanced robots operate in environments which are too hazardous for humans to work in, perform complex tasks which theoretically can be performed by humans or perform tasks requiring close interaction with humans.

2.3   The Limitations Associated with Current Robots

The main limitations associated with current robots which affect the development of advanced robots include:

- Real time sensor data can be uncertain, incomplete and conflicting which results in perception which is slow and unreliable and as a result robot operating procedures at present tend to minimise interaction with people where possible.

- Many different technologies are utilised in an advanced robot, this complicates systems integration and the assessment of robot safety.

- System boundaries are not clear as robots are closely coupled with their environment.

- Safety overrides are often necessary on current robots.

- Current robot behaviour must be predictable at all times to maintain safe operation but future behaviours may be non-intuitive or even, at times, unpredictable.

- Current robot task descriptions are relatively simple, future task descriptions may be complex and of a more generic nature.

Since there is an even wider range of application areas for advanced robots than for current robots, further examination and research into these limitations will be necessary to achieve the wider range of functional capabilities.

3   Safety of Advanced Robots

3.1   Issues Concerning Safety and Advanced Robots

There are three main drivers for considering safety in advanced robots:

- The EEC Machinery Directive means that developers have a mandatory duty to ensure products are safe when used as intended.

- The high technology area, especially in the USA, is becoming more involved in litigation and best safety practice can be viewed as a defence.

- The markets which are the first likely users of advanced robots are ones in which a proven understanding of safety issues and implications is imperative in order to gain sales e.g. nuclear and medical areas.

Having identified the main limitations associated with current robots and the drivers for safety, it is necessary to determine the main safety considerations for advanced robots.

Without a doubt, the most critical aspect of robotic safety is in specifying the robot's environment and identifying the safety related interfaces between the robot and elements of the environment. The more advanced a robot is, the more complex its interaction with the environment will probably be and the more difficult it will be to identify the system boundaries. The robot's environment will therefore need to be analysed in depth. In order to consider safety detailed models of the robot's environment will need to created and the interactions between the robot and the environment examined carefully. These models need to involve the use of techniques such as simulation, and possibly virtual reality, not just functional static models.

Design architectures with proven integrity need to be developed, especially for real time sensor and control data systems, to help overcome the re-use and integration problems such that confidence can be provided in the operation and safety of these areas of the system.

As a result of the uncertainties in the interactions between the robot and the environment it may be difficult to define complete safety requirements which can be verified.

3.2   Applying Current Safety Approaches to Advanced Robots

There are currently no specific standards for considering the safety of advanced robots. The safety life cycles in use today e.g. the IEC safety life cycle [IEC 92], the Def-Stan 00-56 life cycle [MOD 91] are all consistent in use and have similar stages. This safety life cycle is generic in that it can be applied to any application. However the development of advanced robotic systems often involves the use of prototyping, the development of bespoke systems, the development of robots for off-the-shelf use and the use of off-the-shelf components in a robot system. The re-use and integration of existing components is common in the development of advanced robots, i.e. the use of standard design architectures and methods. As a result there may be special considerations and limitations when applying this generic safety life cycle to advanced robots.

Each stage of the generic safety life cycle involves conducting detailed processes e.g. planning, hazard identification and analysis, safety requirements specification, safety risk assessment, safety verification, safety case generation, safety assurance. The major areas where current approaches to safety may be considered weak especially when applying them to advanced robots are as follows:

- Initial hazard identification can be a very subjective process and is dependent on the ability of those performing the analysis and their knowledge and understanding of the robot and its environment.

- Safety requirements are currently specified in a number of different ways, some of which can be difficult to define or interpret in terms relative to the system.

- There can be difficulties in assessing the risk associated with hazards early in the project life cycle when a quantitative approach is used for specifying the probabilities of hazards.

- Application of safety analysis techniques can be subjective and the benefits are dependent on the ability of the analyst and the current state of the system definition.

- It can be difficult to ensure the timely feedback of results of safety analysis into the system evolution process such that the full benefits can be realised.

- It can be difficult to ensure that all aspects relating to safety of a system are considered.

- A safety case is normally compiled at the end of a project rather than built up progressively throughout the project.

- Current approaches to safety analysis only consider the environment informally possibly resulting in an incomplete understanding of the whole system.

Any safety methodology for advanced robots will have to address all, or most of, the points mentioned previously. The ROBUST project is currently developing a safety methodology for advanced robots and the ideas generated to date are discussed in the following sections.

4 Safety Framework Overview

4.1 Background Concepts

The emphasis of the ROBUST project is to develop a rigorous safety methodology to be applied in a variety of development situations, with a bias towards its use for advanced robot applications.

In addition, the application of conventional safety practice to advanced robotics is, in itself, a novel exercise likely to identify a range of problem areas concerning the inadequacies of current techniques. One problem of planning such an exercise is that there are limited examples of advanced robot systems, although some requirements or problem definitions may exist to consider safety issues. As a result, researchers need to sample the robotics field for illustrations and inspiration in order to use known applications to evaluate new development methods or processes, as well as design solutions.

In fact, the feasibility of developing more advanced robot systems partly depends on technological advances, including the existence of an adequate safety methodology. The research problem is to extrapolate the safety methodology findings based on current robot systems to identify the likely safety process requirements for future advanced robots.

The main focus of our research is firmly concentrating on the safety assurance process (referred to as safety process) whose purpose is to provide maximum assurance that a robot system has desirable safe behaviours. These safety processes are required to be separable and collectively define a safety life cycle that will form part of an overall system life cycle. To some extent this separate focus on safety is a part of current practice as embodied in the existing or emerging standards, e.g. [IEC 92] [MOD 91], where safety processes and life cycles are described.

Although the focus is the safety processes, as the safety related systems primarily need to meet operational requirements, any enhanced safety processes must also be closely coupled with all other relevant non-safety related design and development processes. Also, recognition that such design processes consider safety issues, prior to safety assessment activities, is necessary.

In summary, the ROBUST safety methodology is aimed towards reinforcing the separation of the safety process from development processes and the related issues, in addition to concentrating on enhancing the safety process and the associated interactions with other development activities.

4.2   Framework Rationale

Section 3.2 has already raised some of the current difficulties in applying safety techniques to robots. This provides some recognition of the needs and opportunities for safety process improvements. The ROBUST project will advance understanding of the current safety process and develop key components of the revised safety process.

One major thrust has been to acknowledge the generic nature of the safety process across many application domains. Furthermore, the identification of the uniformity in process structure throughout any development life cycle of safety related systems has been a valuable step forward in identifying the core components of the safety methodology.

136

The result is that a generic and uniform development process structure, which emphasises and provide insights into the safety process, has been developed. This process structure is regarded as a basis for developing any part of the safety process. Hence this structure provides the baseline understanding for the ROBUST project activities. This high level process structure is referred as the ROBUST safety framework, or more briefly the framework.

This framework was derived following initial studies on safety issues and techniques, robotics, system modelling, life cycles, formal modelling and process modelling approaches. All of these studies provided valuable insights into the possible structure of a new generation of rigorous safety methodologies.

Prior to the formulation of a framework, it was recognised that the aim of the new methodology is to maximise the utility of existing techniques. The idea is to define a safety methodology which initially links current development techniques with current safety techniques. In addition, the safety process will consist of the enhanced existing techniques as well as incorporating new concepts and techniques which are appropriate to improving system safety. One key area was perceived to involve the significant contribution of safety-oriented modelling techniques.

4.3   Framework Aims

The aims of the framework are:

*   To create a generic and uniform process structure that identifies its constituent elements. This will enable new techniques to be identified and used alongside current best practice for safety and modelling.

*   To emphasise the separation of the safety process from the design and development process reflecting current safety life cycles. This should also provide an appropriate partitioning of the safety process to allow selective use of techniques where beneficial.

*   To define a precise interface between the new safety processes and the existing development and safety processes. These are processes which represent best or current practice by any user.

*   To allow for differences in safety process details across life cycle stages and different technology applications, e.g. for system, hardware and software levels.

*   To be generally directed towards influencing future standards, including being compatible with the major emerging approaches to safety standards development such as the IEC based activities [IEC 92].

- To be flexible so that users of the methodology can easily tailor the process model to their own local life cycles. Any new safety methodology must be generically applicable to attract widespread take up. A generic approach to the system or safety life cycle is desirable so as not to be too prescriptive when users apply the new methodology. In addition, account needs to be taken of any special life cycle features of robotic systems.

- To assist the robot designer in ensuring that the safety requirements are satisfied bearing in mind the need for design trade-offs with other system attributes. Safety should be intrinsic and continuously evolve during the system history. In addition, the safety process should provide standard product guidelines (i.e. acceptable functional architectures) for safe robot systems.

- To define a generic safety case template to be applied throughout the safety life cycle in order to provide the evidence necessary for management to review, and ultimately accept, the current system representation.

- To allow for the use of new, enhanced and existing approaches for considering safety covering a variety of modelling techniques. These processes should allow the system representation to be considered in a formal and systematic manner.

These framework aims are geared towards developing a safety methodology that enhances the current safety techniques for use in advanced robotic applications. This return to basics has been necessary in recognition of the fact that the ROBUST project is developing safety techniques to be applied to advanced robotic systems. Enhancing existing practices involves creating more systematic and formal approaches to applying traditional safety techniques.

Finally, the creation of a framework is considered to be more relevant to the wider safety community in that it is designed to enable safety processes to be enhanced thus creating a means for continuous process improvement.

5 Framework Definition and Structure

5.1 Introduction

This section describes the framework, its relationship with system evolution and life cycles, the main process components relating to a safety related system development activity and the components of the emerging ROBUST safety methodology.

5.2 Definitions

System Evolution is the process of progressively developing (refining or changing) a system through its life history. Hence evolution need not be fixed to any life cycle stage, yet the processes involved may be applied and mapped onto any chosen life cycle. This is an important idea in defining processes to apply through a life history.

System Representation is an expression of all information of a system or component for any stage in its evolution. This will include functional and non-functional specifications including those for each system attribute such as performance, safety, reliability, availability, maintainability, security, etc. This may be in the form of a number of separate documents. For functional elements, the system representation may be presented in structured English or pseudo code, formal or semi-formal languages or in a structured or object-oriented form.

5.3 Generic Approach to System Evolution and Life Cycles

The safety methodology will acknowledge the existence of a range of system and safety life cycles, yet define the safety related system development processes in a generic manner to represent how systems evolve. This may be argued to be the creation of a generic life cycle, although any user of the ROBUST methodology will be able to apply the techniques with a locally defined life cycle.

Figure 1 shows the generic life cycle as a series of connected "process layers" each relating to a defined phase of the system life cycle. The actual layers shown are illustrative, the specific layers used needs to be defined for the local life cycle being used.

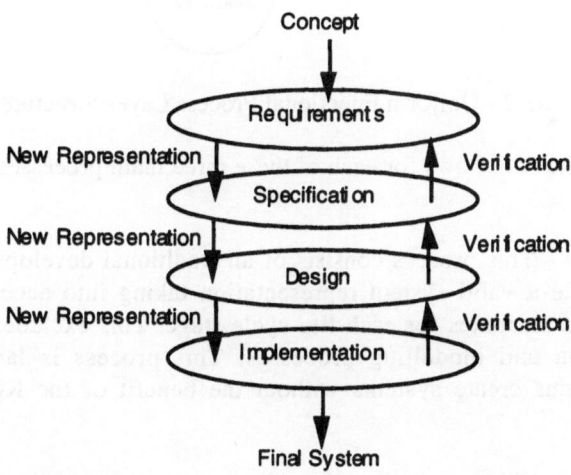

Figure 1 - Illustrative Life Cycle Process Layers

The combined effect of the generic life cycle and the uniform process layers is consistent with known life cycles with waterfall, spiral, assembly, safety, validation and verification characteristics.

5.4　High Level Process Structure

For each process layer, analogous with a life cycle stage, e.g. requirements, the framework initially defines a uniform high level process structure. This consists of three main processes which apply to every stage of system evolution.

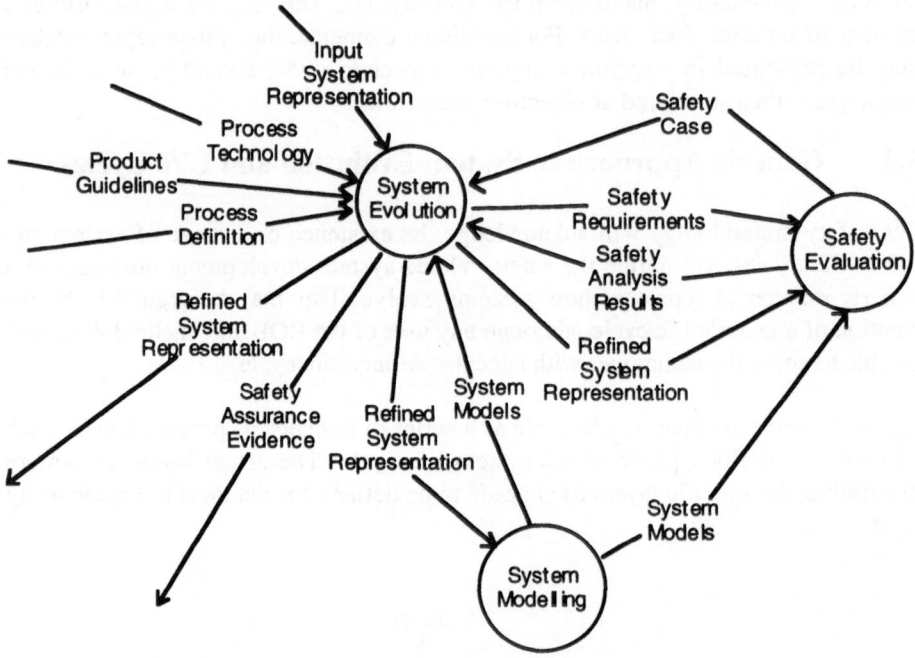

Figure 2 - Uniform Functional Process Layer Structure

Figure 2 shows the data flows for each of these three main processes. Briefly, these processes are:

System Evolution - This process consists of all traditional development processes required to create a valid system representation taking into account all system requirements and attributes for each life cycle stage. This excludes the enhanced safety evaluation and modelling processes. This process is largely how the development teams create systems without the benefit of the ROBUST safety methodology.

System Modelling - This process examines the system representation ensuring that a valid model of the system is generated for use in the safety evaluation process. The system model also feeds into the system evolution process.

Safety Evaluation - This process consists of the full and enhanced safety processes which are used to analyse system representations. The results of the safety evaluation process are used to assist system evolution by providing early analysis results and by generating a safety case. The safety evaluation process uses and refines safety requirements, system models and safety models in generating a safety case for the current system representation. Some of the techniques will be relevant to developers who are not applying the full ROBUST safety methodology.

The system evolution process represents what developers do when they create designs to ensure that the final system meets its intended operational requirement. This activity includes how designers achieve intrinsic safety by following safety guidelines for architectures or safety features.

In contrast, the safety evaluation process is the main focus for defining safety requirements, analysing designs and verifying the results of the analysis against these requirements. This process also involves using modelling techniques, creating a detailed safety specification and compiling a safety case for the system representation.

This safety evaluation process is the core of the new safety methodology in that it will enhance the use of current techniques and make effective use of models and rigorous verification methods. A fuller explanation of how the methodology will enhance the safety life cycle is given in Section 6.

5.5   Major Components of the Framework

The three main process components (system evolution, safety evaluation and system modelling) of the framework have been partially refined to show the major components of the final safety methodology. This level of detail, shown in figures 3 and 4, shows the anticipated interactions and information flows of the major components.

These components are being developed and tested using current studies and trials, and will ultimately be evaluated using a real robotic application. Eventually, the framework with all its method and process details will be documented in a Safety Methodology Handbook.

The major components of the main three processes are now described indicating how the safety methodology is being developed.

5.5.1   System Evolution

Refining system representation - This process progresses the system representation to a validated and accepted system representation for the next process layer. This involves the technical design activities as well as verification, testing, management and quality control activities. This process largely excludes assessing the safety of representations against safety requirements, apart from incorporating safety features and using safe design guidelines. The final safety methodology will enable a generic

141

interface between the safety process and general system evolution process to be developed.

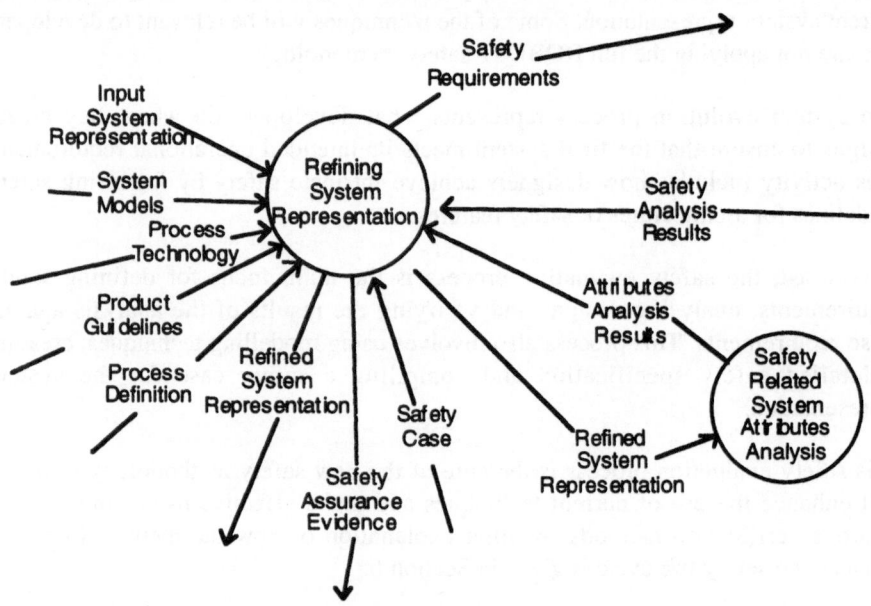

Figure 3 - System Evolution Process Model

Safety related system attributes analysis - This process will identify all system attributes such as reliability, availability, maintainability, performance and examine their relationship with safety in order to provide balanced solutions to any trade-off decisions required. In essence, this process provides a means of reviewing a system representation to ensure that it is suitably balanced in terms of cost, performance etc. as well as being acceptably safe.

5.5.2   System Modelling

This process has not been refined in this section, however it is discussed further in Section 6. The system modelling process examines the system representation ensuring that a valid model of the system is generated. The extent to which designers use modelling techniques may vary and depend on the life cycle stage. These design models are translated into system models for analysis purposes. These models are mathematically oriented using propositional logic expressed in a way that is amenable to tool support. These system models are used as part of the safety evaluation process and the system evolution process.

5.5.3   Safety Evaluation

Safety analysis - This process applies safety analysis techniques to study hazards and their consequences. The safety analysis techniques will apply traditional techniques

142

in a more integrated and formal manner than at present. This will involve the use of modelling to identify and analyse hazards and the creation of a complete and separate safety specification.

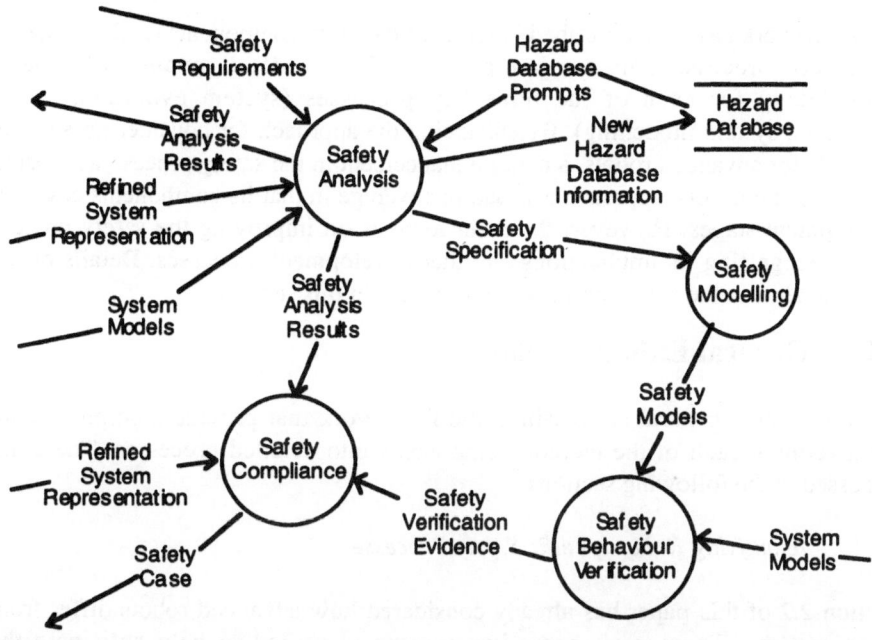

Figure 4 - Safety Evaluation Process Model

Safety modelling - This process creates models that describe the desirable safety behaviour and properties of the system to allow further rigorous verification studies to be undertaken. The form of these models may vary depending on the safety specification but they will formalise what is safe and unsafe. This is so that rigorous examination, using a combination of animation, simulation, formal verification and testing approaches, can be conducted.

Safety behaviour verification - This process generates safety compliance evidence by verifying the behaviour of the system model with the safety model. The safety compliance evidence may consist of many forms of analysis and proofs, each focusing on selected relevant aspects of the system (e.g. logical control) or safety (particular undesirable observations).

Safety compliance - This process collates all the analysis results concerning safety, performs safety risk assessment and prepares a safety case. This safety case is used as the basis for reviewing the acceptability of the current system representation from a safety perspective. The essence of this activity is to identify the approach and criteria for achieving a balanced argument incorporating the evidence on quantitative, qualitative and behavioural aspects of safety.

6   Enhancing the Safety Process

6.1   Introduction

The framework has been described in terms of the main components used to enhance the safety process. This section provides further detail about the general methodology for each of the three key processes (system evolution, safety evaluation, system modelling). By enhancing the approach for considering a safety life cycle for advanced robots, a major enhancement in the safety process will result. Hence the framework approach is broad in coverage in that its components cover all development stages. However, the main focus is on improving the safety process whilst recognising its implications on other development processes. Details of the enhancements to the safety process are discussed in this section.

6.2   General Enhancements

There are some general areas within the framework that provide a common core when refining each of the method components into detailed processes. These are discussed in the following sections.

6.2.1   Identifying Robot-specific Safety Processes

Section 2.2 of this paper has already considered how advanced robots differ from other systems. There were some key aspects identified to help anticipate the problems which will be faced when studying the safety requirements for advanced robots. Therefore there must be special safety processes that will be required to address these special robot characteristics.

6.2.2   Robot Description Techniques

A general requirement for any methodology is that appropriate information about the system and its safety must be formatted and expressed in a usable manner for both system modelling and safety evaluation studies. This information should be available at each stage of the system life cycle. If the robot is described in an inappropriate or incomplete manner then subsequent analysis of the system and its safety properties is likely to be incomplete. Hence the system representation will be progressively more difficult to validate or verify. A standardised approach to specifying robots is being produced as an important input to the ROBUST project. This is imposing a structure on specifications which separates safety aspects and provides the information needed for system and safety modelling.

6.2.3   Hazard Information Database

A prototype information database is being developed for recording and retrieving information about known hazards for robotic applications. This is a computerised checklist of hazard information which assists in the hazard identification activity. As robot systems have many diverse application areas, a comprehensive hazard checklist derived from previous known applications is a valuable source of information. Also as advanced robot systems become more common, maintaining such a hazard information database should be an asset to designers.

6.2.4   Guidance of Design Safe Robots

Guidance material on how to build intrinsic safety into a design is being developed. This guidance will include reference to safe architectures and safety features for systems in general and for robotics in particular. Robot systems appear to created around generic architectures and it is expected that special robot safety features can be embedded into any enhanced generic architecture. If designers deviate from any standard architectures then any subsequent design decisions will require full justification which will form part of the complete safety case for the appropriate system representation.

6.3   System Evolution Enhancements

6.3.1   Life Cycle and Evolution Stages

The method processes will apply throughout the life cycle in that all aspects of system evolution, system modelling and safety evaluation will be employed. In addition, the differences in process details arise when different technologies are used, e.g. hardware, software. The effect is that the safety evaluation and system modelling processes need to be interpreted for the different stages of system evolution. So for software, the specification or the source code may be regarded as the system model. The safety evaluation process needs to be defined to analyse, specify and verify safety properties of the software.

6.3.2   System Evolution Interfaces

The system evolution process needs to be generic, yet adequately detailed, to address the interactions between system evolution and safety evaluation. Furthermore, the process should clearly define how modelling can be of assistance.

6.3.3   Safety Related System Attributes

This process involves systematically relating safety to other system attributes by identifying their influence on safety.

6.3.4   Separation of Development Issues

The framework identifies system evolution, system modelling and safety evaluation as the major processes to be performed. The key aspect is that the system evolution process creates a validated representation which meets the customer requirements. The system modelling process ensures that a valid model is produced and associated analyses are rigorously performed, primarily in support of the safety evaluation activities. The safety evaluation process provides comprehensive evidence in the form of a safety case for the system or any component.

6.4   Safety Evaluation Enhancements

6.4.1   General Enhancements

Enhancements are achievable through process and product improvements by being systematically and formally (e.g. mathematical) rigorous in safety analysis, safety specification, safety modelling and safety verification. The framework demands that safety and development activities are closely linked and that modelling support is provided for safety evaluation. Also current safety techniques are expected to be enhanced through their improved integration based on a greater understanding of what these techniques actually achieve. The consequences of some of these enhancements is that safety evaluation will be more objective in hazard identification, safety specification and safety compliance activities.

In addition, the new approach to safety is to be more rigorous and systematic early in the life cycle especially at the concept stage. This is expected to result in the development of a separate, explicit safety specification document which is discussed later in this section. This safety specification is expected to be the basis of progressive safety compliance assessment at each development stage. The safety specification and subsequent safety compliance analysis require objective understanding of the application, the system and its environment. In addition the safety specification will require a well defined format and structure.

In order to create intrinsic safety, system designs must include adequate safety and protection features. One key aspect is to employ mature and standard architectures requiring strict justification for any deviations.

6.4.2   Uniformity of the Safety Process

The framework has been formulated to highlight the salient processes of a rigorous safety methodology. There is reasonable evidence, from standards in particular, to suggest that a uniform and generic framework is a sound approach to defining a safety process that can be related to any life cycle. Furthermore, this uniformity of the safety process applies between technologies such as systems, hardware and software, and across different applications and sectors.

6.4.3   Understanding the Theory and Practice of Current Safety Techniques

A detailed study of the current approaches to safety has been undertaken. This study considers safety life cycles, constituent processes and compiles a structured and formalised description of widely used safety analysis techniques. This understanding has been essential as the domain of advanced robots is immature particularly where safety evaluation is concerned.

6.4.4   Integrating and Formalising the Safety Process

The effective use of current safety techniques is of interest in the robotics domain. Most techniques are concerned with the review, inspection and analysis of a given system representation which is often a detailed design, e.g. a circuit diagram. This limitation may be a reflection of the difficulty of defining and demonstrating safety, or it may be a limitation of current safety techniques.

One aspect of enhancing the safety process is to consider how best to combine and integrate the use of current techniques. The information generated by applying some safety techniques may be used as inputs to other techniques. In addition, applying different safety techniques to the same design may provide two set of results that can then be shown to be consistent.

Another aspect in attempting to understand and strengthen the meaning of safety techniques, such as Fault Tree Analysis and Sneak Circuit Analysis, is to formalise these techniques based on well formed, perhaps mathematically based, languages. One benefit of this approach is to provide links with current system modelling techniques. In this way, the use of models to assist these safety techniques may be based on a strong foundation. Currently, formalising and integrating current safety techniques is in its infancy with significantly more research being required.

6.4.5   Safety Specification

The aim of this activity is to define and produce a template for a safety specification which provides an improvement on current specification practice. This safety specification is expected to cover all aspects of system safety, e.g. quantitative, qualitative and behavioural aspects. The aim is to ensure that safety requirements are always generated before the corresponding system representation in order to implement a positive approach by defining the levels of safety to be achieved. Safety specifications need to be precise, consistent, complete, traceable, verifiable and capable of being refined throughout system evolution.

6.4.6   Using Models Within Safety Evaluation

This area is receiving a great deal of attention in developing the new safety methodology. Robot models and environment models are being developed and used to enhance the understanding of safety related interactions between the robot and its environment. These models will be used to assist in identifying hazards and be refined to apply to systems, hardware and software. Software based system models will provide a distinct advantage in understanding system safety properties. In addition, safety models will assist in improving the objectivity of the safety compliance assessment. These safety models are descriptions of safe and unsafe

147

properties, or states, including the safety related interactions between the robot and its environment.

6.4.7   Increasing Objectivity in Safety Evaluation

The use of system models and safety models will lead to improved objectivity in hazard identification, risk assessment, assigning and apportioning integrity levels and in applying defined safety criteria.

6.4.8   Developing the Safety Case Concept

Safety cases are required in many sectors of industry, albeit in different forms. A safety case template is being developed which will provide a rigorous argument showing why each representation is considered adequately safe or unsafe. The argument will use information derived from the various detailed safety evaluation processes.

6.4.9   Using Generic Safety Processes

The use of the generic safety process is applicable to a range of projects, sectors and organisations. By being generic, the user may be selective in how the process is implemented. The safety methodology being developed will provide an extended range of safety techniques to be applied in a rigorous and systematic manner. Also, this process will have to take into account the application of safety techniques to new and existing systems. Overall safety is dependent on the integrity of individual components and their integration into the system. These components may be new, bought-in or re-used and may be used with or without modifications. The re-use and bought-in aspects require different safety evaluation approaches from those used for new systems.

6.5   System Modelling Enhancements

6.5.1   Modelling Strategy

A key ingredient of the framework is in the use of models. Models are used to enhance understanding of the intended system and to provide a means of support in studying safety. In addition, another aspect of the modelling that is being used is the idea of a safety model which describes the conditions for the system to remain in a safe state. The safety model is compared with the system model to demonstrate that the safety properties of the system are maintained.

6.5.2   Modelling Aim

The main aim of the system modelling activities is to create models of the robot and its environment so that the safety of the system can be considered in a structured and rigorous way. Two models are created, the first shows the behaviour of the robot and how the robot responds to the inputs presented to it. The second model shows the interactions between the robot and its environment. By using these models, the robot and its interactions with its environment can be better understood.

6.5.3 Modelling Criteria

The models which are being used on the ROBUST project need to be capable of demonstrating safety within the system. In order to be confident that the model demonstrates safety there needs to be some criteria to ensure that safety has been considered in designing the system. For example, one criterion that needs to be considered is to analyse whether the system could ever get into an unsafe state. The modelling techniques being used therefore need to consider the idea of safe states within the system. Another criterion is the interactions between the robot and its environment and the hazards that can arise as a result of the introduction of the robot into the environment. A model is therefore needed which can address the interactions between the robot and its environment.

6.5.4 Safety Criteria

Traditional safety criteria are both quantitative (considering failure probabilities) and qualitative (considering risk categories or system integrity). Further objective criteria are needed for focusing on whether safety requirements for the system have been satisfied. A simple measure of system safety considers whether the system is in a defined safe state which describes the system in terms of its internal variables and its external properties.

Other attempts at measuring safety behaviour may consider logical safety rules that must always be true for safety to be preserved. These rules are expected to form the basis of the ROBUST project's approach to safety modelling. It is intended that these rules can be used to determine if the system can ever enter an unsafe state, given both normal and abnormal operating conditions. This safety model approach can confirm whether the system is acceptably safe thus avoiding the need to evaluate all states of the system.

6.5.5 Elements of Models

The robot model focuses on definitions of *states* which the robot can be in, allowable state transitions, conditions for safe transitions and unsafe states. The robot model also needs to describe the dynamic behaviour of the robot so that safety can still be considered as the state of the robot changes.

In contrast, the environment model focuses on *objects* in the robot's environment. The model describes the life cycles of the objects, the interactions between the objects and the interactions between the objects and the robot. The state of each object in the environment can be examined by considering where it is in its life cycle. The state of the whole environment is therefore defined by the collective state of all the objects in the environment. The interactions between the robot and the environment arise from the relative state of each and the inputs which are presented to the system.

6.5.6   The Modelling Techniques

The main technique for modelling the robot uses the system representation for the current stage of the life cycle. This description of the robot is converted to a state based model of the robot. The state based model considers the main states of the robot and the events which cause the robot to move from one state to another. The model can be built in a hierarchical way which initially only considers the main states of the robot and moves into more detail as the system representation is refined. This state based model is then converted into a mathematical model of the robot using propositional logic to describe the state of the robot. The mathematical model can then be analysed to test the transitions between states, test the conditions for the robot to be in each state and to ensure that the robot cannot enter an unsafe state. However, the accuracy of the analysis results will reflect the accuracy of the system model.

The robot model can be constructed at any stage of the system life cycle from any description of the robot which addresses its functional behaviour. Robot models can be built from informal specifications (e.g. natural language), structured specifications (e.g. Yourdon) and formal specifications (e.g. Z).

The technique for modelling the environment is to consider the objects in the environment. As with the robot model the environment model is considered in a hierarchical manner by first considering the main objects in the environment and then considering the objects which comprise these main objects. In this way the environment can be considered in a structured way and the model can become more detailed as the system representation is refined. Once the objects have been identified the model then examines the states that each object can be in. From this model the interactions between the robot and the different objects in the environment can be considered and this information can be used to identify hazards. The model provides a structured way of considering the interactions and in turn a structured way of identifying and analysing hazards involved with introducing the robot into the environment.

The environment model can be used to consider both functional and non-functional hazards associated with introducing the robot into the environment and can also examine human factors which may affect the safety of the robot.

6.5.7   Use of Models

The system model is used to analyse the dynamic behaviour of the system and compare the behaviour of the system with the description of the safe behaviour of the system contained in the safety model. If the analysis identifies that the system can enter an unsafe state, the models are then used to identify how the system could enter that state. The models can also be used to consider how modifications to the system could overcome any problems.

The environment model can be used to generate a list of hazards which may supplement, or replace, traditional hazard identification activities. Once the hazards have been identified, the model can be used to consider ways in which the hazards could arise.

6.5.8 *Modelling Issues*

The modelling techniques developed only consider certain aspects of the system. The models developed so far consider the dynamic behaviour of the system in a state based model and the interactions between the components of the robot and its environment in an object oriented model. There are other aspects of the system which are currently not modelled, for example the temporal behaviour of the system is not rigorously analysed in the models. The models are currently therefore not a complete reflection of the system representation, but they are useful for considering certain aspects of the system.

One of the main aspects of an advanced robot is the software which controls it. The software could be very complicated, so there needs to be some way that the behaviour of the software can be modelled so that the safety of the system can be assured. Further work is being carried out as part of the ROBUST project to examine ways in which the whole system can be addressed.

The modelling work presumes that the format of the system description is such that a model can be built. So far the state based modelling has been performed on a variety of system descriptions. If there is a system representation which cannot be directly translated into a system model it may be necessary to include an extra step in the modelling process to first convert the system description into a form that can be modelled. This extra translation increases the risk that some information will be lost for the original specification or that errors may be introduced by the translation process.

Further work in the modelling area may lead to more animated models. Already there are several graphical tools which are used for developing robots which use simulation and virtual reality techniques. These tools are very useful for visualising what an intended robot will look like and can be useful for examining how the robot will function in a particular situation. However simulation tools lack the mathematical rigor which is desired for safety analysis. Future development of the modelling work may be able to combine the graphical techniques which are useful for visualising the behaviour of the system with mathematical tools to provide more rigor.

151

7 Conclusions and Future Work

This paper provides an introduction to advanced robotics and identifies the main problems associated with current safety approaches when applying them to advanced robots. The ROBUST project provides a methodology for applying safety processes to advanced robots. This methodology is based on the concept of a safety framework. This framework is a process structure which outlines those processes necessary to ensure that the system safety is addressed in rigorous and systematic manner. The framework outlines the three main generic safety processes which are to be applied at each stage of a system life cycle. These processes comprise the following:

The system evolution process which consists of all traditional development processes required to create a valid system representation taking into account all system requirements and attributes for each life cycle stage. This process is largely how the development teams create systems without the benefit of the ROBUST safety methodology.

The system modelling process which examines the system representation ensuring that a valid model of the system is generated for use in the safety evaluation process. The system model also feeds into the system evolution process.

The safety evaluation process which consists of the safety processes which are used to analyse system representations. The results of the safety evaluation process are used to assist system evolution by providing early analysis results and by generating a safety case. The safety evaluation process uses and refines safety requirements and system models in generating a safety case for the current system representation.

The enhancements to traditional safety processes resulting from the application of these ROBUST processes have been discussed in detail. These enhancements will be tested on trial robotic systems and refined during the course of the ROBUST project. The final results will be presented in the form of a Safety Methodology Handbook.

8 References

[IEC 92]   International Electrotechnical Commission: "Functional Safety of Electrical/Electronic/Programmable Electronic Systems; Generic Aspects: Part 1, General Requirements." SC65A/WG10 Draft Document, Version 5, (IEC Reference: 65A Secretariat 123) January 1992.

[MOD 91]   Ministry of Defence: "Interim Defence Standard 00-56/Issue 1 - Hazard Analysis and Safety Classification of the Computer and Programmable Electronic System Elements of Defence Equipment." MOD, April 1991.

[Pegman 94]   G J Pegman: "Safety and Standards for Advanced Robots", Colloquium on Safety & Reliability of Complex Robotic Systems, IEE Digest Number 1994/085, 1994.

Safe Systems for Mobile Robots
The Safe-SAM project

D.W. Seward, F.W. Margrave
Department of Engineering
I. Sommerville, G. Kotonya
Department of Computing
Lancaster University, UK

Abstract

Development of Large Mobile Robots (LMR) is an area where there are many problems associated with safety. Their requirement to operate in unrestricted, undefined, unstructured environments, and often in relatively close proximity to human beings, present special problems beyond those normally considered in the use of programmable electronics and safety critical software.

The SAFE-SAM project is examining the design of system architectures for use with LMR and the implications of system design, requirements definition and safety and hazard assessment. The concept of an independant safety manager is proposed.

Many existing methods are considered as insufficient for development of LMR and work is investigating new approaches to the design of safety critical systems. Two methods are outlined in this paper, one considers the approach of requirements engineering, the other a safety analysis method. Current practice in these fields often does not begin work early enough in the design process, the authors advocate these issues must be considered as an important part of the initial design phase.

1 Introduction

There is currently a great deal of interest and active research throughout the world in the field of large mobile robots. Applications range from fire-fighting, handling of hazardous materials, nuclear de-commissioning and sub-sea activity to general construction robots. When the time comes for such technologies to reach the marketplace, safety will be a vital issue. Indeed unless considerable research effort is put into addressing the safety issues, it is conceivable that the future exploitation of such robots will be severely handicapped.

In the context of this project "large" means of sufficient size, weight and power to cause significant damage to people, property or the environment. Robots in the above categories differ from conventional industrial robots in four key ways, all of which have very important implications for system safety:-

- Mobility

- Higher power to weight ratios

- More intelligence - to provide autonomy to tackle less well defined problems

- More external sensors - to determine appropriate behaviour in unstructured environments

In addition, the behaviour of these robots must be considered to be **non-deterministic** for the following reasons:-

1. The end-user may need the facility to modify the behaviour of the robot in order to 'train' it to carry out new tasks.

2. The use of heuristic rules is probably essential for flexible operation.

3. They will operate in unpredictable and unstructured environments.

These factors provide a particularly rich setting for the testing and development of advanced safety critical technologies.

2 Background

The issue of the safety requirements of advanced robots is recognised as important, and is the subject of a DTI publication [DTI,90]. However the current state of the art has not yet progressed beyond the recognition of the problem, and the identification of potentially useful background technologies and standards. No specific work has been completed on suitable system architectures for robots which possess a degree of autonomy, are non-deterministic and operate in unstructured environments.

As a result of the above seminar Advanced Robotics Research Limited [Advanced,92] have published a document entitled "Safety and Standards for Advanced Robots - a First Exposition". This excellent report summarises the current situation and clearly exposes the fundamental dilemma facing advanced robot development. It concludes:-

"Certain functions of an advanced robot i.e. its ability to interact with a dynamically changing world, cannot readily be achieved other than by the use of symbolic software representations. To mandate the use of formal methods is in effect to deny this functionality...."

"The issue of artificial intelligence in safety critical applications causes concern and has been side-stepped in the existing standards committees, although they are aware of the problems."

It is the aim of this project to address this dilemma.

The need for explicit safety requirements has been recognised [IEE,89] in the proposed model of the safety life-cycle. However, in the literature concerned with requirements specification, there has been little or no mention of safety issues. The literature on safety critical systems is almost equally sparse on safety requirements achievement. While there has been a great deal of research concerned with safety achievement through formal methods for systems specification (i.e. specification analysis), this is predicated on the assumption that a requirements specification has been formulated and agreed [Atkinson,91], [Bowen,92], [Moser,90].

The construction robotics field is most active in Japan where workers [Gocho, 92] have demonstrated an autonomous hydraulic loader in an asphalt plant. However they have achieved safety by a rigid separation of the vehicle from all manned operations. The vehicle simply stops if any other body enters its enclosure, and this clearly imposes severe operational limits.

3 The Application

For the past three years the Engineering Department at Lancaster has been engaged on a research program in construction robotics. This has concentrated on the automation of the excavation process. An SERC grant was obtained for making site studies of current excavating techniques, and a fifth-scale working model of an excavator arm was constructed to develop automatic strategies. Following on from this work, support from the JCB company enabled the technology to be transferred to a full-sized tracked mini-excavator - LUCIE - the Lancaster University Computerised Intelligent Excavator - see *fig 1*. The following aims were adopted:-

- The project - to concentrate on automating the excavating arm and **not** the mobility of the vehicle

- The task - to excavate a high quality, flat-bottomed trench in variable ground.

- The machine should cope with obstructions in the ground such as boulders.

- It should be self contained with no external power supplies or computers.

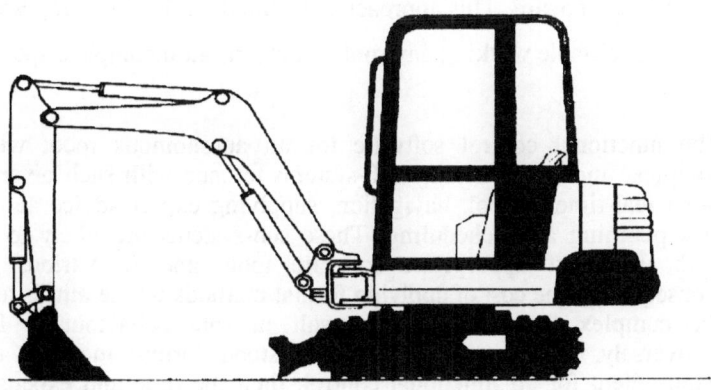

Figure 1. LUCIE - Lancaster University Computerised Intelligent Excavator.

Field trials have shown that these aims have now been substantially achieved. The success of the LUCIE lies in the realisation that the digging strategy must be highly adaptive to cope with the enormous variability and unpredictability of the ground. The conventional robotic approach of a pre-determined path for the tip of the excavator bucket is doomed to failure.

Previous attempts to follow this approach resulted in inadequate filling of the excavator bucket, or worse still, machines of this type are powerful enough to overturn themselves under certain circumstances. Thus the machine must have autonomy to adapt its behaviour to an unknown environment. This clearly presents problems for ensuring safety critical operation.

On the prototype machine, safety is ensured by a tilt sensor and a temporary remote "dead-man's handle" which switches off the machine engine and locks all hydraulic valves. At the present time the excavator must be positioned manually. We are however currently automating the vehicle tracks, and when the excavator becomes fully mobile, a much more rigorous consideration of safety issues is required.

4 The Approach

At the heart of the research is the concept of a safety manager, which it is hypothesised, is the most appropriate approach to safety assurance in large mobile robots. This is conceived as an independent distinct entity, whose job it is to monitor the environment, and give permission for all behaviour which could have a safety critical component. This is a **behaviourist approach** in that it is concerned with achieving safe behaviour, but is not concerned with the processes that determine functional behaviour. Clearly, in the interests of efficiency and reliability, the processes that control functional behaviour should be rigorously designed using the best software engineering practices to maximise safe behaviour. Ultimately the safety manager is, however, responsible, and will block all actions that might create a hazard. This approach is justified on the following grounds:-

1. An unspecifiable working environment means an incomplete specification.

2. The functional control software for an autonomous robot will be very complex, and contain many sub-systems to cope with such diverse tasks as rapid real-time control, navigation, supplying expert advice and long-term task planning and scheduling. These sub-systems are likely to be written with the most appropriate software tools and in various languages. Consequently the cost of applying formal methods whose aim is to show that the complex interactions will result in safe behaviour is formidable. Conversely, if the use of well understood formal methods is made a requirement for all functional control, then the time and expense involved will mean that development of intelligent robots will be significantly retarded.

3. Two schools of thought have recently emerged concerning the most appropriate method of adding intelligence to mobile robots. The conventional approach is to use powerful processors to apply sophisticated reasoning to complex world models. The other school, championed by Brooks [Brooks,86] at MIT, believes in simple reflex actions using such techniques as neural nets. It seems likely that, for tackling real problems, both approaches will need to be integrated. A well designed safety manager is independent, and can cope with either philosophy.

At this stage it is felt that the safety manager should only be concerned with high level behavioural activity, and that the responsibility for safety critical components, such as sensors should be distributed. However the precise nature of this division of responsibility is still a matter for the research.

5 Methodology

The work is divided into six work packages, each of which yields a major deliverable:-

1. **Safety analysis** - The aim is to carry out a thorough assessment of the possible hazards and level of risk associated with large mobile robots, and to assess specific hazards for robot excavators.

2. **Production of a safety requirements specification** - The rules necessary for safe behaviour will be defined in terms of the 'whole system' and not just the software.

3. **Hazard partitioning** - The objective is to identify the functions and components that are relative to the safety argument and to decide whether they are handled centrally, locally or by a mixed approach.

4. **Information requirements of the safety manager** - Consideration must be given to the nature and amount of information required by the safety manager. This applies to information concerning both the internal state of the robot and its environment.

5. **System architecture design** - The aim is to investigate the possible architectures of a safety manager that can implement a safety requirements specification without unduly affecting the functional performance of the robot.

6. **Prototype development** - A real potential hazard arising from the LUCIE excavator will be selected and a demonstrator prototype of the safety manager will be developed and demonstrated in order to prove the system concept.

Work is progressing concurrently on the first three of these tasks. At the same time the functional system hardware for the LUCIE excavator is being improved so that it will form a reliable testbed for the safety manager concept.

Because of the lack of previous work in defining safety requirements, we are extending an existing approach which is based on **viewpoints** as proposed in Kotonya and Sommerville, [Kotonya,92a]. This work is explicitly geared towards making the relationship between non-functional and functional requirements explicit and so is a particularly suitable basis for the work proposed here. An extensive software requirements tool - **VORD** has been developed and extended to accommodate safety analysis. This is described in the next section of this paper.

Also existing methods of hazard and risk analysis have been reviewed and no single technique has been found to be ideal. Consequently a new synthesis has been developed known as **CLASH**, and this will be described in section 5.2

5.1 Requirements Engineering and Safety Analysis

The notion of viewpoints as a means of organising and structuring the requirements engineering activity is now well-known. Viewpoints are implicitly present in SADT [Ross,77] and were first made explicit in the CORE method [Mullery,79] . Since then there have been various other viewpoint-based approaches and proposals [Fickas,91] , [Leite,89]; [Finkelstein,90], [Finkelstein,92] . We have summarised these methods and described our own work on viewpoints for interactive system design elsewhere [Kotonya,92a].

In the initial work, the model which was adopted for viewpoints was a service-oriented model where viewpoints are analogous to clients in a client-server system. The system delivers services to viewpoints and the viewpoints pass control information and associated parameters to the system. Viewpoints map to classes of end-users of a system or with other systems interfaced to it.

This approach can be used to support a user-centred design process [Norman,86]. Like user-centred design it tends to focus the RE process on user issues rather than organisational concerns and this leads to incomplete system requirements. To allow these organisational concerns to be taken into account, we have now extended the concept of viewpoints to consider other inputs apart from direct clients of the system. Viewpoints now fall into two classes:

1. *Direct viewpoints* These correspond directly to clients in that they receive services from the system and send control information and data to the system. Direct viewpoints are either system operators/users or other sub-systems which are interfaced to the system being analysed.

2. *Indirect viewpoints* Indirect viewpoints have an 'interest' in some or all of the services which are delivered by the system but do not interact directly with it. Indirect viewpoints may generate requirements which constrain the services delivered to direct viewpoints.

While the concept of a direct viewpoint is fairly clear, the notion of indirect viewpoints is necessarily diffuse. Indirect viewpoints vary radically from engineering viewpoints (i.e. those concerned with the system design and implementation) through organisational viewpoints (those concerned with the systems influence on the organisation) to external viewpoints (those concerned with the systems influence on the outside environment).

Indirect viewpoints are very important as they often have significant influence within an organisation. If their requirements go unrecognised, they can often decide that the system should be abandoned or significantly changed after delivery. This is particularly true for some classes of safety-related systems which must be certified by an external regulator. If certification requirements are not met, the system will not be allowed to go into service.

5.1.1 Viewpoints Oriented Requirements Definition

Based on this notion of viewpoints, we have developed a method for requirements engineering called VORD (Viewpoint-Oriented Requirements Definition) which covers the requirements engineering process from initial requirements discovery through to detailed system modelling. For the purposes of this paper, the later system modelling stages of the method are not important. The discussion here therefore describes on the first three iterative steps in VORD namely:

1. Viewpoint identification and structuring

2. Viewpoint documentation

3. Viewpoint requirements analysis and specification

The first step, viewpoint identification and structuring, is concerned with identifying relevant viewpoints in problem domain and structuring them. The starting point for viewpoint identification are abstract statements of organisational needs and abstract viewpoint classes. Figure 2 shows some viewpoints identified in relation to an autonomous digger (JCB), using VORD.

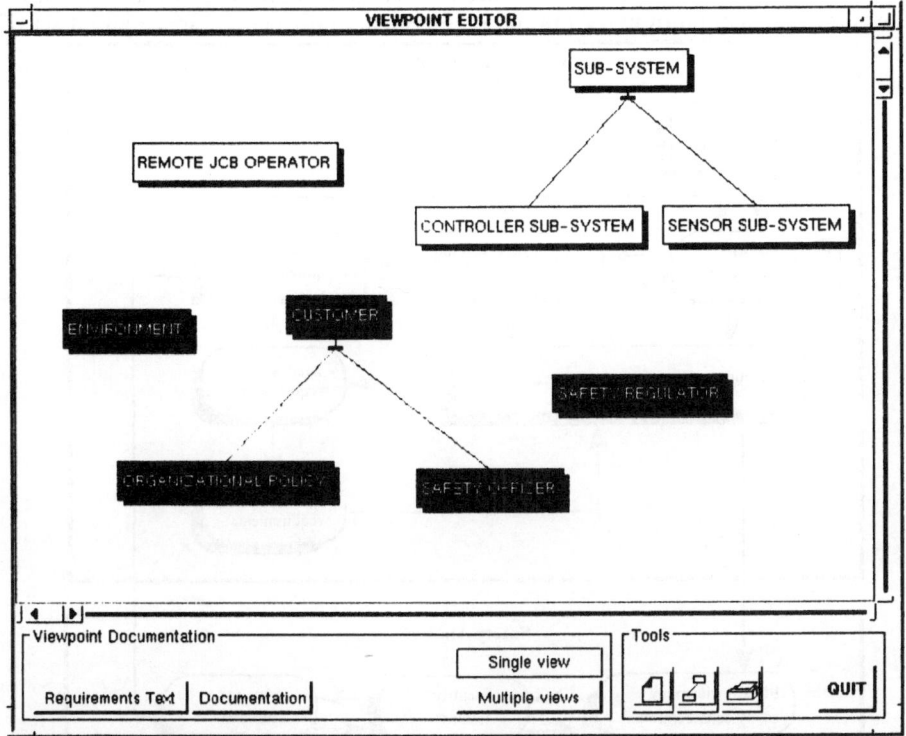

Figure 2. JCB Viewpoints.

The second step is concerned with documenting the viewpoints identified in step 1. Viewpoint documentation consists of documenting the viewpoint name, requirements, constraints on its requirements and its source. Viewpoint requirements include a set of required services, control requirements and set of non-functional requirements.

The final step is concerned with specifying the functional and non-functional viewpoint requirements in an appropriate form. The notation used depends on the viewpoint, the requirement and the requirements sources associated with the viewpoint. Appropriate notations range from natural language (if the requirements source is concerned with non-technical requirements), through equations (if the requirements source is a physicist, say) to system models expressed in formal or structured notations.

Viewpoints and their requirements are collected into a central repository that serves as input to the requirements analysis process. The objective of the analysis process to establish the correctness of the documentation and to expose conflicting requirements across all viewpoints.

5.1.2 Deriving Safe System Requirements

The VORD process has been extended to incorporate an explicit safety analysis activity whose results are used to modify (where necessary) suggested system requirements. This is illustrated in Figure 3. The safety analysis process is based on viewpoint information drawn from the requirements information space and a set of

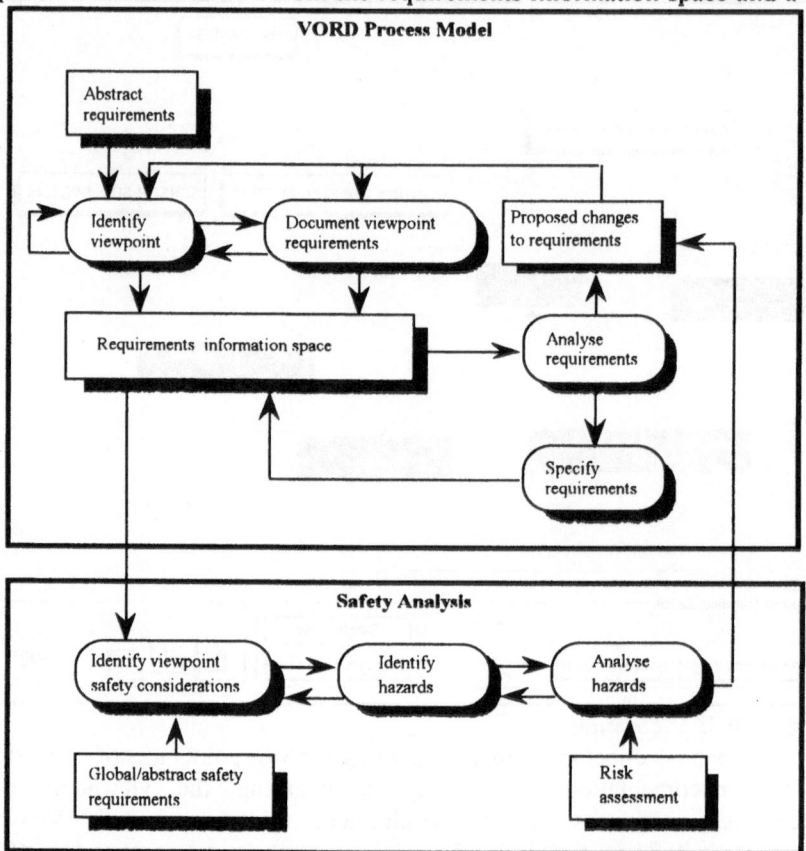

Figure 3 Process Model for integrating safety analysis and requirements engineering.

abstract safety requirements that serve as a reference model for identifying initial safety considerations or concerns relating to each viewpoint. An operator viewpoint, for example, has obvious safety concerns relating to the operation of the machine. The output from the safety analysis process is a set of suggestions and improvements that are fed back into the main requirements process.

The safety analysis process includes the identification of safety considerations, hazard identification hazard analysis, risk analysis and derivation of safety requirements. The hazard and risk analysis stages use any appropriate hazard and risk analysis techniques and are not tied to any particular method. In our example, we have chosen to use fault-trees for hazard analysis and a risk analysis scheme

160

based on accident severity-probability categorisation. We also intend to integrate the hazard analysis technique, CLASH, with VORD at a later date.

We believe this approach provides a sound framework for ensuring that safety is built into the system right from its initial formulation, and a powerful facility for tracing safety related issues and decisions. In the next sections we will describe the components of this process model using the guillotine example discussed above.

The integration of requirements definition and safety analysis is an iterative process. The output from any one stage may be fed back to its preceding stage for review and improvement. The output from the safety analysis, for example, informs the requirements definition process, the result of which acts as the input to the safety analysis process.

5.1.3 Support for Safety Analysis

The process of safety analysis which we have incorporated into VORD includes the following stages:

1. *Safety consideration identification* What are the safety considerations of each system viewpoint?

2. *Hazard identification* What hazards may occur which might lead to an accident.

3. *Hazard analysis* What are the causes of these hazards?

4. *Risk classification and analysis* What is the risk of a hazardous situation occurring and causing an accident which results in injury.

5.1.4 Identifying Safety Considerations

The first step in the safety analysis process involves identifying key safety considerations or concerns associated with viewpoints. A safety consideration is a hazard classification which depends on the system which is being analysed. For the JCB, safety considerations might be operational considerations (i.e. what hazards are associated with the machine operation) and electrical considerations. For other types of system, such as an X-ray machine, radiation considerations would also be identified. Each safety consideration has an associated set of hazards which must be analysed.

Identifying safety considerations serves to limit the number of hazards that can be considered at one time and to place the hazards in context. To help in the process of identifying safety considerations, a set of global safety requirements are used as a starting point. Global safety requirements are derived from general organisational concerns on safety and the need to ensure that safety regulations are met.

5.1.5 Hazard Identification and Analysis

For each identified safety consideration, we identify a list of possible hazards. We then go on to analyse each of these hazards to discover their possible causes. This process of hazard identification largely relies on the judgement and experience of those involved in the process.

The method of hazard analysis which we have initially integrated with VORD is fault-tree analysis. Fault-tree analysis is concerned with discovering the system states which are potentially hazardous. For each identified hazard (condition with potential for causing an accident), a fault-tree is produced which traces back to all possible situations which might cause that hazard. Fault-tree analysis can be applied at different levels from a requirements level through to an analysis of the code of a software system. Fault-trees include and/or operators which allow hazardous conditions to be combined.

We have incorporated a syntax driven graphics editor for fault-trees with the VORD support tools. Figure 4 shows a fault-tree analysis of an operational hazard associated with the environment viewpoint, 'bucket hits obstacle'.

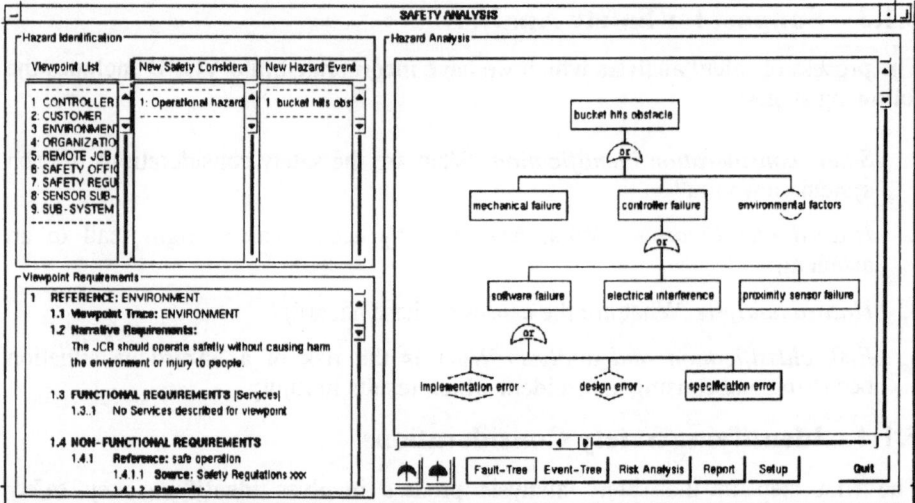

Figure 4. Hazard analysis for the hazard "bucket hits obstacle".

The left side of the form shown in Figure 4 identifies the viewpoint associated with the safety consideration and the identified hazard. The right side shows the hazard analysis which reveals that the obstacle can be hit if there is some controller failure, if the proximity sensor fails, if the is some electrical interference or if there are some external environmental hazards (e.g. an earthquake) which might occur. These are then decomposed into their possible causes.

5.1.6 Risk Classification

Risk classification is concerned with devising a scheme where the risks associated with a system failure can be classified with a view to deciding on whether or not these risks are acceptable. It involves a consideration of the severity of accidents and the probability that a hazard will arise. Note that these will vary depending on the system being developed and on the organisation developing a system. Some organisations may be willing to tolerate a greater level of risk than others because they make an economic judgement that it is cheaper to pay for the consequences of an accident rather than engineer the system so that it will never occur.

We have developed a set of user-definable severity and risk schemes based on the UK Ministry of Defence Standard 00-56 [MOD,93]. The tables comprise a set of accident-severity categories, probability categories and an accident risk classification scheme. This information is used to assist with the later risk analysis activity.

5.1.7 Risk analysis

The accident risk classification scheme can be used to analyse the risks of hazards in the fault-tree. VORD allows the user to select an event within the fault-tree and to open a risk analysis form on that event.

For the top-level hazard, the safety analyst fills in the severity of the event from a menu of possibilities derived from the risk classification. For lower-level events, the analyst makes an estimate of the probability of the event and proposes risk reduction strategies which would reduce that probability if they were applied. The VORD toolset uses the probability of the lower level events to compute the overall hazard probability at the top-level using the event relationships as specified in the fault-tree and fills this in automatically in the risk analysis form.

In Figure 5, the top-level hazard probability has been computed as approximately 0.0199 so is considered as a probable event. As probable events with fatal consequences are intolerable, this implies that some risk reduction strategies must be applied to reduce the likelihood of this hazard occurring or, alternatively, to ensure that the occurrence of the hazard can never result in an accident. This would be documented separately in a system safety case.

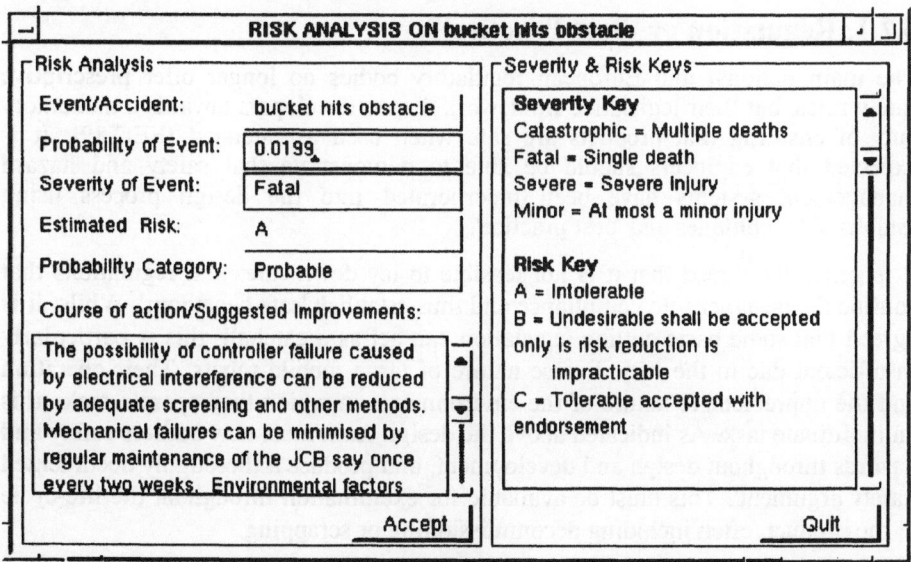

Figure 5. Risk analysis for 'bucket hits obstacle'.

This system may be used as a spreadsheet for safety analysis. Various event probabilities can be proposed and the effect on the overall hazard probability is immediately displayed. Therefore, some acceptable probability for an event can be discovered and system engineering strategies proposed which will ensure that this event probability can be achieved.

The proposed risk reduction strategies for each event are also propagated to the top-level hazard form so that a collected list is available as an input to the requirements analysis process. Each proposed change incorporates a source annotation for tractability.

Suggested changes form the basis for further decisions, in the case of safety analysis they can be used derive more safety requirements.

5.2 Safety & Hazard Analysis

The use of programmable electronics and embedded micro-processor control systems is ever increasing and LUCIE is a typical example of some of the complex problems faced. New problems are faced in establishing safety and reliability of the equipment under control, and its interaction with the environment and its users.

Generally the study of safety and hazards associated with equipment has been undertaken as a retrospective action when a product design is completed or a prototype has been built and is being tested. By their nature large mobile robots present greater difficulties to the engineer than normal, generally there have been more people involved in the design of the equipment.

Software development is carried out by separate groups according to the area of interest and these are only brought together along with their respective hardware at a comparatively late stage in the product development cycle. This is then followed by a great deal of effort to 'get the software working' and finally some work to try and establish that the system will always 'fail safe'.

5.2.1. Regulation or 'Best Practice'?.

The main national and European regulatory bodies no longer offer prescriptive safety rules, but their legislative framework directs developers towards a mandatory duty of ensuring that products are safe when used as intended [EEC,89]. It is expected that engineers should be able to demonstrate that safety and hazard management systems have been incorporated into the design process using established techniques and 'best practice'.

It is generally agreed that it is not feasible to lay down rules and regulations that confine the designer into compliance and thus establish 'safe behaviour'. Whilst it is agreed that some basic outline legislation can act as a catchall, this is particularly insufficient due to the very diverse nature of large mobile robots. Their operation and the unpredictable nature of the environments in which they operate make this an inordinate task. As indicated above the design team must now address safety and hazards throughout design and development, and produce a thoroughly documented 'safety argument'. This must be available for examination throughout the life-cycle of the product, often including decommissioning or scrapping.

The verification and validation of the safety case must also be considered as a necessity both during the development stages. This must also be used as an ongoing check of operational integrity throughout the service life of the product. All of these issues must be addressed for large mobile robots such as LUCIE to be successful and socially accepted, before a system fails and causes their use to be questioned. A further difficulty which is posed for the engineer and other members of a design team is that of establishing a common vocabulary for use in connection with a project, in particular with reference to the safety arguments.

The problem was encountered by two independent committees drafting recommendation for harmonised European Standards[IEC,91a], [IEC,91b] where similar terms were found to have differing interpretations other differing interpretations have also been encountered, in particular with the term software risk. This has been found to posess two extremely different interpretation, firstly the risk associated with the operation of the software and on the other hand the risk associated with developing the software.

A frequently debated question, and one which has no single solution, is 'what is safe?'. The answer always raises many complex responses. A system which is completely safe may never move or operate as its safety case prevents operation. If however it could be accepted as an aim, that a safe system design is one which poses an acceptable level of risk throughout its life cycle, then the only remaining question is 'what is this acceptable level?'. A viewpoint which has been expressed as possibly acceptable by the UK Health and Safety Executive would define "acceptable" as being "at least as safe as an existing comparable system under manual control". Sometimes individuals have attempted to quantify 'acceptable safety' in terms of statistical probabilities, but this still presents problems of quantification where often qualitative or subjective arguments are presented.

5.2.3. The Safety Analysis.

Methods of safety and hazard analysis have been discussed in connection with VORD[Kotonya,92b], there are however drawbacks with these techniques, no one single method is considered suitable for use in isolation. It is proposed that safety and hazard analysis should always be conducted by a team, which should include specialist product knowledge and at least one member with safety critical systems experience, this does not differ greatly from standard practice. These requirements do however lead to a common failing in that apriori knowledge is required as a basis upon which decisions can be made. A prime object of the safety analysis must be:

- To aid identification of all possible accidents which can be foreseen assuming use according to instructions
- To be able to identify the hazards which can cause these accidents
- Conduct a structured assessment of the severity of the accidents
- Categorise the probability of the hazards occurring and their contribution to an accident

Examination of the limitations of each of the major techniques reveals their individual failings for use with large mobile robots. HAZOP[CIA,77],[C&R] was originally conceived by the process industries for hazard analysis where tightly defined circuits and layouts are available. Known conditions can be examined in both stable and unstable operating conditions. Two traditional engineering methods FMEA and FMECA[BS,91] are bottom-up approaches to safety analysis. They employ inductive reasoning to identify levels of criticality and provide indications

165

as to possible methods of reducing them. Finally the last of the most commonly found methods, FTA[IEC,90] utilises a top-down approach or deductive reasoning to establish how a chain of events can be traced from a top or head event. This methods does provide computational power for reliability analysis but is limited by the inability to initially identify accidents and a possible combinational explosion for complex systems. The early development work with VORD has already shown how complex these can become.

Despite their availability and common use, and sometimes misuse, these and a large number of other methods[Bishop,90], the engineer is still faced with a problem in quantifying and categorising risk and hazards, and these are clearly very relevant to the safety argument and the requirements specification. This paper proposes a Consequence Led Analysis of Safety and Hazards (CLASH) technique as a basis for future work in analysis of large mobile robots. This new method attempts to combine some of the best features of each of the major existing techniques, but tailors them more specifically to this area by combining existing methods with a new keyword vocabulary in a structured sequence to establish not only where risks and hazards exist, but also attempt to apply some numerical figures of merit to enable ranking and subsequent redesign and reduction of problems to levels which are 'as low as reasonably practical(ALARP)'[IEC,91b].

6. CLASH - An outline of the proposed procedure.

The development of CLASH [Margrave,94] was instigated through research into development of a safety system manager for autonomous robots[Lanc,93] this has entailed detailed study of current legislation and regulations[HMSO,87] and investigation into the feeling of those currently working in similar areas. Many leading workers feel that the designers must anticipate how an advanced robot might behave in different situations and the associated problems of the technology being employed, only then can the consequential hazards be assessed[DTI,90]. This information and a study of BS EN 292 part 1 and 2 which standardise how safety measures are adopted, and also the work currently being carried out by the European Committees towards harmonised standards for mobile construction equipment[BSprEN 474] have led to the formation of the proposed scheme.

The method advocates a team approach and this should include members as indicated earlier. The team make an initial assessment of areas where problems obviously lie, then using the requirements definition documents, design outlines and briefs, they can begin to apply the CLASH technique. This consists of a series of assessments of each of the areas where consequences might occur. A guide document should be prepared to break down areas into focused studies, e.g. Consequences of Mechanical Actions, Consequences from Electrical Sources, etc. Each of these areas then has a basic standard guide word list, e.g. crushing, shearing, impact, etc. This list can then be augmented by the study team according to the needs of the product design, but all of these should then form part of the final documentation. The design under consideration is then examined in minute detail to ascertain if any of the foreseen consequences are possible, or if, during the examination, any other consequence may be felt possible. Then each member of the team annotates a score to the consequence which they feel represents:

- The probability of occurrence.

- The ease of detection of the occurrence.

- The likely severity of the occurrence.

Each member of the team can arrive at these values by a method which their own knowledge or previous experience can dictate, or use known statistical techniques to obtain values of each category rating. The next stage involves producing a rating of each consequence studied which is achieved from the product of the three values assigned above. The last, and probably most important stage is that each item should be examined and where necessary an action planned to reduce the effect of the consequence.

The use of CLASH throughout the process is seen as paramount to the development of the Safety Analysis, and also as an ongoing tool which contributes to the product design lifecycle. Its use continues throughout as a tool to ensure that a complete understanding of the operation of the design and its associated implications are fully understood. CLASH attempts to check that the output of one stage is completely utilised by the following activity and that the final case which has been developed has actually been implemented as foreseen and tested during earlier stages.

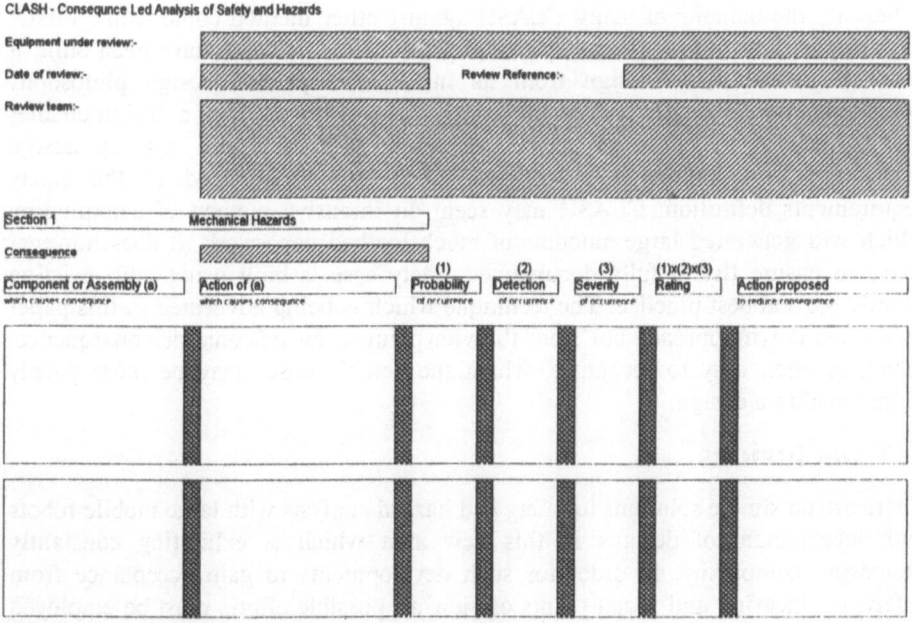

Figure 6. A suggested outline form for CLASH.

Figure 6 shows an outline design of a form for use with CLASH, it is seen that each member of the team would complete their own copy, and that finally all the ratings would then be averaged over the team. this would provide a smoothing effect to take account of the various backgrounds and experience of the individual team members. The documentation for the product would then carry as part of the safety analysis a summary of the findings and actions implemented to reduce each case until it could be agreed that the risks and hazards associated with the design were ALARP.

6.1. Advantages of CLASH.

How then does CLASH differ from other methods of analysis and what particular benefits are gained? The techniques which are employed in CLASH are an attempt to combine some of the best features of other analysis methods into a tool which is both simple to use and comprehensive in its approach. The use of an interdisciplinary team to analyse a design is recognised as an important feature of any investigation, this attempts to ensure that the study has sufficient breadth to ensure complete coverage of as many scenarios as possible. The keyword or guide word technique from HAZOP offers an outline starting point to direct the team, the list is in outline only very basic and it is regarded that the team should augment this as appropriate. The most important feature here is that the process must be fully documented to form a permanent record as part of the product specification documentation. The use of the embedded tree analysis technique from FTA strives to ensure that each situation is followed to a proper conclusion. The final application of numerical values to each of the lines of investigation seeks to ensure that consequences and the actions which led to them are addresses according to the priority of their level of criticality, this technique is partly drawn from FME(C)A.

Where do the benefits of using CLASH against other method come from? Firstly the analysis does not require any physical prototype or model to have been built, it seeks to examine the design from its initial concept and design philosophy documentation. The early start in the examination of the safety of a design enables the engineer to build a comprehensive safety case, and also develop design verification and validation techniques which are closely tied to the safety requirements definition. CLASH may seem an intensive method of examination which will generated large amounts of much loathed paperwork, it does however strive to ensure that a fully documented safety case is built using both existing knowledge and best practice. The technique which is being advocated in this paper uses a top down approach but from the viewpoint of an outcome or consequence, which is often easy to recognise, whilst the actual causes may be more deeply hidden within a design.

7. Conclusions

There are no simple solutions to safety and hazard analysis with large mobile robots and development of designs in this new area which is exhibiting constantly increasing complexity. In order for such developments to gain acceptance from safety, engineering and social points of view all possible efforts must be employed to ensure that issues have been fully addressed. CLASH is just one tool which it is hoped will gain acceptance in this field but should not be regarded as a panacea solution for use in isolation. As everyone is aware, at some stage, something will occur which has been missed, and no matter how small this may be, the critical event which leads to a major disaster. If however it can be demonstrated that all reasonable precautions were taken and the safety case built on sound foundations this will be a major step toward the defence of the safety case for the design.

References

[Advanced,92]   Advanced Robotics Research Limited (1992), 'Safety and Standards for Advanced Robots - A First Exposition', *Report ARRL.92.009*, 1992.

[Atkinson,91]   Atkinson, W. and Cunningham, J., 1991, 'Proving properties of safety-critical systems', *IEE/BCS Software Eng. J.*, 6 (2), 41-50.

[Bishop,90]   Bishop, P G(Ed), Dependability of Critical Computer Systems 3 - Techniques Directory, Elsevier Applied Science, London 1990

[Bowen,92]   Bowen, J and Stavridon, V., (1992) 'Safety-critical systems, formal methods and standards', *PRG-TR-S-92, Programming Research Group*, Oxford university

[Brooks]   Brooks R A, A robust Layered Control System for a Mobile Robot. *IEEE Journal of Robotics*, Vol RA-2, No 1, March 1986, pp 14-23.

[BS,91]   BS 5760: Part 5: 1991, reliability of systems, equipment and components: Guide to failure mode effect and criticality analysis (FMECA and FMEA).

[BSprEN474]   BS prEN 474 parts 1 - 6, Earth Moving Machinery - Safety.

[DTI,90]   DTI, 'Safety and Standards in Advanced Robotics', *Seminar held 10th July 1990*, London.

[CIA,77]   HAZard and OPerability Studies, The Chemical Industries Association, 1977.

[C&R]   Coulson and Richardson, Chemical Engineering, Volume 6, Safety and Loss Prevention.

[EEC,89]   Machinery Directive, 89/392/EEC, 14 June 1989 as amended by 91/368/EEC, 20 June 1991.

[Fickas,91]   Fickas, S., Van Lamsweerde, A., and Dardenne, A. (1991). 'Goal- directed concept acquisition in requirements elicitation'. *Proc. 6th Int.Workshop on Software Specification and Design*, Como, Italy.

[Finklestein,90]   Finkelstein, A., J. Kramer, et al. (1990). 'Viewpoint oriented software development'. *3rd Int. Workshop on Software Engineering and its Applications*, Toulouse, France.

[Finklestein,92]   Finkelstein, A., J. Kramer, et al. (1992). 'Viewpoints: A Framework for Integrating Multiple Perspectives in System Development.' *Int. J. of Software Engineering and Knowledge Engineering* 2(1), 31-58.

[Gocho,92]   Gocho, T., Yamabe, N. and Hamaguchi, T. 1992, 'Automatic Wheel Loader in Asphalt Plant' *9th Int. Symp. on Aut. and Robotics in Const.*, Tokyo

[HMSO,87]   Programmable Electronic Systems in Safety Related Applications, HSE, HMSO 1987.

[IEC,90]   IEC 1025, (1990) Fault Tree Analysis.

[IEC,91a]   IEC 65A Secretariat(122), System Aspects, Software for computers in the application of industrial safety systems. November 1991.

[IEC,91b]   IEC 65A Secretariat(123), System Aspects, Functional safety of electrical/electronic /programmable electronics systems. November 1991.

[IEE,89]   Software in Safety-related Systems, London: Institute of Electrical Engineers.

[Kotonya,92a] Kotonya, G. and Sommerville, I., 1992, 'Viewpoints for Requirements Definition', *IEE/BCS Software Eng. J*, November 1992.

[Kotonya,92b] Kotonya, G O, A Viewpoint-Oriented Method for Requirements Definition, PhD thesis, Lancaster University, 1994, pp133-134

[Lanc,93]   SAFE-SAM: Safe System Architectures for Large Mobile Robots, DTI/SERC Safety Critical Systems Programme Grant, August 1993.

[Leite,89]   Leite, J. C. S. P. (1989). 'Viewpoint analysis: a case study.' *Software Eng. Notes* 14(3), 111-9.

[Margrave,94] Margrave F W, Seward D W, Bradley D A, Hazard Analysis Techniques for Mobile Construction Robots, 11th International Symposium on Automation and Robotics in Construction, Brighton, (May 1994), pp XXX

[MOD,93]   MOD (1993). Safety Management Requirements for Defence Systems Containing Programmable Electronics. Ministry of Defence.

[Moser,90]   Moser, L.E. and Mellior-Smith, P.M., (1990), ' Formal verification of safety critical systems', *Software - Practice and Experience*, 20(8), 799-821.

[Mullery,79] Mullery, G. (1979). ' CORE - A Method for Controlled Requirements Specification.' *Proc. 4th Int. Conf. on Software Engineering* Munich.

[Norman,86]   Norman, D. A. and S. W. Draper (1986). *User-centered System Design*. Hillsdale, N.J: Lawrence Erlbaum.

[Ross,77]   Schoman, K. and D. T. Ross (1977). 'Structured Analysis for Requirements Definition.' *IEEE Trans. on Software Engineering.* SE -3(1), 6-15.

Nuclear Electric's Contributions to the CONTESSE Testing Framework and its Early Application

G. Hughes and D. Pavey, Nuclear Electric

J.H.R. May, P.A.V. Hall, H. Zhu and A.D. Lunn, Open University

Abstract

The various areas of study undertaken by Nuclear Electric for their contributions to the CONTESSE project are briefly listed. One of these areas, methods for statistical software testing, is then reported more fully, after its role in the UK's safety principles for nuclear power plants has been identified. Appropriate techniques for statistical testing of plant protection systems are detailed.

1.0 Introduction

CONTESSE is a DTI/SERC SafeIT project[1] in its final year at the time of writing. The project consortium consists of BAeSEMA, G.P.Elliot Electronic Systems, Lloyd's Register, Lucas Electronics, NEI Control Systems, Nuclear Electric (in collaboration with the Open University), Rolls Royce, Scottish Nuclear, and the University of Warwick. The objectives of CONTESSE were defined as follows:-

1. Assess the contribution of dynamic testing in simulated and final environments to safety arguments.

2. Extend current methods to provide a formal basis for cost effective simulator construction and use.

3. Provide practical and easily assimilated guidance on dynamic test and simulation.

4. Consider how simple the system will have to be before claims could be justified purely on the basis of dynamic testing.

5. Develop a strategy for dynamic testing which will provide a significant degree of confidence in the system before operational use.

This paper will concentrate on the contribution to the CONTESSE project

---

1. DTI project reference: IED4/1/9021

made by Nuclear Electric. The Nuclear Electric work has focused on methods for achieving the nuclear industry safety objectives, with particular reference to the licensing of nuclear power plants.

2.0 UK Safety Principles and their Application to Nuclear Power Plant (NPP) Design and Justification

NPP design and safety assessment in the UK is set within the framework of the Health and Safety Executive's Tolerability of Risk (TOR) from Nuclear Power Stations [HSE 92a]. In this formal context, risk associated with an accident is defined as the product of the frequency of the accident with the consequences (in terms of loss or shortening of lives) of the accident. The TOR document defines a risk of death of 1 in 10^4 per annum for a member of the public as the maximum that should be tolerated from any large industrial plant. It is however insufficient to demonstrate that the plant design will comply with this target; the risk must be demonstrated to be As Low As Reasonably Practicable (ALARP) as is required by UK law [HSW Act 74] until at least a risk of death for the individual of 1 in 10^6 per annum can be justified. These concepts are simply illustrated in Figure 1.

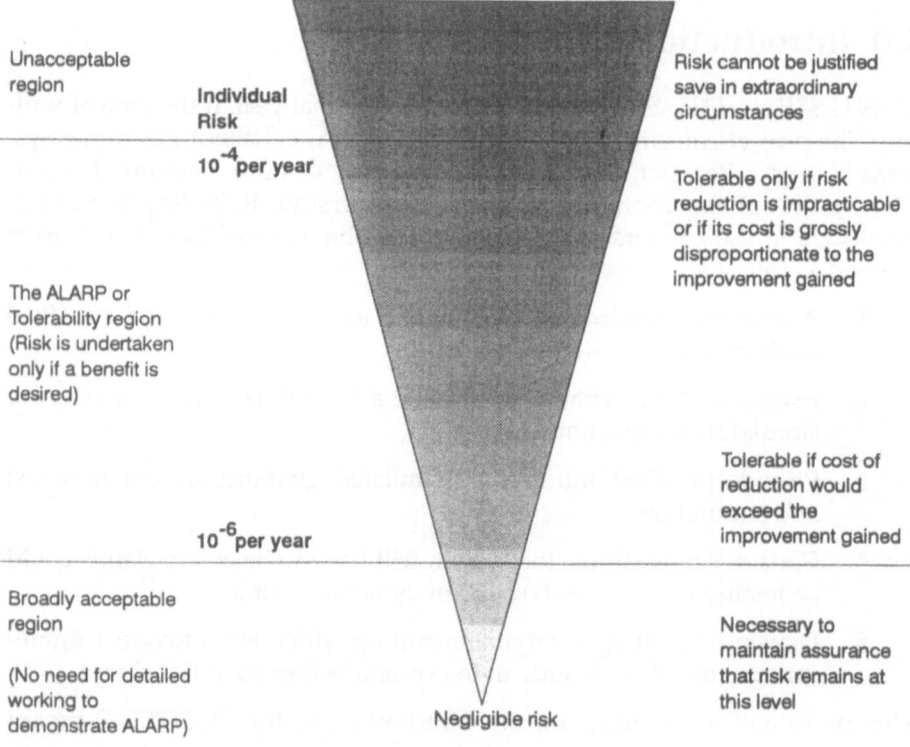

FIGURE 1. Tolerability of Risk and Risk Limitation Concepts

172

It is more difficult to quantify limits for societal risk. Currently TOR requires that it is necessary for the risk associated with a particular plant to be specified.

The TOR requirements are translated into engineering terms by Safety Assessment Principles (SAPs) for NPP, prepared by the Nuclear Installations Inspectorate (NII) on behalf of the HSE [HSE 92b]. The SAPs provide the elements of the structure for the numeric demonstration that the TOR requirements are satisfied for normal plant operation and under normal plant fault conditions. They require that all potential plant faults with the potential to lead to an accident be identified in a systematic way and that the sequences are quantified, using deterministic analysis, Probabilistic Safety Analysis (PSA) and Cost-benefit analysis to show that the individual and societal risks are ALARP. In addition, two pragmatic operational risk measures (frequency of plant damage and of accidental criticality) are introduces to reinforce the requirement for defence-in-depth, i.e. these events need to be prevented regardless of further defences. In practice it is these targets which strongly influence plant and control/protection system design.

The design approach for NPPs has often been described as consisting of three levels: (1) the design for safety in normal operation, providing tolerances for system malfunctions; (2) the assumption that incidents (Design Basis Accidents) will nonetheless occur and the inclusion of safety systems on the plant to minimise damage and protect the public; and (3) the provisions of additional safety systems to protect the public based on the analysis of very unlikely (Beyond Design Basis) accidents. A combination of active and passive features are used to reduce the risks from releases of radioactive material into the environment, as illustrated for a Pressurised Water Reactor (PWR) in Figure 2, with the plant automatic protection and associated Engineering Safety Features (ESFs) playing an important role up to containment failure. The ability to provide accepted probabilistic assessment is fundamental to the ALARP approach, however, in cases where it is not currently considered practicable to demonstrate failure frequencies or reliabilities for plant items critical to safety, the SAPs provide for a 'special case' procedure. The basic idea of the procedure is to construct a robust multi-legged safety case for the plant item in question. These legs could include historical evidence, production standards, analysis of products, testing, inspection, monitoring etc.

Currently a 'special case' procedure is necessary to demonstrate the Incredibility of Failure (IOF) of key pressure vessel components (particularly for rapid failures). A similar approach has been adopted for software, which provides a potential common cause failure mechanism for redundant computer control and protection system hardware channels [Hunns 91].

The current software safety-case, as illustrated in Figure 3, is based on three legs; the quality of the production process, the analysis of the products and testing. The evidence for the three elements is derived from the development process, the suppliers IV&V, independent assessment and customer testing of

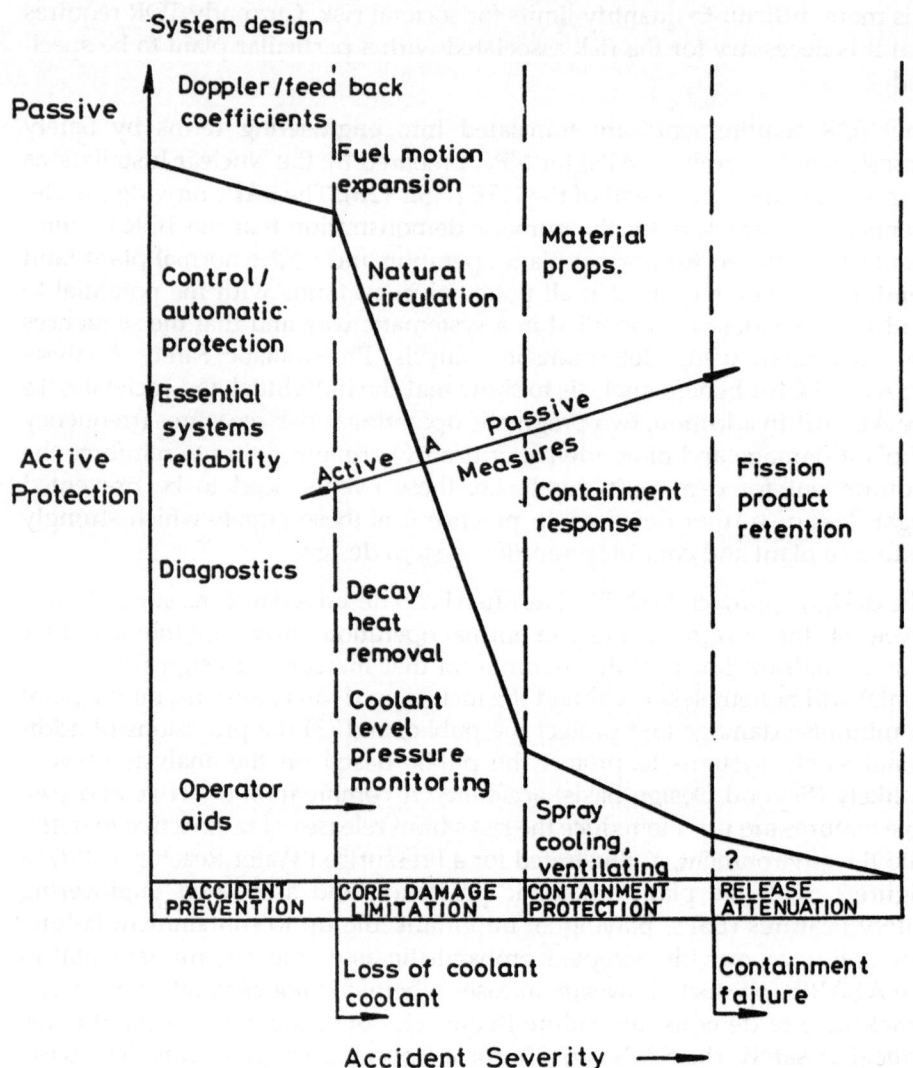

FIGURE 2. Relative Importance of Active and Passive Techniques as a Function of Accident Severity

the final system as illustrated in the Venn diagram of Figure 4. The Nuclear Electric / Open University contributions to CONTESSE have been aimed at strengthening this leg. In particular, the 'black-box' testing of the final system in an environment as close as possible to that in which it is to function is the most direct way of demonstrating that it will perform as desired in the real world. This is facilitated by the earlier testing of components separately, incorporating 'white-box' techniques to exercise every part of the internal structure of the executable software. The ability to simulate the environment

174

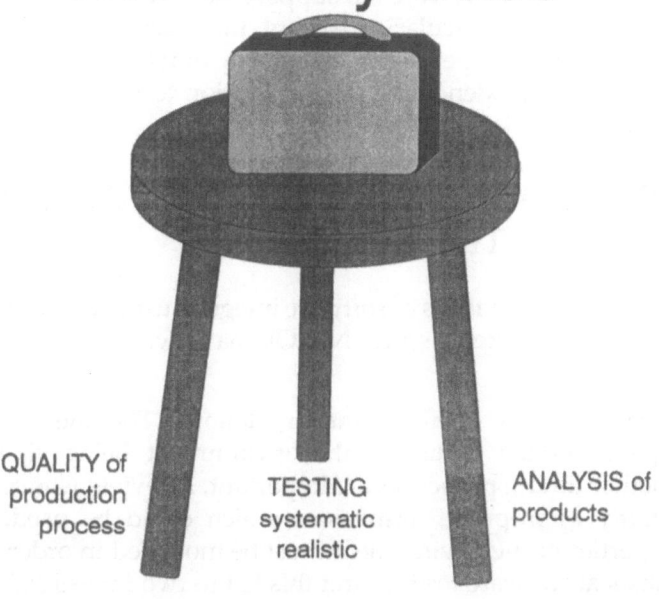

FIGURE 3.   Current Approach to Software Justification

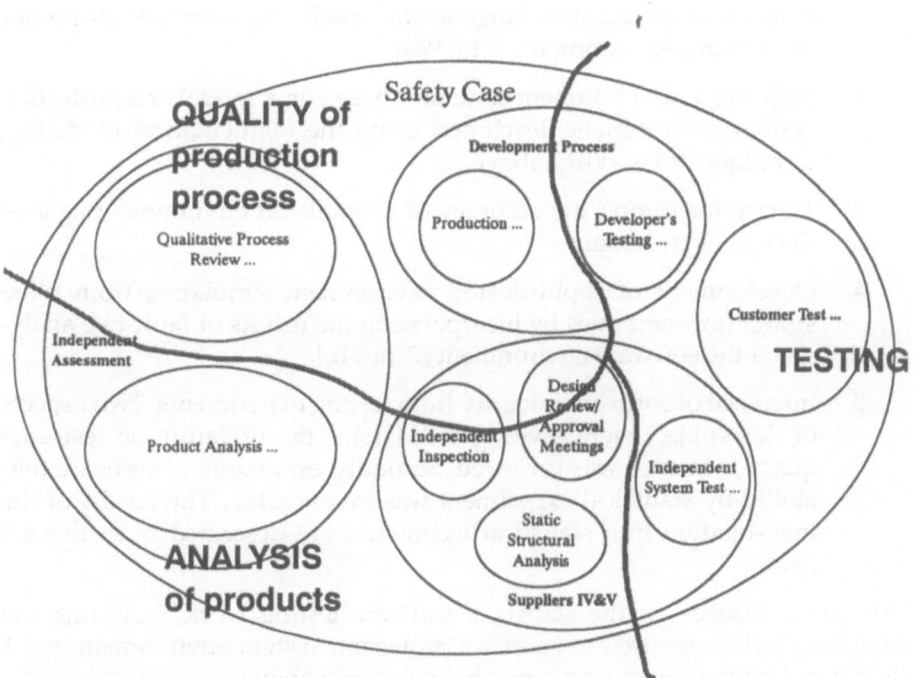

FIGURE 4. Classification of Sources of Safety Case Evidence

and quantify the reliability of the system in a statistically valid way is seen as a primary objective of the research to support the current probabilistic licensing framework. It is particularly relevant for reactor protection systems because, in general, the expected low frequency of risk significant faults limits the ability of even extended system operation to provide empirical evidence of protection system performance.

3.0 Nuclear Electric / Open University Areas of Work within the CONTESSE Project

The goal of quantified analysis of software integrity using environment simulation raises many different issues. NE/OU have worked on the following related problems:-

1. Specification of environment simulations. The methods required depend on the nature of the environment being simulated e.g. whether timing properties are important. A review was conducted to identify appropriate formalisms which could be used. Stochastic properties of the environment must be modelled in order to perform statistical software testing, and this led to two investigations. Firstly, the development of a specific mathematical modelling technique capable of describing the environment for a nuclear plant protection system, including its stochastic properties. This is described further in section 4.2 below. Secondly, the development of a general purpose simulation specification language for specifying stochastic properties of software environments [Zhu 94a].

2. Building an environment simulator based on a model of a protection system environment developed using the mathematical modelling technique devised in 1 above.

3. Theory for testing the accuracy of a simulated environment against the real environment.

4. Development of sophisticated environment simulators from more straightforward ones by incorporating the results of fault-tree analysis on the software environment [Zhu 94b].

5. Inference of software integrity from testing experiments. Two aspects of this subject were investigated. Firstly, the literature on test adequacy measures was reviewed. Secondly, estimation of software reliability by statistical experiment was investigated. The results of the investigation into statistical techniques are described in section 4.0 below.

This paper focuses on the statistical software testing work, including the modelling technique used to specify a protection system environment (see 1 above) and estimation of software reliability (see 5 above).

4.0 An Investigation of Statistical Software Testing (SST) for Estimation of Software Reliability

4.1 Foundations

The particular appeal of software reliability estimation is that it is the only approach to software integrity measurement to provide a quantified measure which is directly useful in current plant PSA. Production of an accepted estimate for software reliability would be a significant step forward for quantified ALARP risk arguments.

The form of statistical software testing reported in this paper could be called demand-based. If the input space of a piece of software is visualised as a sequence of inputs over time, a demand is a time slice of that input space such that the software has a known required behaviour for that pattern of inputs. A demand is thus considered as a single test, and a test set is actually a sequence of demands.

For example, a single demand on code implementing a sine function is a single real floating point number. By contrast, a demand on nuclear plant protection system software is a vector trajectory of readings from the sensors used by the protection system to monitor the plant condition. Thus in this case a demand is any trajectory of the software inputs from plant steady state to plant conditions at which the protection system should react e.g. shut down (trip) the plant. The concept of demand can also be applied to software which is continuously interacting with its environment, for example an aircraft engine controller, but this is beyond the scope of this paper. Note that to the software, the software environment is equivalent to a population of demands, and the means of production of those demands is not important.

The notion of a demand is extremely general. In demand-based testing, the reliability of the software is stated in the form of 'probability of failure on demand' (pfd) which is the probability of failure when a demand is chosen at random from the population of demands forming the software's input space.

The SST problem can be viewed in two parts:-

1. Experimental design - essentially this is the problem of simulating the software's environment, and in particular its stochastic properties. This subject is developed in section 4.2 by reference to an example: a software protection system for a nuclear plant.

2. Statistical inference to estimate software reliability from the test results. This is described in section 4.3.

The work reported here is concerned with the particular situation where all tests are dealt with successfully by the software[1]. Observed faults are not tolerable in safety-critical software. Once failure during testing is diagnosed, the corresponding faults should be corrected, and all testing re-performed on what is considered to be a new system i.e. test results prior to the corrections are ignored. The effects of this correction process on reliability are unpredictable. Indeed, it has been shown empirically that reliability cannot always be expected to increase as a result of attempts to correct faults [Littlewood 88]. Therefore, a testing scenario of particular interest is where no failures are observed during testing, and this is the scenario studied in this chapter. In this scenario, all testing prior to correction work is ignored i.e. assuming that it provides no useful information with regard to the reliability of the newest code version. Furthermore, no other source of prior information is used, the successfully acquitted test set is all that is taken into account.

4.2 Statistical Experiment Design for Protection System Reliability Estimation using Dynamic Testing

Environment simulation for real-time embedded software is currently a difficult and important problem in the field of software testing. When the simulation is to be used in SST for reliability estimation this imposes additional difficulties in assuring the simulation fulfils the requirements of valid statistical experimentation.

If statistical estimation of reliability is to be performed using test results, it is not sufficient for the software to be tested with any sequence of demands which its environment (the plant) would be capable of producing. In addition, the frequencies with which the various possible types of demands would occur in practice must be approximated in a valid statistical fashion. That is, the operational distribution of the protection system input space must be simulated and the protection system software tested with a sample from this distribution of demands.

Demands on a protection system are caused by accidents occurring in the plant surrounding it. The starting point for the simulation is a deterministic model of plant behaviour during accidents. This model will be referred to as the trajectory model (TM). TM produces smooth curves of the response of plant parameters to the accidents of interest. In order to simulate the operational distribution of the software, sources of random variation must be imposed over the output from TM. For brevity a simplified version of the model referred to in 2 of section 3.0 will be described i.e. some sources of random variation will be ignored. Two kinds of random variation are considered. Firstly modelling error in TM. Here we assume this is entirely due to plant noise. Hence let X_{ij} be the j'th software read on the i'th input trajectory, We assume that the actual value appearing at the software will not necessarily be that read from the trajectory produced by TM, but this TM-produced value plus some amount described by a normal random variation whose mean is zero. Secondly, the random variation in occurrence of accident types,

---

1. Although the analysis can be extended to allow estimation in the presence of demand failures.

severities, and other trajectory-influencing factors.

In short, a probability distribution is required which would typically be of the kind shown in (1).

$$p(X_{11}, \ldots, X_{1n_1}, X_{21}, \ldots, X_{2n_2}, \ldots, X_{m1}, \ldots, X_{mn_m}, F, S, G | A) \quad (1)$$

Suppose there are m input ports to the software i.e m individual input trajectories make up a demand. Then X_{ij} is the j'th point on (software read from) the i'th trajectory; n_i is the number of points on the i'th trajectory; F is the accident severity which determines the TM trajectory precisely; S is the accident type; G is the plant start power (i.e. the power output of the plant at the start of the demand); and A is the event that any instance of the accident types of interest occurs.

The modelling technique proceeds by making the assumption that (1) can be factorised to (2) below, using careful assumptions about the dependencies between the random variations.

$$p(X_{11} | F, S, G, A) \ldots p(X_{mn_m} | F, S, G, A) p(F, S, G | A) \quad (2)$$

(2) follows from (1) under the assumption that, conditional on F, S, G, A the X_{ij} are determined except for random noise effects, i.e. they are independent of each other. Implicit in this assumption is that the software reads its inputs at sufficiently large intervals that the noise components for adjacent reads are not dependent.

Furthermore, the algorithm requires that (2) be decomposable to (3).

$$p(X_{11} | F, S, A) \ldots p(X_{mn_m} | F, S, A) p(F | S, A) p(S | G, A) p(G | A) \quad (3)$$

Some assumptions are required so that (3) follows from (2).

Firstly, it must be assumed that, conditional on F and S, X_{ij} is independent of G. This appears reasonable, since F and S determine a deterministic trajectory. This assumption states that once F and S are specified, the only other remaining effects determining the location of the j'th measurement of trajectory i are due to plant noise and are entirely random.

Secondly, it must be assumed that F, given S and A, is independent of G.

Statistical inference using results from direct dynamic testing of a finished software product is a relatively new approach[1]. Consequently, there is little literature on methods for the vital task of simulating operational distributions. A recent paper [Musa 93] has addressed this task based on experience at Bell Labs (US), in particular with telephone switching networks. The Musa approach is based on specifying an operational distribution using a layering

---

1. For reasons which are unclear, it was preceded by the less intuitive and technically more complex approach of reliability growth modelling.

technique which can be compared with the modelling technique described above. However, the two approaches were derived for different types of software system, and hence have different characteristics. Simulation of operational distributions appears to be very application dependent. Many of the layers in the Musa approach are not present in the approach of this paper, since they are not applicable. On the other hand, it has been necessary in this paper to concentrate on one 'Musa layer' in great detail, creating what Musa termed a 'combination operational profile[1].' Musa distinguished explicit, implicit and combination operational profiles, which correspond to different kinds of dependencies between the variations in the software input space. Combination operational profiles are complex to specify and build. Musa did not develop one in his paper. A general method of operational distribution construction is not yet clear. This will emerge as experience with different types of software and software environment are tackled.

In summary, the approach of this report to simulating an operational distribution involves finding a description of that simulation as a composition of simpler 'component' simulations. The key to this approach is an understanding of the dependencies between random variations in the software input space. The technique goes beyond the simple splitting into progressively greater numbers of mutually exclusive alternatives, as explained by Musa. It would be nice to develop a systematic approach to expressing the dependencies between variations. One possible approach might express these dependencies using graphical dependence structures [Lauritzen 88] i.e. operational distributions may turn out to be conveniently formulated as graphical probability models. However, this is a topic for further research, and is not taken further in this report.

Given the formulation in (3), an algorithm to generate a sample from the operational distribution of demands is straightforward provided the distributions for the individual terms in (3) can be simulated. Essentially, production of the full simulation has been broken down into a series of smaller simulation problems corresponding to the individual terms in (3).

4.3 Estimation of Software Reliability from Test Results

In section 4.2 a statistical testing experiment is described in which a plant protection system is subjected to a test set. The test set is of a particular type. It is carefully selected to be consistent with the requisite conditions for statistical testing. Given that such an experiment has been conducted, the next step is to perform valid statistical inference on the experimental test results, and is the subject of this section. Statistical inference must be based on a statistical model. This section describes two statistical models for pfd estimation, both belong to the class known as urn models.

---

1. In the computing literature the word 'profile' is often used instead of 'distribution.'

Urn models are one way to construct probability models and in particular can be used to construct failure probability models for software. Uncertain processes are modelled by analogy with the drawing of objects, usually balls, from an urn (or bag). The population of objects in the urn is the same in some sense as the population of interest, in our case the input space for the software concerned.

These urn models can be divided into two camps. Firstly, those based on testing the input space as a whole; we will refer to these as *global* models. Secondly, those models based on *partitioning* (also called binning) the input space. The estimates from the two types of model can be compared when the partition-based model combines pfd estimates from individual partition subsets to obtain a pfd estimate for the entire demand space. Perhaps the most recent and widely accepted paper which covers these issues is [Miller 92]. Sections 4.3.1 and 4.3.2 present the Miller paper methods.

4.3.1 The Global Model

As stated in section 4.0 the results reported here are based on the situation where all test demands are dealt with successfully by the software[1] and no prior belief is brought to the experiment regarding the software's reliability. Quantifying such prior beliefs remains problematic [Miller 92].

The current use of the global urn model in SST can be described by the following statistical thought experiment [Miller 92]. Each demand in a program's demand space has a (natural) number of balls associated with it, which are placed in the urn. A demand either contributes balls which are all black, or contributes balls which are all white, a demand cannot be associated with both black and white balls. Balls are black if the software fails on that demand, and white if the software succeeds on the demand. The number of balls associated with a demand is in proportion to its likelihood of occurrence i.e. in accordance with the software's operational distribution of demands. A precise procedure for achieving this population of balls is described by Miller et al. Then random drawing of balls from the urn is a model for the testing of the software with demands sampled from the software's operational distribution. We will call this the single urn model (SUM) of software failure. It can be convincingly argued [Miller 92] that the probability model for the number of black balls encountered after n draws from the urn[2] is a binomial distribution [Casella 90]. This distribution can be used to construct a statistical estimator for θ, the proportion of black objects (balls in this case) in the urn[3].

If the experiment described above is conducted with replacement of drawn

---

1. Although the analysis in Miller can be generalised, allowing estimation in the presence of demand failures.
2. By analogy, the number of failures encountered when sampling demands from the software's operational distribution.
3. by analogy, the probability of failure on demand of the software

balls and N draws, Miller et al. argue that the θ statistic "black balls/all balls," conditioned on the event that n black balls were drawn (written $\theta_{SUM} \equiv \theta| \, (N, n) \,)$, has the probability density function of (4) when n=0, and also the properties in (5) and (6) when n=0, where E denotes expected value, and σ^2 denotes variance.

$$f_{SUM}(\theta| N, 0) \; = \; (1+N)\,(1-\theta)^N \tag{4}$$

$$E\,(\underline{\theta}_{SUM}) \; = \; \frac{1}{2+N} \tag{5}$$

$$\sigma^2\,(\underline{\theta}_{SUM}) \; = \; \frac{N+1}{(N+2)^2\,(N+3)} \tag{6}$$

Applying the above to the software reliability estimation problem, the distribution (probability density function) for the statistic θ the 'proportion of all demands which fail' (i.e. the probability of failure on demand) is given in (7), in which '|x=0' is a reminder that the validity of this distribution for θ is conditional on the number of observed demand failures x being zero, and t is the number of successful demands executed. The explicit representation of the fact that this distribution is conditional on t is dropped here - it is obvious since t appears in the expression for the density. θ is the protection system pfd which we wish to estimate i.e. the pfd as determined by the protection system itself, its desired behaviour, and its operational input population.

$$f(\theta| x=0) \; = \; (1+t)\,(1-\theta)^t, \, 0 \le \theta \le 1 \tag{7}$$

We wish to estimate θ based on a finite sample from the operational input population. Accordingly, the expected value of θ (i.e. a 'best' estimate), denoted $\hat{\theta}$, is given in (8).

$$\hat{\theta} \; = \; \frac{1}{2+t} \tag{8}$$

A measure of confidence in this estimate, the variance denoted σ^2, is given in (9).

$$\sigma^2 \; = \; \frac{t+2}{(t+2)^2\,(t+3)} \tag{9}$$

An alternative way to express the confidence is using a confidence interval. In this case, solving (10) gives an interval $[0, \theta_c]$ in which we are $c \times 100\%$ confident that the true value of θ lies.

$$c \; = \; 1-(1-\theta_c)^{t+1} \tag{10}$$

Note that (10) implies that at 99% confidence, 5000 tests justify a pfd of less than 10^{-3}, and 50000 tests justify a pfd of less than 10^{-4}.

The above results in this section are based purely on the test results. Miller also discusses how to incorporate prior knowledge into the estimation model using Bayesian techniques. However, there is some uncertainty and lack of formality involved in converting prior knowledge (e.g. of the software production methods used) into the quantitative form required. For this reason, this paper does not include any detail on this fusion of priors with the experimental data.

4.3.2 The Binning Model (with Independent Bins)

Miller et al. also describe a partition-based estimator which uses input space partitioning. Similar partition-based estimators have been proposed elsewhere [Thayer 78]. The global model is applied within the individual partition subsets ('bins'), to produce pfd estimates $\hat{\theta}_i$ for these subsets. An overall system pfd estimate $\hat{\theta}$ is computed using 11:

$$\hat{\theta} = \sum_{i=1}^{M} p(i)\hat{\theta}_i \qquad (11)$$

where there are M partition subsets, and $p(i)$ denotes the probability of a randomly-picked[1] demand falling in the ith partition subset. A crucial assumption is that the results of demands in different bins are completely irrelevant (i.e. probabilistically independent) to each other.

4.3.3 Conclusions regarding the discussed statistical models for pfd estimation

Note that in the case where no prior knowledge of the software reliability is used, and given that both estimators were conservative (i.e. if biased then biased to err on overestimation of the pfd), there would be no reason to use Miller's partition-based estimator. This is because the expected reliability value of the partition-based estimator is always lower than that of the SUM estimator, given the same test results. The strength of the binning model is its ability to inject more detailed prior evidence; the binning approach allows the incorporation of prior information for individual partition subsets. In the absence of such evidence the global model is always preferable. Given the informal state of the current approach to prior evidence, its use must be regarded with some scepticism (see section 4.3.1) and consequently this paper does not consider its use.

It is interesting that the global and binning models result in different pfd estimates. Miller et al. seem to imply that this difference is unimportant. However, we consider a study of the discrepancy between the models to be an

---

1. From the operational input population.

important line of future research. Some new results in this field are reported in [May 95]. An interesting development from this line of research at Nuclear Electric and the Open University is new statistical estimators based on software structure. One such estimator is discussed below in section 4.3.4.

4.3.4 An Enhanced Estimation Model

The key feature of the enhanced model proposed here is that it brings some white-box [Myers 79] considerations into statistical software testing theory, traditionally a wholly black-box [Myers 79] activity. It is not clear how the assumptions on which the global model is founded relate to some aspects of software testing. For example, the notion of complexity or length of code appears to be completely irrelevant to the global model; a certain number of tests produces the same pfd estimate for any program. This section introduces a theory which is a first attempt to build a statistical theory based on physical assumptions which relate to software.

The enhanced model is a partition-based model. It is assumed that the input space of the software can be partitioned into subsets such that within a subset all demands execute the same area of code. There is evidence to suggest that the number of failures increase with code area i.e. "lines of code" [Mills 87]. This observation inspired the Open University to investigate the fundamental assumption of the enhanced model: that the evidential worth of a demand is commensurate with the amount of code it exercises. That is, provided testing has revealed no failures and there is no other evidence to suppose otherwise, if the i'th scenario exercises a proportion α_i of code, the assumption is that P(failure on demand caused by scenario i) is $\alpha_i \theta$.

A further assumption assumes a state of ignorance regarding the whereabouts of faults within the code. This is an interesting assumption, which is also made for global model. In particular, when testing within a bin it is not assumed that faults are more likely in 'complex' areas of code (unless code area is considered a complexity measure), nor is it assumed that faults are more likely at any kind of boundary such as a partition boundary or a boundary which exists in the specification of the software. However, the enhanced model could be said to model complexity in a wider sense, 'outside' of the bins. This is because the number of bins required is a type of complexity measure in itself, and the pfd estimate increases with the number of bins (i.e. with complexity, as would be expected).

In [May 93] it is shown that enhancing the binomial model of failure using the above considerations leads to a particular statistical pfd estimator[1]. It is shown analytically that an approximation to this estimator is the right hand side of (12):

---

1. In the case of this report, this is an estimator for the pfd of the whole protection system.

$$\hat{\theta} \approx \frac{1}{\displaystyle\sum_{i=1}^{m} n_i \alpha_i} \qquad (12)$$

where n_i is the number of demands passed in bin i, and m is the number of bins. The integral expression specifying the variance of this estimator does not appear to have a readily derivable analytic solution. However, recent developments in numerical statistical methods mean that it is a relatively simple matter to compute the variance numerically. The OU statistics dept. has recently acquired the relevant software, so the answers should be available soon.

Table 1 below compares the estimators from the enhanced model with the currently available global and binning models.

Model	pfd estimate
Global	1.8e-5
Miller Binning (10 bins)	1.8e-4
Binning - each bin exercises 100% of code	1.8e-5
Binning - each bin exercises 80% of code	2.3e-5
Binning - each bin exercises 50% of code	3.6e-5
Binning - each bin exercises 20% of code	9.1e-5
Binning - each bin exercises (100/no. of bins)% of code, (10 bins)	1.8e-4

TABLE 1. Comparing pfd estimates on 55000 tests

The results in Table 1 are for 55000 demands. For convenience, the examples chosen for the enhanced model are the cases where all bins exercise the same proportion of code. It can be seen that, at its extremes, the enhanced model reduces to the particular cases of the global model and Miller et al Binning model. In addition the following points are noteworthy:-

i.   If each bin exercises (100/no. of bins)% of the code, there can be no code sharing between bins. This is why the estimate in the last row of table 1 agrees with the Miller et al binning model estimate which assumes independence between the results of demands in different bins.

ii.   Most rows do not refer to the number of bins. This is because the type of estimates given in these rows do not depend on the number of bins: 1000 bins each exercising x% of the code produce the same estimate as 10 bins exercising x% of the code. However, for a given

185

piece of code, the pfd estimate will increase with the number of bins, since a higher number of bins will result in each bin executing less code.

iii. It is not necessary for all bins to exercise the same proportion of the total code. This particular type of binning theory calculation was performed for convenience.

5.0 Conclusions

Sections 4.2 and 4.3 present a method of demand-based statistical software testing which can be applied to plant protection systems. With suitable definition of demands, it appears that the approach could be applied to continuous feedback control software also.

A significant problem in any statistical software testing to estimate reliability concerns the production of the operational distribution for the software. No general methods exist for this task which appears to be very application dependent. A modelling technique has been developed and briefly described in this paper which is appropriate for plant protection systems.

The suggested method of statistical inference to estimate reliability is taken from the current literature on the subject [Miller 92].

Further research, not reported above, is being conducted into additional statistical estimators capable of utilising information regarding software structure, and also into methods of validating software reliability estimators.

References

[Casella 90] Casella,G. & Berger,R.L. *Statistical Inference*, Wadsworth and Brooks/Cole, Belmont Calif. 1990

[HSE 92a] Health and Safety Executive: *The Tolerability of Risk from Nuclear Power Stations (TOR)*. HMSO, London 1992

[HSE 92b] Health and Safety Executive: *Safety Assessment Principles for Nuclear Plants*. HMSO, London 1992

[HSW Act 74] *Health and Safety at Work etc. Act*, HMSO, London 1974

[Hunns 91] Hunns,D.M. and Wainwright,N. "Software-based protection for Sizewell B: the Regulator's Perspective," Nuclear Engineering International, Sept 1991

[Lauritzen 88] Lauritzen,S.L. and Spiegelhalter,D.J. "Local Computations with Probabilities on Graphical Structures and Their Application to Expert Systems," J. Royal Statistical Society B, v50 n2 1988

[Littlewood 88] Littlewood,B. "Forecasting Software Reliability," in *Software Reliability Modelling and Identification*, Ed Bittanti,S., Springer-Verlag, Berlin 1988

[May 93] May,J.H.R. and Lunn,A.D. "A Model of Code Sharing for Estimating Software Failure on Demand Probabilities," Technical report, Dept. of Computing, The Open University 1993

[May 95] May,J.H.R. and Lunn,A.D. "New Statistics for Demand-based Software Testing," to appear in Information Processing Letters 1995.

[Miller 92] Miller,W.M., Morell,L.J., Noonan,R.E., Park,S.K., Nicol,D.M., Murrill,B.W., and Voas,J.M., "Estimating the probability of failure when testing reveals no failures," IEEE Trans. on Software Engineering v18 n1 1992

[Mills 87] Mills,H.D., Dyer,M. and Linger,R.C., "Cleanroom software engineering," IEEE Software, Sept 1987

[Musa 93] Musa,J.D., "Operational Profiles in Software Reliability Engineering," IEEE Software, March 1993.

[Myers 79] Myers,G. *The Art of Software Testing*, Wiley, New York 1979

[Thayer 78] Thayer,R., Lipow,M. and Nelson,E. *Software Reliability*, North-Holland, Amsterdam 1978

[Zhu 94a] Zhu,H., "Software Testing via Environment Simulation," CONTESSE Task 4 DTI deliverable, rep. no. OU?Task4/006. July 1994

[Zhu 94b] Zhu,H. and Hall,P.A.V., "Testing Protection System Software through Simulation of Failures," Proceedings of the Real-Time Systems Conference (2nd Ed.), Palais des Congrés Porte Maillot, Paris Jan 11th-14th 1994

Current Practice in Verification, Validation and Licensing of Safety Critical Systems
- The Assessor's Point of View -

Günter Glöe, Gerhard Rabe
TÜV Nord e.V
Hamburg, Germany

Abstract

Dealing with the verification and validation of safety critical systems we have to consider two different approaches: the type approval which is independent of any application and is targeted to the certification of components. On the other hand in an application dependent approval it has to be proven that all requirements for a specific application are really met. Both procedures and relevant standards will be described and explained on the basis of concrete examples.

1   Introduction

At SSS'93 a presentation was given with general information about TÜV as an inspection organisation and about principle procedures of certification. National and international standards which form the basis of the verification and validation of safety critical systems were highlighted. Amongst these the German standard [DIN94] can be considered as a basic element in the procedures of verifying and validating systems with safety relevance.

Dealing with safety relevant systems we have to consider two aspects:

a) Concentration on individual <u>components</u> which are implemented in a system. The question is whether these components meet the requirements.

b) Individual components are being implemented in a <u>system</u> for a special application. The question here is whether the system meets the requirements for the process / application.

Both these aspects lead to two different v&v procedures. The procedure targeted to a) is the application independent <u>type approval</u> which is a prerequisite for certification of individual components. Validation for b), the <u>application dependent approval,</u> on the other hand is a necessary step to get the final approval by licensing authorities for operation of the system.

In the following chapter 2 the principle procedures of a type approval will be described. A detailed explanation will be given with reference to a digital multiprocessor system. The particulars of relevant standards as well as used validation tools will be entered.

Chapter 3 describes the principle relation between type approval and application dependent approval.

The application dependent approval is dealt with in chapter 4 where we come back to basic standards like [DIN94], [IEC91] and [IEC92]. The classification of safety critical systems and eventual problems are decribed with reference to two industrial projects including programmable logic controller (PLC) based safety systems.

In chapter 5 we will give a view on future developments of verification and validation techniques and tools.

2 Type Approval

In the area of type approval we are engaged mainly in systems for use in nuclear power plant applications, eg. reactor protection systems, radiation monitoring or control rod control systems. Besides this we do type approval for control systems of medical equipment, e.g. breathing machines, for theaters, some chemical plants and railway use, e.g. control of locomotives.

Object of approval are software components as well as hardware components and complete control channels, e.g. for neutron flux measurement.

Besides approval of safety related control systems - computerised or non computerised - we are working on 'normal' systems as man-machine-interface-generators or project organisation tools.

May be different from other countries type approval of safety related control systems in Germany is not a task strictly attached to one and only one governmental selected institution but there is some sort of market with a lot of competition. This means that quality of type approval is not only jugded on thoroughness but also in terms of costs and adherence to schedules.

2.1 Standards Applied

Standards we use for approval of *software components* mainly are KTA3503 [KTA86], IEC880 [IEC86], IEEE 829 'IEEE Standard for Software Test Documentation' (1983), IEEE 830 'IEEE Guide to Software Requirements Specifications' (1984), DIN V VDE 0801 [DIN90], ISO 9000 Part 3 and IEC 65A (Sec) 122 [IEC91].

Standards we use for approval of *hardware components* are IEC 68 (part2), IEC801, IEC987 'Programmed digital computers important to safety of nuclear power stations' (1989), IEC1131, IEEE323 'IEEE Standard for Qualifying Class 1E Equipment for Nuclear Power Generating Stations' (1983), IEEE344, EN55011 and KTA3503 [KTA86]

Standards we use for approval of *complete control channels* mainly are KTA3503 [KTA86], DIN V VDE 0801 [DIN90], IEEE308 'IEEE Standard Criteria for Class 1E Power Systems for Nuclear Power Generating Stations' (1980) and IEEE323 .

2.2 Basic Scheme

For safety critical control on the one hand testing must be very thorough and on the other hand it shall be as efficient as possible. To reach these two goals during *hardware type approval* the scheme shown in figure 1 is applied. In this scheme the boxes represent documents. The lines between the boxes represent the testing activities. 'Static papers' are intended to be created just once during the whole test procedure. 'Dynamic papers' are intended to be updated according to work progress.

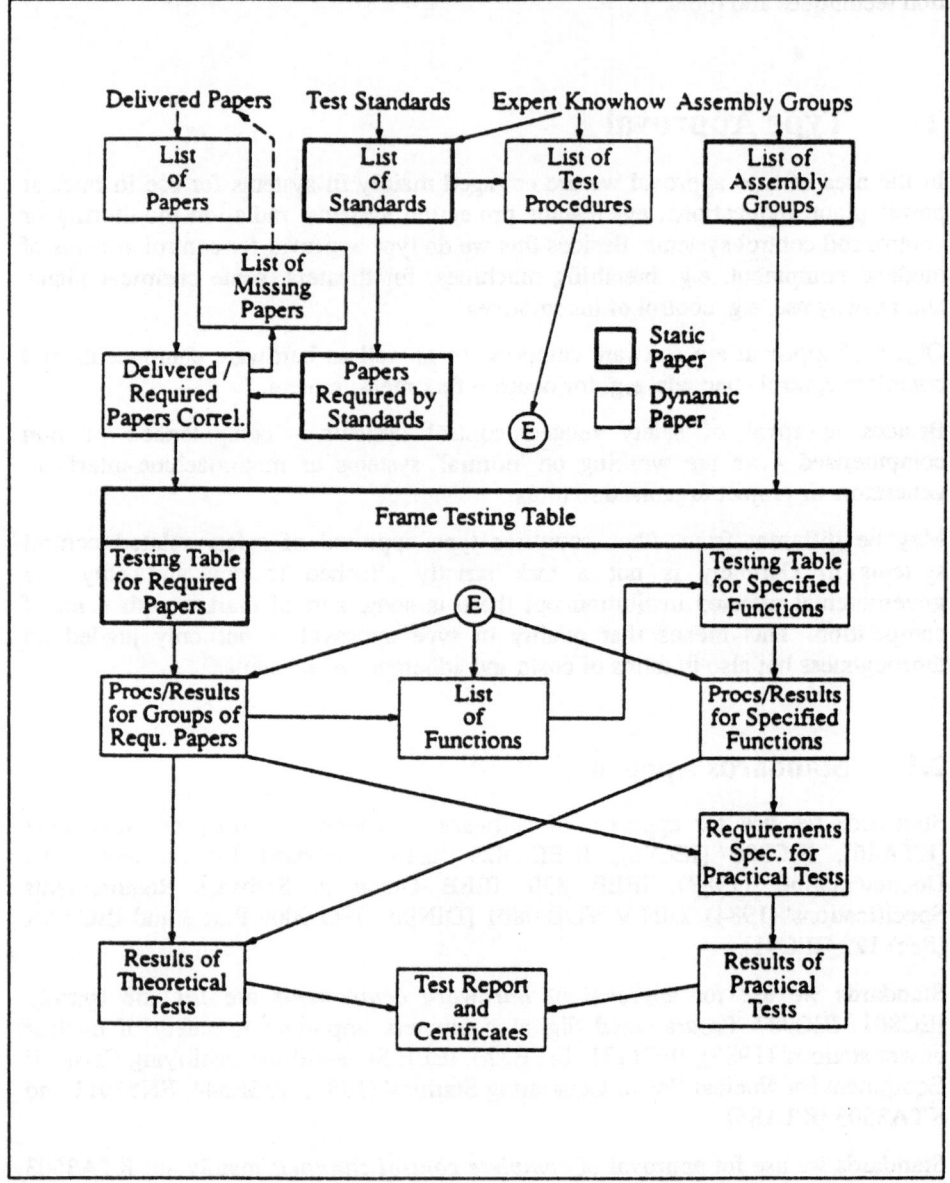

Figure 1: Basic scheme for type approval of safety critical hardware

For the more pretentious testing activities beside the basic scheme there are checklists describing the activity in more detail. The most voluminous checklist has about 90 pages. An overview on the set of checklists used for type approval in TÜV Nord is shown as table 1.

Test Object	Checklist
General	General Preliminaries on Checklists
	Preconditions for Testing
	Instructions for Building Test Frames
Requirements Specification	Preliminaries on Requirements Specification Checking
	Formal Check of Requirements Specification
	Functional Check of Requirements Specification
	Generation of Test Data from Requirements Specification
System Design	Preliminaries on System Design Checking
	Formal Check of System Design
	Functional Check of System Design
	Generation of Test Data from System Design
Hardware Realization	Preliminaries on Hardware Realization Checking
	Check of Hardware Realization
Software Realization	Preliminaries on Software Realization Checking
	Check of Software Realization
Whole Integrated System	Preliminaries on Whole Integrated System Checking
	Check of the Whole Integrated System
	Self Test Procedures of Single-Channel Computer and Microprocessor Systems

Table 1: Overview of set of checklists used in TÜV Nord for type approval of safety critical control systems

On a certain type test we have been working according to the basic scheme for type approval of safety critical hardware components (figure 1) for about a year. From company internal experience as well as from discussions with colleagues from industries and other TÜVs we could learn that the approach provides a clear and reproducible way to determine the characteristics of a testee. Furthermore it supports to a sufficient extent control of budgets and schedules.

The degree of organisation using this basic scheme is such that it is possible to do parts of the testing in another company (another TÜV) in another town some hundred kilometers away without significant problems. Interface documents are the "Requirements Specifications for Practical Tests" and the "Test Reports on the Practical Tests".

For *type approval of software components* a similar basic scheme is applied. Overview on the checklists used for software is part of table 1 as well.

Normunterlagen	KTA 3503	IEC 987 k.A.>> IEC780 IEC880	IEC 1131 >> Teil2	IEEE 323	EN 55011 (k.A.)	WB 26	Experten-wissen	IEC 880 Suppl.	
Stand der Norm	11/89	1989		1983	03/91	10/86		08/93	
Unterlagenverzeichnis	4.1.2								A
Anforderungsspezifikation				6.1		3A1		4.2.4	N
Einsatzumfeld(Umg.bed.)			S.137	6.1.5					F
Sicherheitsfunktionen			S.141	6.1.4					
Beschreibung Meßaufgabe						3A1			E
Funktionsbeschreibung	4.1.3					3A1			N
Schnittstellenbeschreibung			S.137	6.1.2					T
Blockschaltbild			S.139			3A1			W
Technische Beschreibung			S.139	6.1.1					
Aufbauzeichnung						3A1			
Beschriftungsvorlage							X		F
Stromlaufplan	4.1.6					3A1			E
Stückliste	4.1.7					3A1			R
Lageplan Bauelemente	4.1.8					3A1			T
Druckvorlage	4.1.8						X		I
Gebrauchsanweisung	4.1.5		S.139			3A1			G
Datenblatt	4.1.4		S.137			3A1			
Ausfalleffektanalyse	4.2							4.7.1.3	
Ausfallratenanalyse	4.2		S.139	6.1.3					Z
Grenzbelastungsanalyse	4.3								U
Security Nachweis								4.3.10	V
Wartbarkeitsstudie			S.139					4.1.1.5	
Prüfspezifikation	4.4					3A1			T
Qualitätssicherung						3A1			E
Testsoftware			S.139				X		S
Tool-Qualifikation								4.3.2.2	T

Table 2: 'Papers required by Standards' for hardware type approval of a family of assembly groups

2.3   An Example

For type approval still in progress of a family of some twenty assembly groups until now more than 400 different documents filling 20 files have been presented. For this type approval Table 2 shows the 'Papers required by Standards' table mentioned in the basic scheme of figure 1.

Table 3 shows the 'Frame Testing Table'.

Prüfobjekt - Gruppen	Prüfumfang theoretisch			Prüfumfang praktisch
	formal	inhaltlich	Anforderungs-spezifikation für prakt. Prüfung	
Baugruppen (*)	–	–	–	KTA 3503 (u.a) Prüfspezifikation
Normunterlagen	verständlich vollständig konsistent	vollständig konsistent korrekt eindeutig Funktionen spezifizieren	Randbedingungen Testverfahren	–
Funktionen (*)	–	vollständig konsistent korrekt eindeutig realisierbar	Randbedingungen Testverfahren	–

Table 3:   'Frame Testing Table' for hardware type approval of a family of assembly groups

193

2.4 Tools to Support Quality Models

European Standard EN 45011 [EN89b] defines certification of conformity as "Action by a third party, demonstrating that adequate confidence is provided that a duly identified product, ... is in conformity with a specific standard or other normative document."

Prerequisite for demonstrating that a safety critical control system or its components are in conformity with its requirements is the break down of requirements (also known as desired quality characteristics) into testable or measurable characteristics. That is just what quality models provide.

2.4.1 TASQUE

A tool supporting this derivation process is the Tool for Assisting Software QUality Evaluation (TASQUE; see Figure 2). It is result of EUREKA-Project EU 240 which

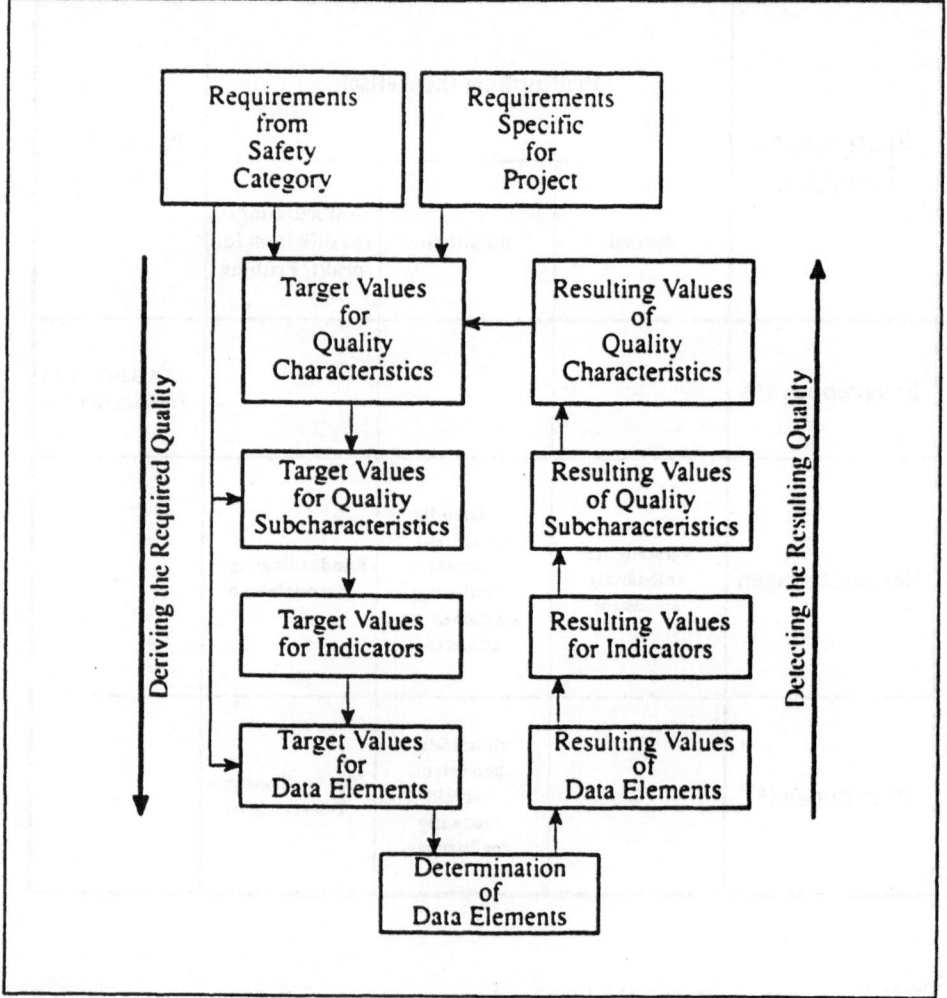

Figure 2: Concept of TASQUE

besides others is sponsored by the German Federal Ministry of Research and Technology as Projects 13RG8812 and 12R9309. Partners in this project have been CEP (Paris), ENEA (Rom), ETNOTEAM (Milan) and ISMES (Bergamo). TASQUE enables in a systematic and efficient way to derive the detailled characteristics for a control system assigned to a certain category (e.g. from IEC 1226 'Nuclear power plants - Instrumentation and control systemes important for safety - Classification' (1993)). The work started within TASQUE will serve among others as a basis for ESPRIT III project Software QUality In Development.(SQUID; P8436).

2.4.2   SQUID

ISO 9001 [ISO90] as well as ISO/IEC 9126 [ISO91] emphazise that product and process quality must be auditable and preferably measurable. The goal of the SQUID (Software Quality in the Development Process, ESPRIT project no. 8436) project is to provide models, methods and tools to assist quantification and auditing of process and product quality.

For a specific project the SQUID toolset will assist to apply the SQUID models and methods to predict end product qualities from product and process characteristics observed during development. This will ensure early identification of process and product quality problems. The tools will allow a project manager to investigate different solutions to identified problems as and when they arise. For a development organization and the assessor the methods and tools will provide information needed to
- show conformance with ISO 9001 [ISO90] and ISO/IEC 9126 [ISO91]
- identify process inefficiencies and monitor the effects of change to the to the process
- increase the organization's CMM (Capability Maturity Model) level by the use of objectively founded process improvement.

2.5   Tools for Testing

2.5.1   CATS

In order to achieve a sufficient depth of examination with acceptable costs within reasonable time and to get well documented reproducible results in the quality assurance of software for embedded controllers in addition to our extensive checklists we developed CATS (Code Analyzer Tool Set). The concept of CATS is shown in Figures 3.

The processor specific disassembler DisCAT acts as a front processor. It operates program path related and produces the source data base for the CATS-Analyzers and some global metric data about the code processed.

ProCAT is provided for the analysis of the global controlflow structure. The tool analyzes the composition of the code in terms of programs, interrupt-routines and subroutines. For these components the relocation and the interaction by call or code-sharing is evaluated.

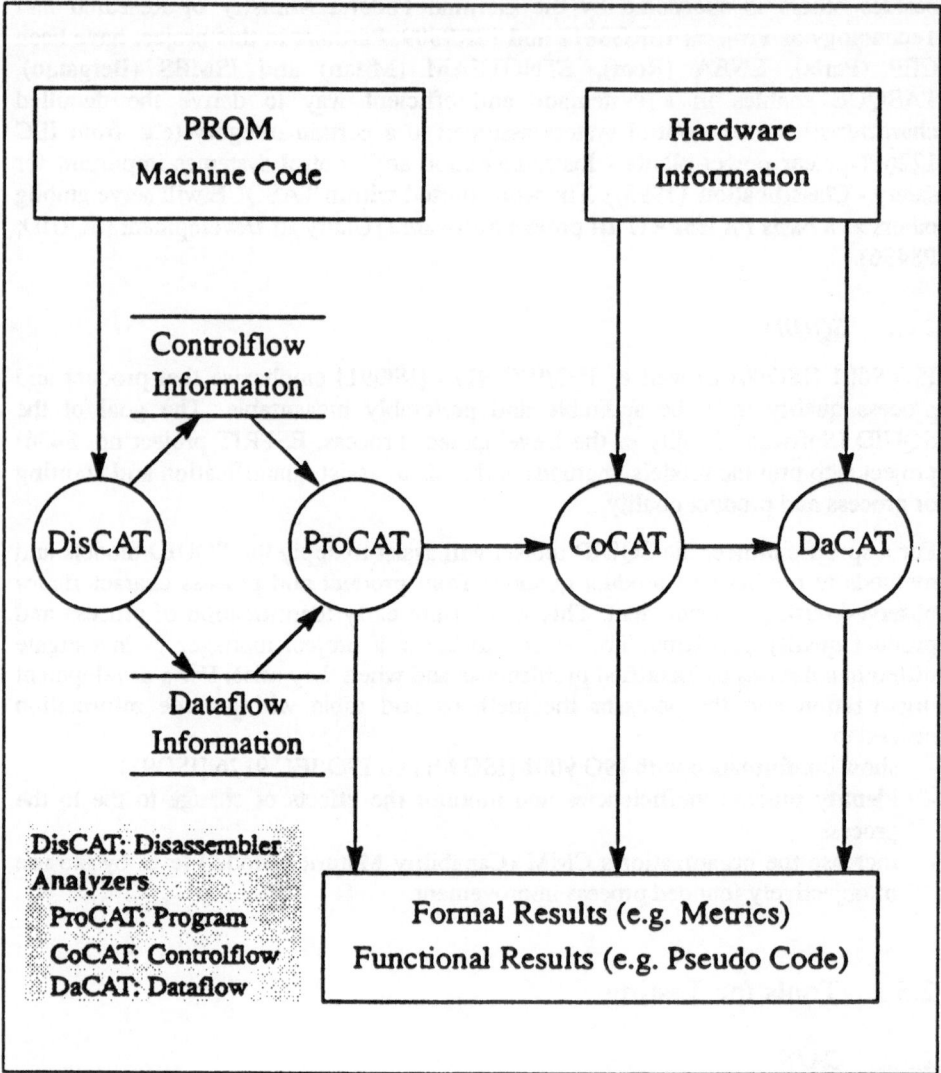

Figure 3: Concept of CATS

CoCAT does the controlflow analysis of routines defined by ProCAT. It provides well documented and reproducible quantitative and qualitative informations on formal quality criteria as well as on realized functions.

Accordingly DaCAT analyzes the dataflow of the whole program and of routines defined by ProCAT, providing informations in the same manner.

Besides these main CATS-Tools there are two further tools: RealCAT, which analyzes the runtime amount for programs, interrupt-routines and subroutines, and MeCAT, which computes metrics and arranges statistical data gained by controlflow and dataflow analysis.

2.5.2 Commercially Available Tools

Simple tools

For more simple validations of object code (e.g. verification of code changes, calculation of module size) we used tools included in operating systems, like
- editors
- search functions
- 'differences'.

SoftDoc

SoftDoc (version 2.02) is a static analysis and documentation tool for C object code with the purpose to give following information:
- calling hierarchy of modules and functions
- static information on object code (metrics)
- use / unuse of declared types, macros, constants, variables and functions
- lists of references for types, macros, constants, variables and functions.

DEC SCA

('Source Code Analyzer' of Digital Equipment Corp., V4.0-93/VAX_VMS) This analyzer is a static interactive tool targeted to different programming languages based on object code. It was used for the purpose of checking following characteristics:
- use / unuse of declared types, macros, constants, variables, functions and procedures
- initialisation / non-initialisation and read / non-read of variables
- compatibility / non-compatibility of types.

2.5.3 Tools in Trial Use

From National Computing Center (NCC), Manchester, we started to apply the 'Data Collection and Storage System' M-Base to learn about the benefits during type approval.

From Daimler Benz an improved method for derivation of black box test cases - called Klassifikationsbaum-Methode (classification tree method) [Gri92] - has been suggested. A tool provided by a major German company to support this method will be applied during type approval of hardware, software and integral system of a locomotive control system. And again this is to find possibillities to improve the ratio from thoroughness of testing to time and effort needed.

2.6 Accreditations

Not so very fast but step by step purchasers start to ask for the qualification of the testing labs they want to entrust with type approval. So testing labs have to submit themselves to evaluation procedures from accreditation bodies. TÜV Nord has got accreditation to act as testing lab according to EN45001 (General criteria for the operation of testing laboratories) from Deutsche Koordinierungsstelle für IT-Normenkonformitätsprüfung und -zertifizierung (DEKITZ) for some 80 standards as e.g. IEC 65A (Sec) 122 [IEC91], IEC 65A (Sec) 123 [IEC92] or IEC 880 [IEC86]. Furthermore we have got accreditation by Gütegemeinschaft Software (GGS), a special organisation concerned with DIN66285 (which is an early german version of ISO/IEC 12119 'Software packages - Quality requirements and testing' and will be superseeded by it). For testing of programmable logic controllers we have got accreditration from PLCopen. In progress is accredidation according to the regulations of Bundesamt für Sicherheit in der Informationstechnik (BSI) in security area.

2.7 Certification

Of course not the test lab but another body of TÜV Nord is authorised to act as certification body for control systems, computerised or not, according to EN45011 'General criteria for certification bodies operating product certification' [EN89b]. So based on the test report from our testing lab in addition customers of TÜV Nord have the opportunity to get a certificate on their product and its characteristics.

3 Relation Between Type Approval and Application Dependent Approval

As we can see in the following figure the procedures for type approval and the application dependent approval are similar. They differ amongst others in standards

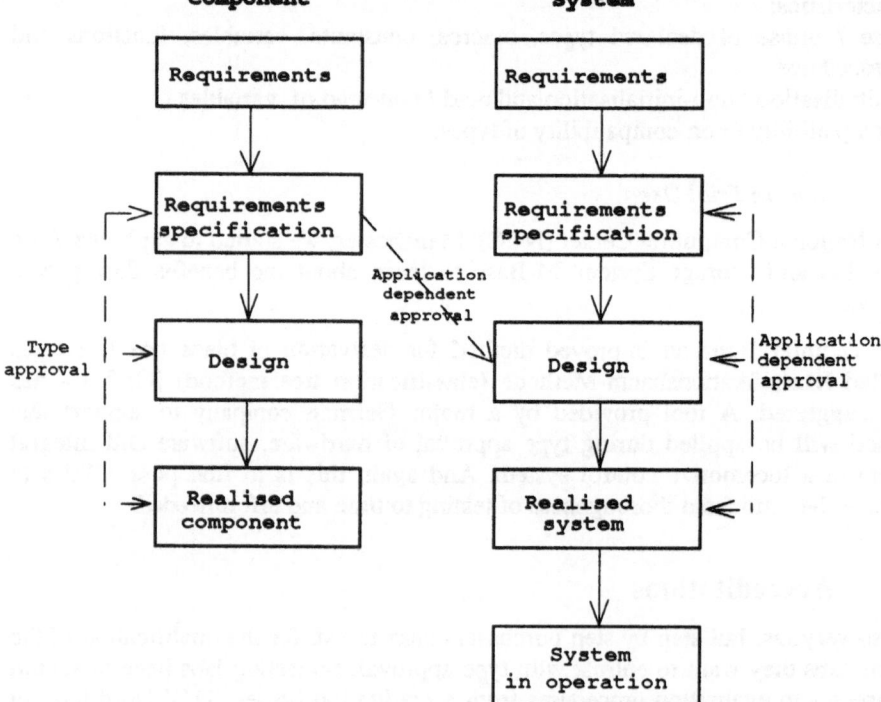

Figure 4: Relation type approval / application dependent approval

being applied and in requirements to be me by the component resp. system. Whereas a type approval can be conducted without any connection to concrete applications the application dependent approval in most cases takes into account the results of type approvals. The results in form of test reports and certificates are then included in the validation procedures.

4 Application Dependent Approval

4.1 General Remarks / Standards

As already mentioned in the introduction the main issue of an application dependent approval is the proof that the system - hardware as well as software - meets all requirements related to a specific application. These requirements can be derived only partly from application specific standards (if they exist). E.g. the German standard [DIN89] includes concrete procedures for the validation and prescribes which issues have to be considered (e.g. a system has to be safe against primary failures, up to two undetected failures have to be considered). As this standard can not be applied to all systems in different applications we have to come back to general standards like [DIN94], [DIN93], [DIN90], [DIN94a], [IEC91] and [IEC92]. We also have to take into account that in most cases of safety critical applications type approved components are being used.

An application dependent approval is divided into

- an analytical validation   and

- functional tests on-site.

4.2 Analytical Approval

A lot of questions concerning system structures, time behaviour and system behaviour in the occurance of failures have to be considered. The following list is an extract of all these questions:

- Is a safety philosophy described for all safety functions ? Can each individual safety function be specified by means of logic diagram ?
- Which hardware / software system is the best one to meet the requirements ?
- Does the safety function include redundancy and / or diversity ?
- Is the safety system fail-safe ? Does there exist a prescription of a safety philosophy or failure calculation in the application dependent standards ?
- What are the timing constraints for individual safety functions ?
- Is the application software separable into safety related and non-safety related parts ?
- Where in the application are fail-safe, where non-fail-safe actuators in operation?
- Which control signals are static, which ones are dynamic ?
- What kind of strategies are applied to find passive failures ?
- Is there a link / communication between fail-safe and non-fail-safe systems in the application ?

4.2.1 Validation of the Safety Concept / Risk Evaluation

Starting point for a validation of the safety concept is a risk estimation which is conducted by application engineers in co-operation with control systems engineers. Based on the risk parameters
- C   consequence
- F   frequency and exposure time
- P   possibility of avoiding hazards   and
- W   probability of the unwanted occurance

and using the risk graph in Fig. 5 a certain <u>requirements class</u> for the application under control is defined.

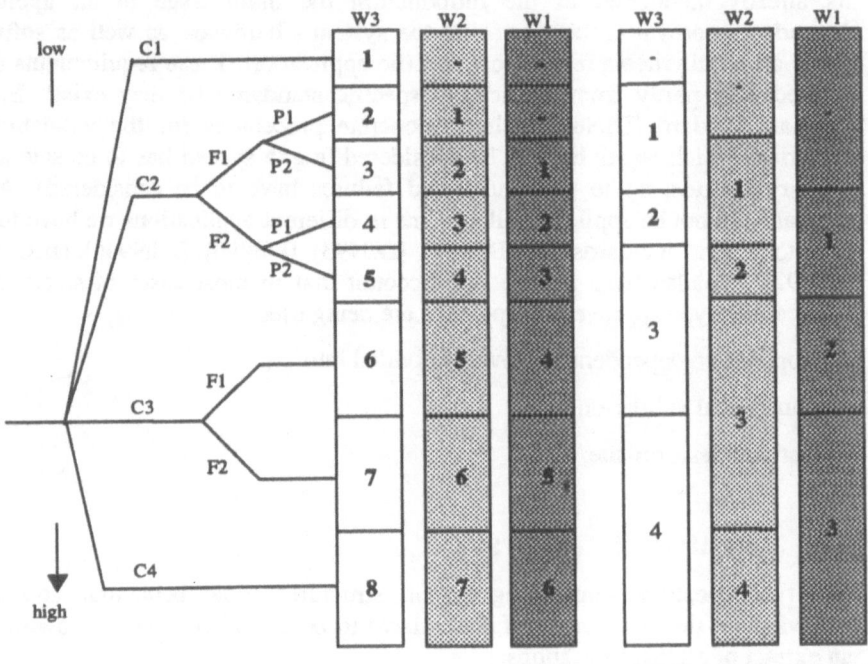

DIN 19250 requirement classes   IEC 65 system integrity level

Figure 5: Risk graphs

<u>Extent of damage</u>

C1 minor injury to one person,
    minor damage to the environment
C2 serious permanent injury to one or
    more persons, death of one person /
    temporary major damage to the
    environment
C3 death of several persons or permanent
    major damage to the environment
C4 catastrophic consequence

<u>Probability of unwanted event</u>

W1 very low
W2 low
W3 relatively high

200

Frequency of exposure to hazard	Avoiding the hazards
F1 seldom to quite often	**P1** possible under specific conditions
F2 permanent exposure	**P2** almost impossible

It is important to mention here that all risk estimations must not be done targeting the whole system (application) under control. It should only be conducted taking into account individual events / risks as a consequence of failures in the control system. [DIN94] and [IEC92] introduce requirements classes resp. system integrity levels as to be seen in Fig. 5. The risk parameters of [DIN94] include personal as well as environmental damages whereas [IEC92] - and even the expected new draft - concentrates only on persons' damage.

In addition to the qualitative risk estimation as described here [IEC92] offers a quantitative one to estimate system integrity levels. But this approach is only feasible if

- tolerable risk can be specified numerically (e.g. a consequence of a failure must not occur more than once per thousand years)
- numerical targets have been specified for the system integrity levels.

The measures that have to be taken to meet the requirements for a specified requirements class are defined in [DIN93], [DIN90] and [DIN94a]. The required efficiency of these measures is shown in the following diagram:

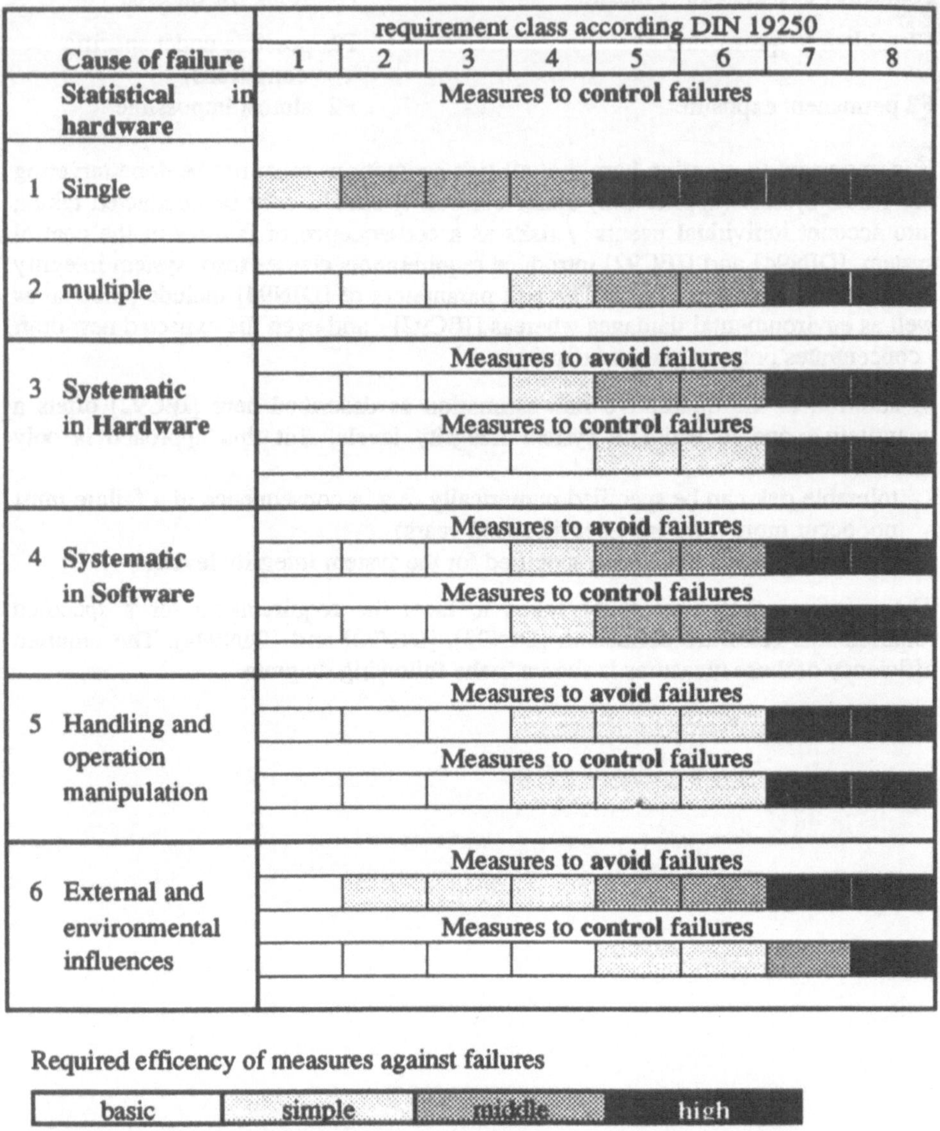

	requirement class according DIN 19250							
Cause of failure	1	2	3	4	5	6	7	8
Statistical in hardware	Measures to control failures							
1 Single								
2 multiple								
3 Systematic in Hardware	Measures to avoid failures							
	Measures to control failures							
4 Systematic in Software	Measures to avoid failures							
	Measures to control failures							
5 Handling and operation manipulation	Measures to avoid failures							
	Measures to control failures							
6 External and environmental influences	Measures to avoid failures							
	Measures to control failures							

Required efficency of measures against failures

basic	simple	middle	high

Figure 6: Required efficiency of measures

Based on these estimates and the delivered documents (requirements specifications) it is validated whether the intended system and the safety concept meet just these requirements.

4.2.2 Verification of logic diagrams

The operational procedures for the application are transferred to logic diagrams which form the basis for the following realisation (coding). These diagrams are validated mainly by process engineers with respect to the requirements specification.

4.2.3 Verification of the application software

The application software is verified with respect to the requirements specifications and the logic diagrams. The following steps are conducted:
- extraction of all safety critical parts (moduls, subroutines) of the software
- control flow and data flow analysis
- verification of all specified functions
- if possible simulation of all functions and time behaviour under normal and erroneous conditions on a simulator. This is recommended but not prescribed.

4.2.4 Verification of hardware design and installation documents

These documents are the basis for the installation of all components in the system on-site. They include all mechanical, electrical and environmental information. This verification again is done in a close co-operation between electrical and process control engineers.

4.3 Functional Tests On-site

The first step in this phase is to control the physical installation of the hardware and the application software on the target system. It is validated whether both the hardware system as well as the application software are in accordance with the pre-validated documents.

With respect to the hardware the following tests are conducted:
- test of the installation with respect to
  * field wiring, e.g. separate installation of redundant wiring
  * protective and functional earthing
  * noise and transient suppression measures
    # separation of cables for inputs, outputs and power circuits
    # correct length of wiring
    # separation of the field wiring from internal I/O cabling and from bus lines
    # control of mechanical contacts which are in series with inductive loads
- test of compliance with the actual service and environmental conditions, e.g. temperature, shock and vibration, electromagnetic influence
- loop checks (interaction between control system and process periphery)
  * check of binary and digital input signals to ensure that physical states of sensors comply with signal latches in the control system
  * check of analog input signals to ensure equivalence of physical value and data received by the control system
  * check of binary outputs including check that no forced binary and digital outputs are set

* check of analog outputs
* check of supervised inputs and outputs with repect to the detection of opens and shorts.

The second step is to test all <u>system functions</u>. During these tests the functional behaviour of the system with respect to the specified functions is validated. All functions are tested under operational conditions.

In a third step extensive <u>fault simulations</u> are conducted based on a pre-defined list of failures. According to our experience the majority of failures can be located in I/O and other periphery. Therefore failures are simulated with respect to sensors, contacts, actuators, inputs, outputs, field wiring and interlocks. During failure simulations it must be validated that the system - in the case of failures - is being brought into a safe state. The kind of this state is depending on the application. It can be a de-energized state of all outputs or the shutdown of certain parts of the application.

In the final step it is being assessed whether all <u>restrictions</u> fixed in the type approval report (if a certified PLC is used) are fufilled.

4.4 Examples

4.4.1 Refinery / Crude Oil Loading-Unloading of Tankers

A double channel PLC system which has been certified for requirements class 6 was to be implemented in a system to control loading and unloading of tankers. The classification of this application resulted via C3, F2 and W2 (see chapt. 4.2.1) to class 6. So far all requirements could be met.

But when we went through the restrictions of the type approval for this PLC (see last step in chapter 4.3) - and this was done just at the beginning of the validation of the safety concept - we realised the following:

One restriction for class 6 is that the PLC system has to be shut down within one hour in the occurance of a failure in one channel. Due to process necessities such an immediate shutdown would not be feasible.

To solve this problem a new classification was done resulting via C3, F1 and W2 in class 5. The type approval allows for this class in case of failure in one channel an ongoing operation up to 72 hours. This was acceptable since the process could be shut down in a defined way without problems.

4.4.2 Gas pipeline

With the following example we want to demonstrate that risk estimations according to [DIN94], [DIN93], [DIN 90] and [DIN94a] always have to consider individual events / risks and <u>not</u> a whole system:

The - unintentionally wrong - estimation of the control system of a gas pipeline receiving station led to a classification in class 5 wereas other systems in comparable applications have been classified in class 4. The whole hardware system was designed on the basis of class 4 requirements. So a change up to class 5 would have had major consequences with respect to system design and costs.

A solution was found with a new risk estimation including all active parts in the safety chain since it was realised that the PLC system was only one system amongst others. In addition electrical, mechanical and hydraulic devices have to be considered. By this the requirements for class 5 can be met because of the combination of all parts although the different parts may be classified in a lower class.

5   Future Trends

Testing software years ago we felt that software production was some sort of an artistic discipline. It seems to arrive now to be an engineering discipline which means that the most important basis for succesfull verification and validation will be met.

Complexity has an important influence on correctness. Increasing processing capabilities tempt to increase complexity as well. So it may tend to exceed acceptable limits but it should be limited. This is not only a software problem but a problem about the demands on actual safety critical systems.

Verification and validation very often splits up into hardware and software v&v. Doing this - at least in Germany - people believe that we can show the hardware to be free of design errors. For software- sometimes doing just the same job as hardware- we just assume the contrary: software can not be shown to be free of design errors. But what we are really interested in is not hardware or software but reliability of integrated systems hardware and software included of course. To manage this job we have to synchronize and integrate our hardware and software v&v activities. Especially for hardware this will mean that we have to assume design errors.

For products in general and such as well for safety related systems European standardisation introduced a quite clear distinction between testing (EN 45001 [EN89a]) and certification (EN45011 [EN89b]). We will have to make fit together concepts of verification and validation on the one hand and testing and certification on the other hand.

Progress in control systems technologie is very fast, much faster then progress in v&v. If we do not succeed in following advances in control systems technologie by v&v we will not be capable to benefit for our safety from improved technology! Because the market for control systems is much more promising than the market for v&v the gap will increase and we will need substantial support in development of v&v technics and tools to solve the problems.

And vice versa to make potentials of modern control systems accessible for safety purposes in a market - and it is a market not a monastery - we will have to increase ratio from thoroughness of testing to time and effort needed. Besides others this means to increase the usage of checklists and tools.

Acknowledgements

We would like to thank our collegues Dr. D. Haake, E.-U. Mainka, G. Pillmann and R. Westhäußer for their contributions which are used in this paper.

References

[DIN89] DIN VDE 0116, Elektrische Ausrüstung von Feuerungsanlagen (Electrical equipment of in firing stations), Oct. 1989

[DIN90] DIN V VDE 801, Grundsätze für Rechner in Systemen mit Sicherheitsaufgaben (Principles for computers in safety-related systems), Jan. 1990

[DIN93] DIN 19251, MSR-Schutzeinrichtungen; Anforderungen und Maßnahmen zur gesicherten Funktion (MC-protection equipment; requirements and measures for safeguarded function), Dec. 1993

[DIN94] DIN V 19250, Grundlegende Sicherheitsbetrachtungen für MSR-Schutzeinrichtungen (Fundamental safety aspects to be considered for measurement and contol equipment), May 1994

[DIN94a] DIN V VDE 801 A1 Grundsätze für Rechner in Systemen mit Sicherheitsaufgaben (Principles for computers in safety-related systems), 1994

[EN89a] EN 45001, General criteria for the operation of testing laboratories, 1989

[EN89b] EN 45011, General criteria for certification bodies operating product certification, 1989

[Gri92] Grimm, K,. Grochtmann, M., Automatischer Softwaretest, Design& Elektronik, 25, 01.12.92

[IEC86] IEC880, Software for Computers in the Safety Systems of Nuclear Power Stations, 1986

[IEC91] IEC 65A (Sec)122 Software for computers in the application of industrial safety-related systems Nov. 1991

[IEC92] IEC 65A (Sec)123, Draft Standard: Functional safety of programmable electronic systems, 1992

[ISO90] ISO 9001, Quality systems - Model for quality assurance in design / development, production, installation and servicing, 05.1990

[ISO91] ISO/IEC 9126, Information technology - Software product evaluation - Quality characteristics and guidelines for their use, 1991

[KTA86] KTA3503, Typprüfung von elektrischen Baugruppen des Reaktorschutzsystems, 1986

A Code of Practice for the Development of Safe PLC Software

Stephen Clarke, Gerald Moran
ERA Technology Ltd
Leatherhead, UK

Peter Faulkner
Servelec Ltd
Sheffield, UK

David Hedley
LDRA Ltd
Liverpool, UK

Des Maisey
ICL
Reading, UK

Stuart Pegler
British Gas plc
Loughborough, UK

Abstract

The DTI/SERC sponsored collaborative project[1] entitled Software Engineering Methods for Safe Programmable Logic Controllers (SEMSPLC) has produced a Code of Practice for developing safe PLC application software. The Code of Practice is based on the joint experience of both the software engineering community and the PLC industry. Currently the project is applying the Code of Practice on demonstrator projects. This paper introduces the Code of Practice and examines the early results of its application.

---

[1] SEMSPLC collaborators also include University of York, York Software Engineering Ltd, Cegelec Ltd, ICI Ltd, IDEC Ltd, Nestle Ltd, Nuclear Electric plc, the HSE ,and Cincinnati Milacron Ltd.

1   Introduction

At the safety critical systems symposium in 1993, at Bristol, Audrey Canning presented an overview of the SEMSPLC project after the first year of work [CAN 93]. The project has now completed three years of its four year timescale, and is now looking towards completion at the end of 1995. In the presentation in 1993 the initial challenges to members of the project were identified through differences between the conventional views of software engineers and the approach used by PLC practitioners.

The original ideas behind the project were based upon the wide use of PLCs for safety interlocks because of their use of a language, ladder logic, which was developed to allow the implementation of configurable relay systems. PLCs are also 'ruggedised' which make them suitable for operation in industrial environments. However, the problem which the project sought to address was that typical methods for development and design of PLC application software were not considered to be of an appropriate qualitative level for safety system engineering.

PLC programs often exhibit problems in testing and debugging code especially where subtle timing properties are concerned. Further, PLCs suffer from primitive development and analysis environments in comparison to the software engineering industry and the level of support is not available to allow developers of PLC application software to cope with the increasing complexity of the systems they implement.

The objectives of the project are to:

(i)   establish a method suitable for development of safety-critical PLC based programs in an application language suitable to industrial control

(ii)   define conformance for automated tools to support the application of the method

(iii) identify metrics suitable to evaluate the benefits to product safety arising from the work programme

(iv)   evaluate the benefits in software reuse in safety-critical applications.

This paper will concentrate on work carried out to meet objectives (i) and (iii). In section 2 the Code of Practice and the reasoning behind it will be described and contained in section 3 is a description of the metrics work and the demonstrator projects. The plans for the remaining workplan and a discussion of some of the questions still to be fully addressed by the project are contained in section 4.

2   The SEMSPLC Code of Practice

2.1   Production

The workplan of the first year of the project was to investigate practices in PLC programming, industrial process control applications, relevant software engineering techniques, safety requirements, and hazard analysis techniques. The objective of the first phase was to identify relevant techniques from both existing PLC practice and conventional software engineering for combination in a method for development of safe PLC application software. The initial body of work highlighted the differences between the software engineering fraternity and actual practice in the PLC industry and led to the challenges reported in 1993.

The second phase of work drew on the work of the first year to form the Code of Practice. The information gathered during the first part of the project was classified into three groups:

(i)   current practices in PLC software development

(ii)   problems in PLC software development, in terms of either limitations of current practice in satisfaction of safety requirements, or particular PLC related constraints

(iii)   proposed solutions to the problems of (ii).

Initially, a process model of a typical PLC application software development lifecycle was built up from the practices of the collaborators in the project who were actively involved in PLCs. The process model also included the information classified as (i) above. There were a number of issues which were raised as a result of the production of a typical PLC lifecycle model.

It was noted that the contractual relationship between parties involved in the software development influenced the lifecycle, often leading to a blurring of requirements capture activities and design. The requirements are often written for the users by the suppliers, and interpreted in a functional specification, which was seen as the first level of design. Consequently, the functional specification becomes a key contractual document which is signed off by the users and is the basis for acceptance of the system during testing. The acceptance test itself can also be written by the supplier and authorised by the user. Often, the user will add extra tests to those specified during the factory acceptance tests or site tests, but the control the supplier has over the requirements and his ability to determine the satisfaction of the requirements is not ideal from a theoretical or practical point of view. Further, the lack of analysis of requirements and the difficulty of determining the level of testing required to validate safety and reliability requirements were seen as areas which could be improved for safe software development.

The process model of current practice in PLC software development was used as a basis for an initial version of the Code of Practice. In its initial form, the SESMPLC Code of Practice consists of 86 pages which contain guidance on the development of safe PLC application software. In the development of the Code of Practice it was recognised that there was a large gap between what was considered to be 'ideal' in terms of the development of safe software, and the current practices in PLC software development. To some extent, the theoretically ideal approach of developing demonstrably safe software[2] was agreed by members of the project to be unattainable with the current technology, especially since the size and complexity of most software systems means that there are so large a number of internal states that their behaviour cannot be considered predictable until many years of operation has passed. To make some impact on the PLC industry, adoption of the theoretical approach could be counter-productive in its aim of improving safety. The project took the stance that a Code of Practice which could improve the status quo, and which moved the status quo closer towards the ideal of demonstrably safe software, would be more acceptable to the industry. The difference in the pragmatic and theoretical viewpoints was described more fully in a paper which appeared in the IEE review [CAN 94].

2.2   Precepts

The Code of Practice was developed to embody a number of precepts for quality and safe software development which are stated initially, and these are:

(i)   there is little evidence to support the use of one software development method above another; however certain notations and techniques can produce software which is easier to understand and analyse

(ii)   a shift of emphasis towards requirements and design and away from coding and testing can be cost effective and improve safety and reliability

(iii)   phases of development do not run in strict sequence; ordering of activities is based on the control of the configuration management system

(iv)   there is no absolutely safe software system: only levels of risk against which each system must be judged

(v)   the existence of a 'safety culture' where safety requirements are seen as the priority is assumed within user and supplier organisations

---

[2] software which can be shown to be safe before operation commences.

(vi)   measurement of the development process is an essential part of continually improving the quality, safety and reliability of software products.

2.3   Scope

The scope of the Code of Practice is limited to the development of the PLC application software   programs which execute on one or more PLCs, and also includes the I/O configuration.

The input to the Code of Practice is assumed to be a PLC system requirements specification from which the software requirements are developed during a software requirements phase.   The main body of the Code of Practice completes with validation of the software and system requirements.

The Code of Practice is designed to fit into the framework defined in the draft IEC standards for Programmable Electronic Systems [IEC92] [IEC91].   The Code of Practice could provide an application specific interpretation of the more generic Programmable Electronic System safety standards.

2.4   Contents

The Code of Practice embodies a traditional V-lifecycle for software development. Each phase of requirements/design has a corresponding test/validation phase to confirm that the development has met the requirements.

The Code of Practice is concerned mainly with technical issues of what is required to complete a successful software development programme in terms of satisfying requirements and achieving software which is maintainable.   Concentration on the development process alone deflects the project from the aim of investigating methods specific for PLC software development.   However, the process is required to provide a framework within which specific techniques can be used, and which also provides reasons for the application of those techniques.   For instance, the use of a particular analysis technique is not appropriate unless the process of which the analysis techniques forms a part, contains a way for exploiting the results of application of the technique in order to improve the design.

There are however aspects of development which are not specific to PLC software engineering, but which are necessarily part of all development processes, for example project management and quality management.   It was considered that the remit of the project was not to provide a quality manual, or to provide general advice on project management.   The Code of Practice does provide some guidance on quality and project management,   but only to the   extent   that a   quality standard such as ISO 9000-3 is recommended as a minimum in safety applications, and that effective project management is assumed.

211

Similarly, system development aspects such as commissioning and site testing which are outside the scope of a software development Code of Practice are assumed to be driven by a system lifecycle, although it is recognised that software requirements validation activities may need to be carried out during these activities. Operation and maintenance are system issues, of which software operation and maintenance form a part.

There are two further aspects of a software development lifecycle which have not yet been discussed, but both are relevant to the SEMSPLC Code of Practice. These are configuration management, and safety.

Configuration management, although relevant to all development activities, is particularly relevant to software safety and has some specific implications for PLCs. The importance of providing an audit trail of activities is stressed as a reason for configuration management in many quality standards, and is pertinent for safety systems. Software configuration control is seen as an area which requires careful consideration. In particular, the complexity of software and its configurations, and the ease with which changes to software can be made can cause problems.

For PLCs, configuration management of documentation is assumed to be similar, and all development products require control. The configuration control of actual PLC application software requires definition in the terminology of PLCs. For instance it is not adequate to say that PLC software entities will have unique version numbers and build status, rather that each software entity will have a unique version number and configuration information such as I/O card and rack definitions, scan time information, programmer version number and versions of function blocks from libraries which may be used.

One of the major differences between the Code of Practice and the process model representing current practice is the emphasis on safety. It is assumed that input to the software development process defined in the Code of Practice is a system safety requirement together with a list of plant hazards. In line with safety standards, the Code of Practice advocates:

(i)   the clear definition of software safety requirements

(ii)   the identification of a target safety integrity level for the application software

(iii)   a safety plan for meeting the safety requirements

(iv)   analysis of the design against the safety requirements using hazard analysis techniques and the safety plan at each stage of the development process.

With a similar idea to that advocated in the Interim Defence safety standard 00-56 [MOD 91], it is suggested that a list of hazards can form the basis of a coherent safety effort. Further a hazard log can be used to record design decisions, reasons and comments on safety aspects throughout the software development. With the safety plan, the hazard log can be used as the basis for a safety case. A complete lifecycle for the Code of Practice which contains safety and configuration management activities is contained in figure 1. All activities which are not covered fully in the Code of Practice as they are wider issues not specifically concerned with PLCs are shaded.

The applicability of the Code of Practice is aimed at a wide range of safety application solutions. For this reason, notations are not mandated in the Code of Practice, although some are suggested. In the early phases of software development, more generic notations are recommended, such as data flow diagrams. Design structuring is mandated. As the software design progresses then the notations can become targeted towards the PLC specific language: for instance logic diagrams for ladder logic, state machines for sequential function charts.

PLC idiosyncrasies are highlighted for consideration at the lower levels of design, such as the effect the scanning algorithm can have on scan times, the need to ensure that global memory access in not abused, and problems which may be encountered with analogue to digital conversion functions.

The choice of PLC language in a software development is not mandated, and it is recognised that the choice of PLC will often dictate the language to be used. A selection of characteristics of safe languages are included for consideration by the developers.

It is noted that current PLC languages do not satisfy all the desirable characteristics for safety, and that the developer can produce safer software by structuring the code simply, limiting the interaction between functions via global data and omitting the use of ill-defined or complicated functions. Hence the Code of Practice recommends the use of an in-house, language specific, coding standards to enforce safe characteristics.

Each design phase in the development lifecycle develops a software test specification, for performance of tests once the software code has been written and the system is being integrated and installed. The testing techniques recommended in the Code of Practice consist of both 'black-box' functional testing methods, and 'white-box' techniques as appropriate.

As well as activities during the testing phases designed to validate the software, review and inspection activities are recommended during design phases. Additionally, modelling and simulation is encouraged in analysis of the software requirements, and in particular, involvement of the eventual operators in the software requirements process is highlighted.

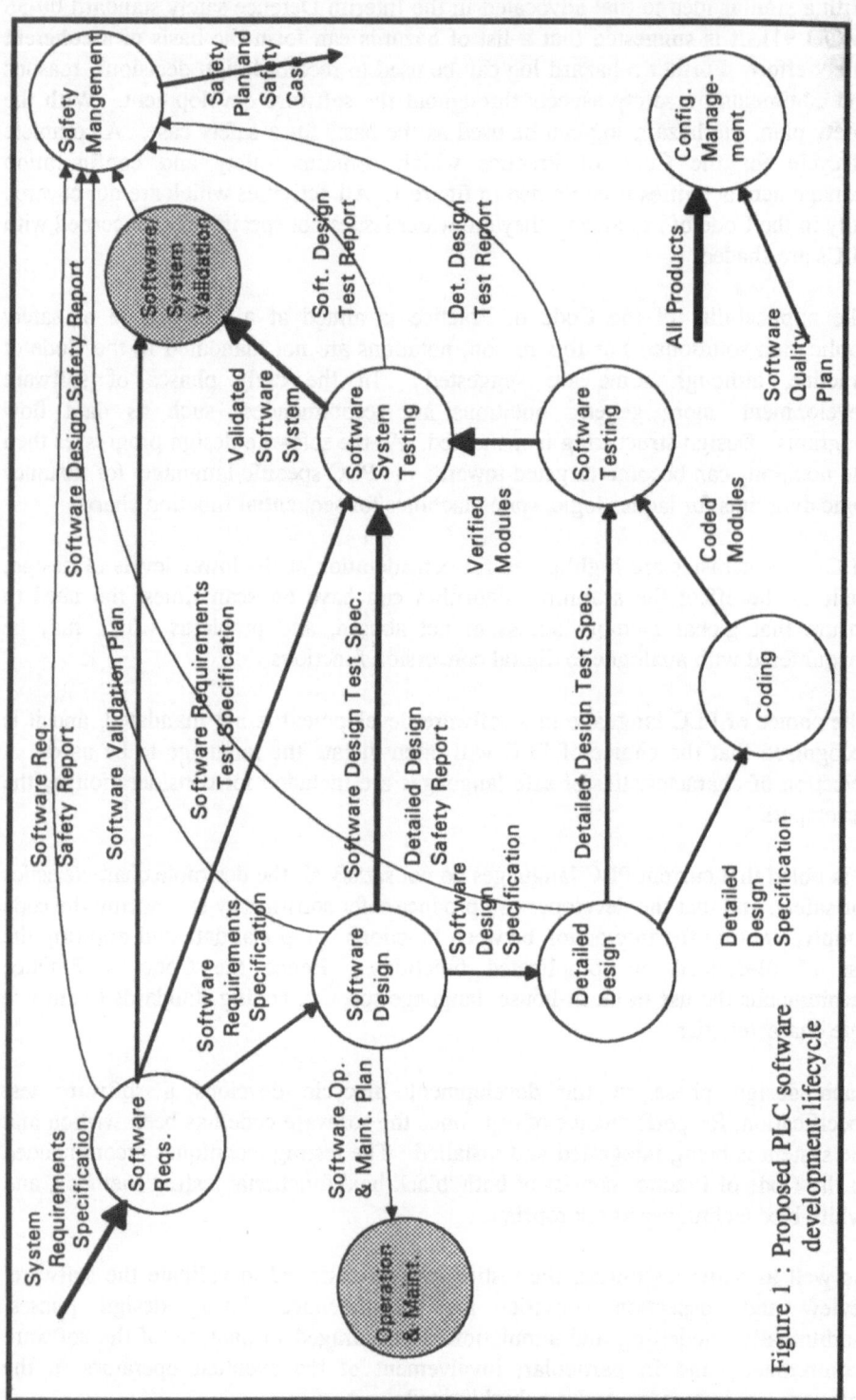

Figure 1 : Proposed PLC software development lifecycle

3   SEMSPLC Demonstration Projects

To evaluate the effectiveness of the SEMSPLC Code of Practice, demonstrator projects were established to gain experience of its application, and to compare the results with typical PLC software development practices. Further, experience gained during the demonstrator projects can be used to improve the Code of Practice. The metrics used to measure the results of demonstrator projects, and the initial results from the demonstrator projects are described in the following sections.

3.1   Metrics and Measurement

As part of the overall objective of producing a Code of Practice for safety-critical industrial PLC based programs, it was planned to investigate and establish metrics for the measurement of software product safety. The metrics developed are being used to measure the effect on software product safety of using the Code of Practice as compared to the safety of software developed using traditional practices.

For the purposes of the SEMSPLC project, metrics to measure software product safety, or software safety integrity needed to be identified. All entities produced during the software development process are open to measurement, and these include processes, products and resources. Producing a metric to directly measure software safety integrity is infeasible since safety is a characteristic which is open to qualitative measures during development, and can only be quantified after a period of operation. So the 'primary goal' of software safety integrity is broken down into second level quality factors and third level attributes (see Table 1).

Each phase in the software development lifecycle will produce products which can be measured. Unfortunately the measurement which is produced is based on the internal attributes of the product and an attribute such as software safety integrity is an external attribute. However, the splitting of the primary goal of software safety integrity into 2nd level factors and 3rd level attributes is to enable the measurement of software products in terms of internal (and measurable) attributes. It can be argued that inferences can be made about the external attribute of interest from the measurements collected on the 3rd level attributes.

One of the important product assessment measures which evaluation of the Code of Practice is based on is the 'defect occurrence' attribute. The number of defects discovered in software products are being collected, together with the phase in which they were discovered. Further, each defect is being assessed to determine if the implications of its non-discovery would have a bearing on the safety of the system. Thus, a metric for the discovery of safety related defects raised during use of the Code of Practice against the number raised through use of traditional PLC software design methods will be identified.

215

Process assessment metrics are based on the requirement that a safety software supplier should, as a minimum meet the requirements of ISO 9000-3. It is recognised that the pass/fail criteria against which ISO 9000-3 certification is based may not be sufficiently sophisticated a scale to judge a safety-critical software development process. It is intended to use the Software Engineering Institute's Capability Maturity Model [PAU 93] to assess a typical PLC software development process, and to compare this with the process maturity of the process implied through use of the SEMSPLC Code of Practice.

Table 1 : Breakdown of primary goal into quality factors and attributes

Primary Goal	2nd Level Factor	3rd Level Attribute
Evaluate safety integrity	Reliability	Defect occurrence Change occurrence Complexity Modularity
	Maintainability	Modularity Complexity Size
	Testability	Degree of testing Test effort Complexity Modularity Integration granularity Size
	Understandability	Self descriptiveness Complexity Modularity Size Readability
	Completeness	Traceability of requirements Degree of testing
	Correctness	Verification of phase outputs
	Process maturity	SEI Maturity Model
	Quality assurance	ISO 9000-3
	Analysis techniques and notations	IEC 65a Secretariat 122

Resource assessment metrics on elapsed time, man days of effort, and the amount of tool support used are being collected to determine the cost of applying the Code of Practice in comparison to the cost of applying traditional methods.

3   SEMSPLC Demonstration Projects

To evaluate the effectiveness of the SEMSPLC Code of Practice, demonstrator projects were established to gain experience of its application, and to compare the results with typical PLC software development practices. Further, experience gained during the demonstrator projects can be used to improve the Code of Practice. The metrics used to measure the results of demonstrator projects, and the initial results from the demonstrator projects are described in the following sections.

3.1   Metrics and Measurement

As part of the overall objective of producing a Code of Practice for safety-critical industrial PLC based programs, it was planned to investigate and establish metrics for the measurement of software product safety. The metrics developed are being used to measure the effect on software product safety of using the Code of Practice as compared to the safety of software developed using traditional practices.

For the purposes of the SEMSPLC project, metrics to measure software product safety, or software safety integrity needed to be identified. All entities produced during the software development process are open to measurement, and these include processes, products and resources. Producing a metric to directly measure software safety integrity is infeasible since safety is a characteristic which is open to qualitative measures during development, and can only be quantified after a period of operation. So the 'primary goal' of software safety integrity is broken down into second level quality factors and third level attributes (see Table 1).

Each phase in the software development lifecycle will produce products which can be measured. Unfortunately the measurement which is produced is based on the internal attributes of the product and an attribute such as software safety integrity is an external attribute. However, the splitting of the primary goal of software safety integrity into 2nd level factors and 3rd level attributes is to enable the measurement of software products in terms of internal (and measurable) attributes. It can be argued that inferences can be made about the external attribute of interest from the measurements collected on the 3rd level attributes.

One of the important product assessment measures which evaluation of the Code of Practice is based on is the 'defect occurrence' attribute. The number of defects discovered in software products are being collected, together with the phase in which they were discovered. Further, each defect is being assessed to determine if the implications of its non-discovery would have a bearing on the safety of the system. Thus, a metric for the discovery of safety related defects raised during use of the Code of Practice against the number raised through use of traditional PLC software design methods will be identified.

As well as metrics collection during the development of the demonstrator projects, each body of software arising from the development using the Code of Practice, and from the traditional methods, is being independently tested. Once acceptance testing have been completed on the software development, and all aspects of software validation has been completed to the satisfaction of the client and the supplier, the PLC software code will be tested by one of the partners on the project, through analysis of unusual structures, based on experience of conventional software testing methods, and relating these back to their effect on the real world.

3.2   Demonstrator Projects

At the time of writing there were currently two demonstrator projects in operation: one to implement a fire and gas detection and protection system, and the other to implement a CHP (Combined Heat and Power) simulation display system. The first of the demonstrators will be described in this paper.

The fire and gas system demonstrator project consists of two parallel development processes: one using the SEMSPLC Code of Practice as the development method, and the other using existing in-house PLC software development procedures. The requirements are based on realistic fire and gas systems although, in practice, there is no client willing to pay money for the product. The project has a number of functions which cover both continuous monitoring and sequential control. A "dummy" client was selected, with experience in fire and gas systems, but with little knowledge of PLCs or software. The client would formulate the requirements, provide clarifications, and finally accept the completed product from the suppliers.

The size of the project can be considered small but realistic, with about 150 I/O and software of around 600 rungs of ladder logic. Different people were chosen to perform the development using the Code of Practice, than were chosen to use the traditional methods. The personnel are considered to be of a similar experience and ability, and representative of a mature PLC software engineer.

Both of the development teams were issued with the same client requirements consisting of: the requirements specification, an I/O schedule and an identification of plant hazards.   A standard PLC, a test rig/simulator along with the programming manuals were also issued. All personnel involved in either of the projects were asked to fill in a log book of subjective comments on their project. Independently from both of the development teams, metrics were collected on the development processes as they proceeded and software products as they were produced. Neither of the development teams were aware of the metrics being collected.

3.3   Initial Results

At the time of writing, only the software requirements phase of the software development had completed on both fire and gas detection projects. The SEMSPLC Code of Practice development had nearly completed the design phase, and the development using traditional methods had completed the coding phase.

The initial results are mixed. From the metrics collected during the first phase it can be seen that the amount of effort to produce the software products associated with the Code of Practice was greater by a factor of three, 40 days to 12 days, than the effort required using the traditional methods. However, the number of defects discovered in the client requirements was greater also by the same factor, 32 to 9. This also included a proportional increase in the number of defects which were assessed as being 'unsafe' (7 to 2). There is quite a remarkable correlation between the ratios of effort and defects discovered.

Reading the comments in the log book filled out by the engineers who carried out the first phase of development using the Code of Practice provided useful feedback. The subjective comments in the log book are many and varied! For the purposes of this paper, the comments can be collected into two main points.

The first is that application of the Code of Practice is seen as boring. Particularly, this is believed to be because there are many paper products associated with each development phase and that production of the documentation is tedious. The objective of each of the software products could be more clearly stated in the Code of Practice to enable their purpose to be better understood.

The second point is related to the first point. There are comments about the detail of the level and type of information required during each phase of development. In particular, there are problems as to what information should be produced in one phase or the next, which sometimes led to duplication of information.

Some of the comments in the log book could be related to the lack of experience the engineers have in some of the techniques advocated in the Code of Practice. Further, comments may also be related to the lack of help contained in the Code of Practice on instructing engineers how to design, and more specifically how to apply a generic process to a specific application project. It was noted by the developers using the SEMSPLC Code of Practice, that they believed they would produce a better end product. However, the next revision of the SEMSPLC Code of Practice will aim to improve the detail in the structuring of the method, and strengthen the reasoning behind the requirements for production of software products and for recommending certain analysis and design techniques.

4   Conclusions

As originally planned, the SEMSPLC project is to revise the Code of Practice as a result of the findings of the demonstrator projects, especially addressing those issues raised in the previous section.

There are still answers to questions which the project has to finalise which could greatly affect the form of the Code of Practice, i.e.:

(i)   What should the final presentation of the Code of Practice look like?

(ii)   Who should the Code of Practice be targeted towards?

(iii)   How should the Code of Practice be used?

(iv)   Why should anyone want to use the Code of Practice?

(v)   What techniques should be mandated for what types of applications?

(vi)   How much tool support is required?

The first three questions are all interrelated, and it seems that there are at least three possible options open for the Code of Practice:

(i)   a set of work instructions targeted at individual PLC software engineers, for use during development

(ii)   a guideline for the PLC industry in general on how to develop software, which can be used as a basis for the formation of in-house work instructions

(iii)   a standard to be adopted by legislative bodies which must be adhered to, and against which assessments can be made.

If the project were ambitious it could try and produce all three forms. Clearly though, this fundamental question will have a great impact on the language and format of the Code of Practice.

The fourth question of why should anyone want to use the Code of Practice is related to safety. The early indications are that the demonstrator projects are providing evidence that application of the Code of Practice will have an impact on the safety of PLC software, though probably at a cost in effort.

In answer to the fifth question above, although the Code of Practice was aimed at being generic for production of all types of safe PLC software, some of the

220

techniques appropriate for continuous process control applications are clearly different from those appropriate for shutdown logic, and those appropriate for sequential control. The current Code of Practice based demonstrator projects are struggling with the choice of appropriate notations because of the generic stance taken. This problem seems similar to that caused by the IEC software safety standard [IEC 91] because, although a wide range of techniques is suggested in the tables in the standards, it still does not provide specific help to the engineer as to how to choose from a range of techniques and to fit them into the current application to achieve safety integrity levels. If the Code of Practice is to take the form of work instructions then it may be that a separate set of work instructions will have to be created for each application area.

In response to the sixth question, the use of tools required to support the Code of Practice may relieve some of the boredom reported by the engineer in producing the software products demanded by the Code of Practice. There has been a lower level of tool support used in the PLC industry in comparison to other areas of software engineering and although production of large integrated project support environments would be useful if some of the problems of integration could be solved, the use of a high level of automated support would be overkill when considering the size of the typical PLC project. The final year of the project aims to produce a number of prototype tools to support the Code of Practice, and the level of support required for a PLC safety development process is being investigated.

References

[CAN 93] Canning A. Software engineering methods for industrial safety related applications, Directions in safety-critical systems: proceedings of the safety-critical systems symposium, Bristol 1993, Ed. Felix Redmill & Tom Anderson, Springer-Verlag, 1993, 96-102

[CAN 94] Canning A, Moran G, Clarke S, Maisey D, Pegler S, Hedley D. Sharing ideas the SEMSPLC project, IEE review software for engineers supplement, IEE March 1994, S-23 - S-26

[IEC 91] International Electrotechnical Commission. Software for computers in the application of safety related systems, IEC 65A(Secretariat) 122, Draft Standard, IEC, Geneva, 1991

[IEC 92] International Electrotechnical Commission. Functional safety of electrical/electronic/programmable electronic systems: generic aspects, Part 1: general requirements, IEC 65A(Secretariat) 123, Draft Standard, IEC, Geneva, 1992

[MOD 91]   Ministry of Defence.   Interim defence standard 00-56/Issue 1 - hazard analysis and safety classification of the computer and programmable electronic systems of defence equipment, MOD, April 1991

[PAU 93]   Paulk M.. et al. Capability Maturity Model, version 1.1, Technical report CMU/SEI-93-TR-25, Software Engineering Institute, Carnegie-Mellon University, Pittsburg, 1993

USING INCIDENT ANALYSIS TO DERIVE A METHODOLOGY FOR ASSESSING SAFETY IN PROGRAMMABLE SYSTEMS

Eamon J. Broomfield and Paul W. H. Chung
Loughborough University of Technology
Loughborough, U.K.

Abstract

This paper describes the development of a generic methodology for assessing safety in programmable systems. A functional model, a new graphical technique and Method Study are used to analyse incidents. The results of the analysis form the basis of the methodology. A case study is used to show how the methodology can be applied.

1 Introduction

Much research work has been done on developing and assessing the safety of programmable systems. One of the major problems is in relating the embedded software to the system as a whole. As Redmill remarks *"Safety does not depend arbitrarily on the quality of software but on a deeper study of potential hazards ..."* [Redmill 93]. There are a number of well established techniques for analysing these *potential hazards*, such as HAZard and OPerability Study (HAZOP), Failure Mode Effect Analysis (FMEA), Fault Tree Analysis (FTA) and Checklists. Although the software analyst has a number of techniques for developing systems such as formal methods, structured methods, and static and dynamic testing, no established technique is available for relating software faults to system hazards. Some researchers have attempted to overcome this problem by modifying traditional hazard identification techniques, including combining them with software techniques. Taylor suggests an integrated approach which considers the interaction between hardware, software and the operator [Taylor 94]. He proposes functional analysis for the system specification, he uses HAZOP, FMEA, simulation and sneak analysis for the system model and he suggests Human Reliability Analysis (HRA) for operating procedures. Chudleigh & Clare state *"... a common representation is necessary"* for the total system [Chudleigh 93]. They use Dataflow diagrams for modelling the system and a modified version of HAZOP for analysis. Fink et al. analyzed data management in a clinical laboratory information system [Fink 93]. Their aim was to identify and analyse potential failures of the system. They combined HAZOP and Failure Mode Effect and Criticality Analysis (FMECA). The Health and Safety Executive (HSE) provide guidelines including Checklists on safety requirements specification, hardware, software, installation, testing, operations and maintenance modification [HSE 87]. Leveson used a modified version of FTA for analyzing system requirements and software code [Leveson 83]. Bowen & Stavridou state that there is a gap between the safety level required and that which is obtainable with current practice [Bowen 93] and they suggest a possible way of narrowing this gap is to use formal methods in conjunction with other techniques.

As shown above, although there are many possible ways of assessing the safety of a system, there is no established methodology uniquely suited to analyzing the total system.

2. Safety and Faults

Hazards do not occur in software, however, *faults* in software can contribute to hazards. Therefore, it is these faults in software which must be considered when assessing the safety of programmable systems. In attempting to achieve a safe system, we can employ *fault prevention, fault removal* and *fault tolerance* concepts. Laprie has defined these concepts in relation to dependable systems [Laprie 93]. For this work, we view *fault prevention* as a systematic methodology which can help prevent the introduction or occurrence of faults, *fault removal* as the application of this methodology to help remove faults and, finally, *fault tolerance* as the application of this methodology to minimize the effects of any faults remaining in the system.

In order to use these concepts, we must first find a mechanism for analyzing faults. We can categorise faults according to their type [Kershaw 93], random failures of components; systematic errors in design; errors in the interface to the outside world; maintenance errors; and common mode errors. We can categorise faults in terms of where they occur in the development of a system. There may be faults in the requirements, design, coding etc. These faults could be further subdivided into type e.g. in the design phase we could have hardware interface, software interface, user interface and functional description [Grady 93]. Yet another approach is to analyse incidents and develop a mechanism for identifying faults. Incidents are very useful because they are representative of the 'real world', they allow us to investigate how faults occur and how these faults propagate in a system i.e. incidents can be used for more than just a classification system. Brazendale & Jeffs have analyzed a number of incidents and have highlighted where faults occur and what precautions are needed to prevent recurrence [Brazendale 94]. These faults include, specification error; inadequate specification of safety integrity; inadequate design and implementation; inadequate installation and commissioning; inadequate operation; and maintenance.

The approach taken in this work is different in that a generic model of an incident is proposed. This model, in combination with a new graphical technique, an Event Time Diagram (ETD) [Broomfield 94a] is used to analyse each incident, the results of the analysis form the basis for a safety assessment methodology. The technique used for analyzing incidents is Method Study, a popular technique for problem solving. The basis of this technique is to examine a problem by asking the questions What?, Why?, How?, and When? Elliott & Owen's pioneering work in analyzing chemical plant design using Method Study [Elliott 68] has evolved into the well established HAZOP technique used worldwide by the chemical industry. One of the major problems with Method Study is the amount and complexity of the information that can be generated. In this work, the Method Study is constrained by asking questions related to *fault prevention, fault removal* and *fault tolerance* and by grouping the questions in a 'keyword' structure.

3. A Functional Model for Incidents

Over 300 incidents were provided by two major organisations, one involved in the process industry, the other in avionics. The format of the information received is shown in Tables 1 and 2. No user requirements, functional specifications, architecture diagrams or software code were provided for any of the incidents. This might first appear to be a disadvantage, however, it forces one to take a 'real world' view of a system and to decompose the system in a general manner thereby preventing the methods developed from becoming application-specific.

A/C Type	Flight Phase	Location	Date	Occnum	Permpub
BXXXXXX	CRUISE	LXXXXX	XX XXX XX	XXXXXXXF	P

FMS malfunction in cruise at FL350 A/C nosed over lost 600ft in 5sec ---- departing altitude due to the loss of air data reference power caused by a faulty one amp circuit breaker

Table 1: An example incident from the AVIONICS INDUSTRY

Although industries record incidents, there is no established method of using these incidents to identify hazards at a generic level in re-engineering old systems or developing new systems i.e. each incident is dealt with individually following its occurrence.

OBSERVED EFFECT	ROOT CAUSE	CAUSE CATEGORY
SEQ. STARTED PREMATURELY	SEQ. PERMISSIVES INCOMPLETE	APP.S/W

Table 2: An example incident from the PROCESS INDUSTRY

The initial objective was to construct a generic model for incidents i.e. a model which could be used to represent incidents similar to those shown in Tables 1 & 2. The basic components of a generalised system include:

- computer
- output comms link
- human input device
- actuator
- input comms link
- sensor
- display
- operator

Each incident which occurs involves one or more of these components and some result from complex interactions between the components. We can associated with each component a functional behaviour (e.g. an operator intervenes; a sensor provides input).

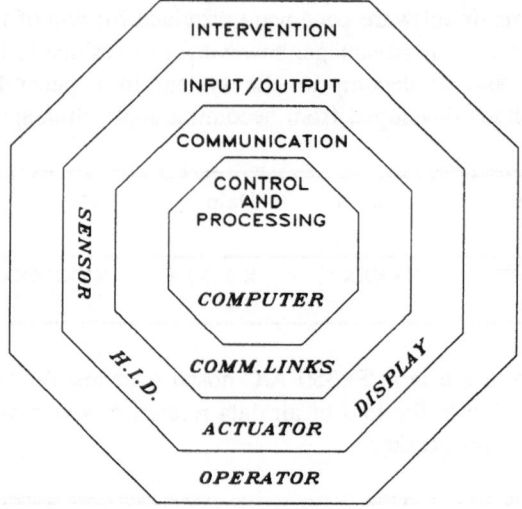

Figure 1: Model showing basic components and functional levels

A model (Figure 1) showing the basic components and associated functional levels was constructed. This model shows the hierarchy of the functional levels and illustrates their interdependency. All components associated with the intervention level that interact with the computer, must use some component at the I/O level. Similarly, all components associated with the I/O level that interact with the computer, must involve some component at communication level. It can be used to establish 'what if' scenarios. For example, if an operator inputs incorrect data, what happens if the error propagates through to the inner levels? In the worst case, the operator error causes a failure at the control and processing level.

4. Analyzing Incidents

An incident may be viewed as a sequence of events. There are many structured design methods which can be used to analyse events [Edwards 93], however none were suitable to map directly onto the Functional Model developed above. Instead, a new graphical technique, called an Event Time Diagram (ETD), is proposed for viewing and analyzing events. This technique is found to be useful for focusing on the safety aspects of the 'total' system. An ETD is used to model system behaviour in terms of events, time, control and data flow, entities and associated functional levels. Figure 2 is the ETD for the incident described in Table 1. An ETD may be viewed as a polar diagram where the angle represents time, the distance from the centre gives the functional level, and the arrows give direction of flow of information (either control or data). A node within an ETD may be described by two coordinates (c_x, r_x) where c_x represents the entity, and r_x represents the functional level. The ETD shown in Figure 2 is an 'interpretation' of the incident in Table 1. It has to be creative in that assumptions are made such as *the computer made a calculation based on erroneous data* (E4). It is seen that a number of events (E1, E2, E3 and E4) lead to the final consequence (E5). E1 is the root cause, however, any of the events (E2, E3, E4) could be root causes under different circumstances. For example, incorrect computation (E4) may result in E5 occurring irrespective of E1, E2 and E3. In

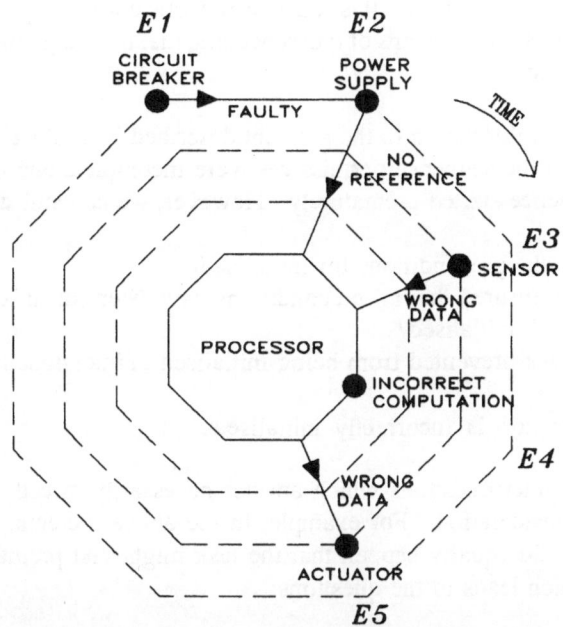

Figure 2: ETD for the incident described in Table 1

addition, we could prevent E4 by preventing the computer from acting on erroneous information, however, this would not be sufficient as it is quite likely that the Actuator requires continual updating, therefore, even if the computer recognizes erroneous data, there must be some recovery mechanism. This indicates the importance of considering all events. The ETD also allows us to analyse an event on an individual basis. Even if E1, E2, E3 and E4 did not occur, E5 could still occur if, for instance, the communication link between the processor and actuator failed. The ETD provides a very effective mechanism for analyzing incidents.

Having represented the events related to the incident, the next step is to pose a series of questions which, had they been asked in the early design stage, would have prevented this and similar incidents from occurring. The questions are:

- What hardware reliability data is available for hardware components?
- How is failure of power detected?
- If power fails, how is the system placed in a safe state?
- How is the sensor to be calibrated?
- What is the range of the expected input values?
- Are multiple sensors required?
- Is a continuous self-test sequence required (e.g. to detect dramatic changes in input)?
- Is there any method of verifying or correlating output data to detect out-of-range values?

At this stage we also construct a general descriptor which best describes the failure mechanism of the incident; in this case 'erroneous/corrupt operation'. General descriptors provide a useful means of crosschecking that the main failure mechanisms have been identified.

There is very little information in the incident described in Table 2. The root cause in this case is that the sequence permissives were incomplete and the consequence was that the sequence started prematurely. However, we can still derive questions:

- What are the preconditions for initialisation?
- How is it ensured that all preconditions have been identified?
- How is task initialised?
- How is task prevented from being initialised unintentionally?

The general descriptor is 'incorrectly initialised'.

Incidents trigger other questions which are not necessarily directly relevant to the incident under consideration. For example, in the above incident, the task started prematurely, it could equally happen that the task might end prematurely or might not end at all which leads to the questions:

- What are the sustaining conditions for this task?
- What are the postconditions for this task?

5. Framework Based on Analysis of Incidents

Using the above method for analyzing incidents, we can see that we rapidly build up a library of questions. We need to structure these questions and general descriptors so that they can be used in a logical and effective manner when assessing the safety of a system. The questions and general descriptors were grouped as follows:

Specification:	Questions related to understanding what is involved in the particular task, such as objective and timing/control.
Associative:	Questions related to what other tasks or devices are associated with this task, useful for cross referencing or traceability.
Implementation:	Questions related to what should be considered when implementing this task.
Protective Measures:	Questions related to what protective measures can be taken for this task.
Failure Modes:	All general descriptors derived from incidents. This group provides a means of checking that the main failure modes have been considered.

The next stage was to create *keywords* within the above categories. e.g. in the Protective Measures Group, we can have the *keyword* 'failure detection'. Any assessment has to be creative in that there is no way all possible questions could be asked. The idea of the *keywords* is to prompt the user to think of other questions which may be important. It also helps to make the questions less subjective. Table 3 shows the *keywords* within their groupings. Finally, we introduce the concept of a *question-type* associated with a *task-type*. This *task-type* is based on the functional model described above e.g. a Display Task, a Human Task etc. For any given task the questions asked would be all the general questions, together with those specifically related to that task (i.e those specifically related to other *task-types* would be ignored). Table 4 shows how the incident described in Table 1 is mapped into this framework. The framework provides a systematic way

Specification	Protective Measures
definition	failure detection
objective	interlocks
options	trips
inputs/outputs	security
timing/control	fault recovery
operational modes	procedures
	verification
Associative	**Failure modes**
tasks	not initialized
devices	incorrectly initialized
	incorrectly executed
Implementation	not terminated
selection	incorrectly terminated
installation	erroneous/corrupt
testing	operation
maintenance	no input/output
environment	incorrect input/output
utilities	lockup
	too fast
	too slow
	defective hardware
	failure not detected

Table 3: *Keywords* and their Group

of applying the questions and is also a useful means of auditing the development process, since the results may highlight weaknesses in certain areas.

QUESTIONS (reading clockwise around the ETD; Fig. 2)	GROUP	KEYWORD	TYPE
What hardware reliability data is available for hardware components?	Implementation	testing	General
How is failure of power detected?	Protective Measures	failure detection	Intervention
If power fails, how is the system placed in a safe state?	Protective Measures	fault recovery	Intervention
How is the sensor to be calibrated?	Implementation	installation	Sensor
What is the range of the expected input values?	Specification	I/O	Sensor
Are multiple sensors required?	Specification	timing/ control	Sensor
Is a continuous self-test sequence required (e.g. to detect dramatic changes in input)?	Protective Measures	failure detection	General
Is there any method of verifying or correlating output data to detect out-of-range values?	Protective Measures	failure detection	Actuator
General Descriptor for this incident	Failure Modes	erroneous/ corrupt operation	

Table 4: This table shows how the incident described in Table 1 is mapped into the framework

6. Applying the Methodology

The methodology developed is generic so, to apply it to a specific system, we must first of all identify the top level hazards, then specify the safety requirements for the system and, finally, determine the associated requirements. The methodology is then applied to each requirement as follows:

- Decompose requirement into tasks
- Draw ETD showing tasks
- Identify critical tasks
- For critical tasks, apply *keywords* and associated questions

To illustrate the methodology, a case study follows where two safety requirements are selected for analysis.

6.1 Case Study

The case study is based on a rotary screen line printing machine [Broomfield 94b]. The development of such a machine requires a multi-disciplinary approach, bringing together expertise in process engineering, logic control, variable speed drives and mechanical engineering. The rotary screen printing press consists of a number of print stations. Each print station prints a different pattern. Material flow is shown in Figure 3: paper passes to a print head where it is coated with plastisol, it then passes through a drying cabinet, before passing to the next print station. Three different types of computer systems are used. One computer controls the running of the line. Each print head has a single board computer to control the printing. Each burner has an associated proportional, integral and derivative controller.

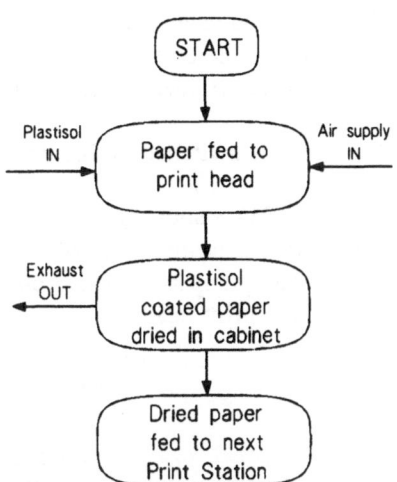

Figure 3 Material Flow through print machine

The use and control of flammable solvents in printing inks and plastisols (a type of paste) is a well known source of hazards in the printing industry. Each drying cabinet in the machine has explosion panels, however, the machine user is responsible for ensuring that flammable solvent levels in the circulating air and exhaust air associated with each cabinet are safe as defined in the HSE regulations [HSE 71].

From the above, one safety requirement is to ensure that the flammable solvent level must not exceed that specified for the machine while the machine is running. There are a number of requirements associated with this safety requirement but only two of these are selected here for analysis. This analysis is performed by applying the methodology as described above. For each Requirement, only one task (the one thought to be the most useful in demonstrating the methodology) is selected for further analysis. Only some *keywords* and questions are selected, depending on the particular task. Finally a list of actions is given including comments, warnings and design constraints.

6.1.1 Requirement 1

The requirement is to pump the correct quantity of plastisol into the print head when automatic mode is chosen. A pump at each print station fills the print head at intervals, the time between intervals is dependent on the required coat weight. If the pump overfills the print head, the solvent concentration in the dryer will increase and may increase beyond the specified limit.

231

Task Decomposition:
Operator selects automatic mode for pump operation
Line sensor indicates line is running
Nip sensor indicates that nip is closed (nip is the gap between the printing rollers)
Level sensor indicates low plastisol
Processor determines pump 'on time' t, based on line speed and required coat weight
The pump is switched on for time t

ETD: The ETD is shown in Figure 4. The pump can operate in automatic or manual mode. Automatic mode is only operative when the line is running and the nip is closed. The pump is switched on when the level detector detects a low plastisol level in the print head.

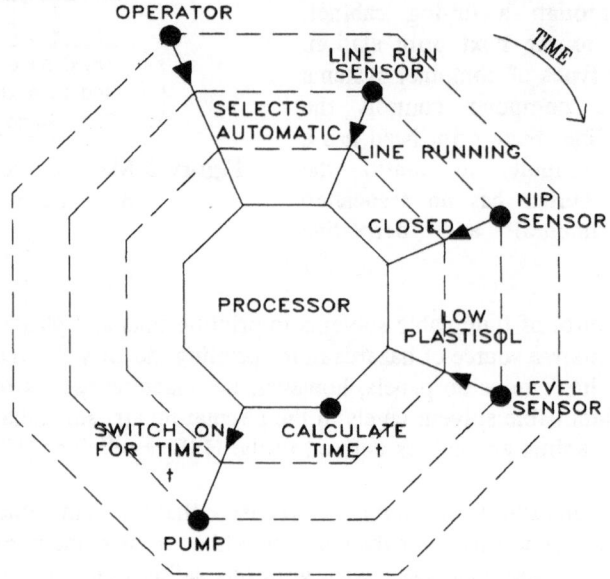

Figure 4: ETD for Requirement 1

Identify critical task for further analysis: The safety aspect of this requirement ultimately depends on the switching on of pump for time 't'.

Analysis of PUMP SWITCH ON task: A selection of keywords, generalised questions and questions specific to an actuator task are used

Specification

Definition: *Q: What action is required?*
               A: To turn on pump for a 'set' time
Objective: *Q: Why is this action required?*
               A: To fill print head with required amount of plastisol

232

Options:	Q: *What other way could this task be accomplished?*
	A: Rather than using the level sensor to trigger switch on of pump, a detector which directly measures coat weight could be used.
Inputs/ Outputs:	Q: *What are inputs/outputs for this task?*
	A: The pump solenoid receives an energise signal from the main system
	Q: *What is the range of this signal?*
	A: On/off only. No associated range.
	Q: *What parameters are associated with this task?*
	A: 'Line speed' and 'pump factor'. 'Pump factor' is set by operator. It is dependent on coat weight required.
Timing/ Control:	Q: *How is this task to be initialised?*
	A: Level probe senses that the plastisol level is low
	Q: *How is this task to be terminated?*
	A: The task is terminated after a time 't' set by the main programmable system which de-energises the pump solenoid.
Operational modes:	Q: *What relationship does this task have to manual mode?*
	A: In manual mode the pump action is interlocked only to the level sensor, the pump action is independent of line speed and status of nip.

Implementation

Selection:	Q: *What actuator will be used?*
	A: A solenoid valve
Installation:	Q: *How will this device be interfaced to the system*
	A: The solenoid valve is energised via a signal from an electromechanical relay. One input of the relay is connected to programmable system via a digital output
Testing:	Q: *How will the implementation be tested?*
	A: The switch on time of pump is dependent on coat weights. The coat weight is related to the pattern being printed on the paper. Trials are required.
	Q: *What reliability data is available on hardware items?*
	A: There is no reliability data available at present.

Protective Measures

Failure detection:	Q: *How will the system know if any of the hardware devices associated with this task are functioning incorrectly/defective?*
	A: There are no directly associated detection mechanisms as there is no automated feedback from the coating operation. Detection is dependent on the operator
Interlocks:	Q: *What are preconditions for initialization of any associated tasks?*

Trips:

A: Line running, nip closed, level sensor indicating low plastisol

Q: *What trips are associated with this task?*

A: There are no trips associated with this task

Security:

Q: *What parameters associated with this task can be modified by the operator?*

A: The 'pump factor'

Q: *Why can these parameters be modified by the operator?*

A: There are many different coat weights required and there are always inconsistencies within batches

Failure modes

Lockup:

Q: *How could this task lockup?*

A: No signal to deactivate pump so plastisol would overfill print head, resulting in excessive coating. To prevent this, incorporate in software a maximum on-time for pump which, if exceeded, would trip power to pump.

Actions:

1. It would be worth considering the use of on-line coat weight monitor as there would be immediate feedback on excessive coating. This would be useful for safety, from an economic point of view and for better consistency in coating. The difficulty of using such a device is that the coat weight profile varies across the paper for a given pattern and there are also variations due to inconsistencies in batch production (i.e. continual recalibration would be required).

2. Check calibration of level sensor as the operation of the pump is dependent on the sensitivity of the level sensor.

3. Trials are required to determine maximum and minimum pump on-times.

4. Obtain hardware reliability data on pump and associated devices.

5. Consider implementing trip in case of pump remaining on unintentionally.

6. If operator selects and sets up pump on time, he/she may need training on how to do this.

6.1.2   Requirement 2

If no solvent vapour is being exhausted from the drying cabinet while the line is running, the machine must be shut down and placed in a safe state.

Task decomposition:

Flow sensor indicates no exhaust flow
Line sensor detects line is running
Burners are switched off
Nips are forced to open
Line is kept running
Alarm is activated
Warning is shown on Display

ETD: The ETD is shown in Figure 5. The burner is switched off to remove heat source from drying cabinet. The nip is forced open to stop coating of paper and the line is kept running to remove paper from the oven and prevent it from igniting. Siren is activated and warning is displayed.

Identify critical task(s) for further analysis: The operation of the flow switch is critical. Incorrect operation or logic associated with this device could lead to catastrophic events. This is the only task chosen from this requirement.

Analysis of FLOW SWITCH task: A selection of keywords, generalised questions and questions specific to a sensor task are used

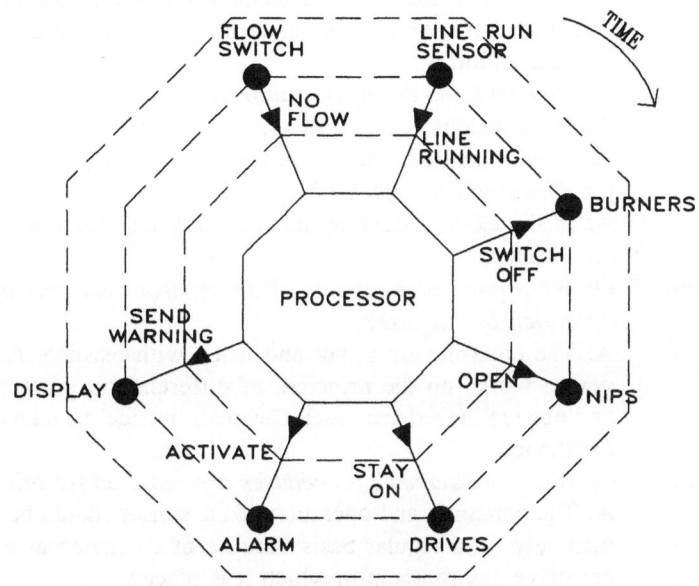

Figure 5: ETD for Requirement 2

Specification

Definition:	*Q: What state is to be monitored?*
	A: The exhaust flow
Objective:	*Q: Why does this state need to be monitored?*
	A: To prevent build up of solvent concentration in drying cabinet
Options:	*Q: What other way could this task be accomplished?*
	A: A sensor which directly measures solvent concentration or one which measures airflow
Inputs/	*Q: What are the input(s)/output(s) for this task?*
Outputs:	A: Output signal from flow sensor indicating no exhaust flow
	Q: Over what range is the signal to be monitored?
	A: Output signal on/off, no associated range
Timing/	*Q: How often does this state have to be scanned?*
Control:	A: Scan time should be less than a second.

| Operational Modes: | Q: *What relationship does this task have to startup?*
A: The flow switch is associated with a purge sequence which must complete before startup. |

Implementation

Selection:	Q: *What sensor will be used?* A: An air flow failure detector which has an embedded mercury switch.
Installation:	Q: *How will this device be interfaced to the system* A: The mercury switch is connected to a electromechanical relay. One output of the relay is connected to programmable system via a digital input. Q: *How will this device be calibrated?* A: The device is supplied by the manufacturer calibrated, it will be tuned when the machine is being commissioned Q: *Where will this device be positioned?* A: In the exhaust ducting, not near any disturbances e.g. fresh air dampers
Environment:	Q: *What particular aspects of the environment may affect the operation of this task?* A: The environment is hot and laden with plastisol fume. The device works on the principle of difference in pressure inside and outside the exhaust duct. Clean air outside the duct activates the device.
Maintenance:	Q: *What maintenance procedures are required for this task?* A: The condition and operation of the sensor should be checked manually on a regular basis because of its importance and the corrosive environment in which it is placed.

Protective Measures

| Failure Detection: | Q: *What alarms are associated with this task?*
A: A siren and alarm message on a display
Q: *What is the purpose of this alarm(s)?*
A: To warn operator that the sensor has indicated no flow in the exhaust, and to check that the machine is shutdown safely.
Q: *How will the system know if the sensor is performing OK?*
A: There will be a test sequence pre-startup which will check the sensor indicates 'off' state with the exhaust fan off indicating no exhaust flow and on state with the fan on to indicate there is an exhaust flow |
| Interlocks: | Q: *What are the preconditions for initialisation of any associated tasks?*
A: Flow sensor indicates no flow and line sensor indicates line running. |

	Q: How are postconditions of completion of any associated tasks checked?
Trips:	A: Must be checked manually by operator. Q: What trips are associated with this task? A: Nip is forced off and burner is switched off Q: Why are these trips required? A: The nip is opened to stop coating, the burner is switched off to remove heat source
Fault Recovery:	Q: What fault recovery procedures are associated with this task? A: Manually test operation of exhaust fan and sensor, purge dryer and measure concentration of solvent in dryer.

Failure Modes

| Incorrectly terminated: | Q: How could task be incorrectly terminated?
A: If operator presses emergency stop, this would leave coated material in dryer which may ignite. To prevent this, the siren should be unique to this task, indicating that no operator intervention is required. |
| Too slow: | Q: How could task response be too slow
A: If the response of the sensor was too slow, this would cause a build up of solvent vapour in dryer and possible subsequent explosion. Verify response of sensor is adequate. |

Actions:
1. It is worth considering on-stream solvent monitoring rather than monitoring the exhaust flow . Firstly, it measures the required parameter directly and, secondly, continuous feedback from on-stream monitoring could detect a unpredicted rise in solvent concentration. The major disadvantages of this are cost and number required.
2. A good maintenance procedure for the airflow switch is essential, because of its importance and the 'dirty' environment. The device should be easily accessible for the same reasons.
3. Hardware reliability information is required to make sure the airflow switch is robust enough for the number of operations and environment.
4. Correct calibration and response time of the airflow switch is critical.
5. The dependence on a single device for monitoring exhaust flow is questionable.
6. The alarm siren should be readily identifiable as being associated with 'dangerous' level of solvent vapour.

7. Conclusions

A functional model was used, in conjunction with an ETD, to represent the events leading to a hazard. Incidents were analyzed and a library of questions and general descriptors compiled. The questions were categorised into groups and *keywords*.

The ETD provides a very effective mechanism for analyzing incidents and system requirements. In applying this generic methodology, one major problem is the identification of the top level hazards. A possible way of overcoming this would be to enhance the methodology by adding sets of *keywords* specific to different industrial sectors. Applying the methodology to a Case Study showed that it is systematic and effective, however, it also proved to be time consuming. It would be worth considering what aspects of the methodology could be automated. This would have the added advantage of providing a documented record for the user which could prove useful in the design of new systems or re-engineering of old ones.

Acknowledgements

The research reported in this paper is supported by the Engineering and Physical Sciences Research Council grant J.18217. The authors are grateful to ACM for permission to include part of the case study previously described in [Broomfield 94b]. Paul Chung is also grateful to British Gas and the Royal Academy of Engineering for financial support through a Senior Research Fellowship.

8. References

[Bowen 93] Bowen J, Stavridou V: "Safety-critical systems, formal methods and standards". Software Engineering Journal 8:189-209.

[Brazendale 94] Brazendale J, Jeffs AR: "Out of control: failures involving control systems". High Integrity Systems 1:67-72.

[Broomfield 94a] Broomfield EJ, Chung PWH: "A hazard identification methodology for programmable systems". In: "Risk Management and Critical Protective Systems", Safety & Reliability Conference, Altrincham, Cheshire, 12-13 October 1994. SaRS Ltd.

[Broomfield 94b] Broomfield EJ, Chung PWH: "Hazard Identification in Programmable Systems - A Methodology and Case Study". Applied Computing Review 2:7-14.

[Chudleigh 93] Chudleigh MF, Clare JN: "The benefits of SUSI: safety analysis of user system interaction". In: "Safecomp 93" - 12th International Conference on Computer Safety, Reliability and Security (Ed. Gorski J), Springer-Verlag, London, pp.219-229.

[Edwards 93] Edwards K: "Real-time Structured Methods - Systems Analysis", John Wiley & Sons, Chichester, UK, 554pp.

[Elliott 68] Elliott DM, Owen JM: "Critical examination in process design". The Chemical Engineer, November:377-383.

[Fink 93] Fink R, Oppert S, Coolison P, Cooke G, Dhanjal S, Lesan H, Shaw R: "Data management in clinical laboratory information systems". In: "Directions in Safety-Critical Systems" (Eds: Redmill F, Anderson T), Springer-Verlag, London, pp.84-95.

[Grady 93] Grady R: "Practical results from measuring software quality". Communications of the ACM 36:62-68.

[HSE 71] Health & Safety Executive: "Evaporating and other ovens", HSW 46, HMSO, London.

[HSE 87] Health & Safety Executive: "Programmable electronic systems for safety-related applications". No.2 Technical Guidelines, HMSO.

[Kershaw 93] Kershaw J: "The special problems of military systems". Microprocessors and Microsystems 17:25-30.

[Laprie 93] Laprie J-C: "Dependability: from concepts to limits". In: "Safecomp 93" - 12th International Conference on Computer Safety, Reliability and Security (Ed. Gorski J), Springer-Verlag, London, pp.157-168.

[Leveson 83] Leveson NG, Harvey PR: "Analyzing software safety". IEEE Transactions of Software Engineering SE-9: 569-579.

[Redmill 93] Redmill F: "Software in safety-critical applications - a review of current issues". In: "Safety critical systems: current issues, techniques and standards" (Eds. Redmill F, Anderson T), Chapman & Hall, London.

[Taylor 94] Taylor JR: "Developing safety cases for command and control systems". In: "Technology and Assessment of Safety-critical Systems" - Proceedings of the 2nd Safety-critical Systems Symposium, Birmingham, UK, 8-10 February 1994 (Eds. Redmill F, Anderson T), Springer-Verlag, London, pp.69-78

PROCESS SYSTEMS APPLICATIONS OF ARTIFICIAL NEURAL NETWORKS

Gary Montague, Julian Morris and Paul Turner

Department of Chemical and Process Engineering

University of Newcastle

Newcastle upon Tyne, NE1 7RU, England

email: julian.morris@newcastle.ac.uk

Abstract

Since the mid 1980's interest in two major areas of strategic research have emerged - that of Artificial Neural Networks and Multivariate Statistical Data Interpretation. The much wider availability and power of computing systems, together with new research studies, is resulting in expanding areas of application. This paper examines both of these very important developments and assess the contribution that both technologies can make in process supervision and control. Feedforward networks with sigmoidal activation functions, radial basis function networks and autoassociative networks are reviewed and studied using data from industrial processes.

Two applications of artificial neural network model-based predictive control (MBPC) are discussed. The first introduces the concepts of neural network MBPC by application to a comprehensive simulation of a binary distillation column; and the second to a C3 Splitter operating at an ICI site in Australia. In the second study, four control algorithms are compared - two PI regulators, a linear MBPC and a neural network MBPC. The neural network MBPC is shown to significantly improve the control performance of the column.

Situated above the process control loops are Statistical Process Control (SPC) procedures responding to operational deviations and process malfunctions. SPC schemes are now receiving significant attention in response to increasing demands for improved process performance and improved product reproducibility. In particular the multivariate statistical approaches of partial least squares and principal component analysis have been found to be useful to provide for effective predictive model development from highly dimensioned and ill conditioned monitored process data. The strong links between the multivariate statistical methods of Principal Component Analysis and Projection to Latent Structures with neural network techniques is discussed with particular reference to a continuous polymerisation reactor condition monitoring.

Finally, the safety critical aspects that may well be met during the implementation of artificial neural networks in process systems applications is discussed.

Introduction

System representation, modelling and identification are fundamental to process engineering where it is often required to approximate a real system with an appropriate model given a set of input-output data. The widespread interest in input-output mapping has been given impetus by the increasing attention being focussed on pattern processing techniques and their potential application to a vast range of complex and demanding real-world problems. These include, for example, image and speech processing, inexact knowledge processing, natural language processing, event classification and feature detection, sensor data processing, control, forecasting and optimisation. The model structure needs to have sufficient representation ability to enable the underling system characteristics to be approximated with an acceptable accuracy and in many cases the model also needs to retain simplicity. For linear time-invariant systems these problems have been well studied and the literature abounds with many useful methods, algorithms and application studies [LYUNG 83]. Widely used structures are the Autoregressive Moving Average (ARMA), the Autoregressive with Exogeneous Variables (ARX) and the Autoregressive Moving Average with Exogeneous Variables (ARMAX) representations.

In practice most systems encountered in industry are non-linear to some extent and in many applications non-linear models are required to provide acceptable representations. Non-linear system identification is, however, much more complex and difficult, although the Nonlinear Autoregressive Moving Average with Exogeneous Variables (NARMAX) or Nonlinear Autoregressive with Exogeneous Variables (NARX) descriptions have been shown to provide useful unified representations for a wide class of non-linear systems. Efficient parameter identification procedures are particularly important with non-linear systems so that parsimonious model structures can be selected. The work of Billings and colleagues [BILLINGS 83], [BILLINGS 86], [BILLINGS 91], [CHEN 90a], [CHEN 90b] provide seminal work in nonlinear system studies.

The problem of identifying, or estimating, a model structure and its associated parameters can be related to the problem of learning a mapping between a known input and output space. A classical framework for this problem can be found in *approximation theory*. Almost all approximation, or identification, schemes can be mapped into, ie. expressed as, a network. For example, the well known ARMAX model can be represented as a single layer network with inputs comprising of lagged system input-output data and prediction errors. In this context a network can be viewed as a function represented by the conjunction of a number of *basic* functions.

Another strategically important area of process control is that associated with Statistical Process Control (SPC). Here the aim is to monitor process operation and performance in order to be able to detect the occurrence of off-specification production, important process disturbances and process malfunctions (faults). The early detection of process malfunctions, followed by the location of their causes, can lead to significant improvements in process operability. Process Condition monitoring and SPC requires the efficient handling of large amounts of monitored plant data which often is subject to measurement errors, is ill conditioned, highly correlated and collinear. The effective treatment and characterisation of complex data has been the subject of much attention, particularly in the chemometrics field. Powerful statistical methods such as Principal Component Analysis (PCA) and Projection to Latent Structures, or Partial Least Squares, (PLS) have been developed leading to Principle Component Regression (PCR) modelling procedures. These approaches allow the development of models relating process input/output measurements to the prediction of future 'quality' variables which can be very difficult to measure continuously. An approach suggested here is to make use of the nonlinear mapping capabilities of artificial neural networks. In particular the feature extraction properties of autoassociative neural networks can be exploited for nonlinear process data characterisation and fault signature extraction.

Artificial neural network representations

The problem of identifying the parameters of a model structure essentially reduces to the determination (or learning) of a set of parameters, **w**, that provide the best possible approximation to the real system (according to *a-priori* chosen criteria) when trained on an 'example' data set. One of the major obstacles to the widespread use of advanced modelling and control techniques is the cost of model development and validation. The utility of neural networks in providing viable process models was demonstrated some years ago [BHAT 89] where the technique was used to successfully characterise two non-linear chemical systems as well as interpret biosensor data. Since that time studies in neural networks theory and applications have flourished [IEE 88-90], [HUNT 92], [THIBAULT 91].

The sigmoidal feedforward network

The basic feedforward network performs a non-linear transformation of input data in order to approximate output data. Figure 1 shows a feedforward network structure. The number of input and output nodes is determined by the nature of the modelling problem being tackled, the manner of input data representation, and the form of network output required. In developing a model which is representative of a systems behaviour, it is the topology of the network, together with the neuron processing function, which determines the accuracy and degree of representation. A number of papers have shown that a feedforward network has the potential to approximate any non-linear function. It has been shown that the two-layered feedforward network can uniformly approximate any continuous function to an arbitrary degree of exactness - providing that the hidden layer(s) contains a sufficient number of nodes [CYBENCO 89], [WANG 92], [WANG 93]. The ability of the network to approximate non-linear functions is dependent upon the presence of hidden layers. Without these only linear combinations of functions can be fitted. The number of nodes in the hidden layer(s) can be as small or large as required and is related to the complexity of the system being modelled and to the resolution of the data fit required.

The input layer to the network does not perform processing but merely acts as a means by which scaled data is introduced to the network. The data from the input neurons is propagated through the network via the interconnections. The interconnections within the network are such that every neuron in a layer is connected to every neuron in adjacent layers. It is the hidden layer structures which essentially define the topology of a feedforward network. Each interconnection has associated with it a scalar weight (w) which acts to modify the strength of the signal. The neurons within the hidden layer perform two tasks; they sum the weighted inputs to the neuron and then pass the resulting summation through a non-linear processing, or activation, function. In addition to the weighted inputs to the neuron, a bias is included in order to shift the space of the nonliterary. For example, if the information from the ith neuron in the $j\text{-}1^{th}$ layer, to the kth neuron in the jth layer is $I_{j-1,i}$, then the total input to the kth neuron in the jth layer is given by:

$$S_{j,k} = \sum_{i=1}^{n_H} (b_{j,k} + w_{j-1,i,k} I_{j-1,i})$$

where n_H is the number of neurons in the hidden layer, and $b_{j,k}$ is a bias term which is associated with each interconnection and determines the co-ordinate space of the nonlinearity.

243

The output of each node is obtained by passing the weighted sum, $S_{j,k}$, through a nonlinear operator. The most widely applied nonliterary is the sigmoidal function in the interval (0,+1 or -1,+1). The sigmoidal function has the mathematical description:

$$O_{j,k} = 1/(1 + \exp[-S_{j,k}])$$

Although this function has been widely adopted, in principle any function with a bounded derivative could be employed [RUMELHART 87]. It is, however, interesting to note that a sigmoidal nonlinearity has also been observed in human neuron behaviour [HOLDEN 76]. Within a neural network, this function provides the network with the ability to represent nonlinear relationships.

Feedforward network training

In network training one method adopted is to split the data, randomly, into training and test data sets. If the network approximation is adequate then the squared error between the training data outputs and network predicted outputs should be relatively small and, more importantly, should be uncorrelated with all combinations of past inputs and outputs. Comparison of the average error achieved on the training set with that achieved on the test data set can be used to indicate the adequacy of the model (the network). For example, if the test data set error is significantly larger, then an over-dimensioned network is indicated.

The problem of determining the network parameters (weights) can be considered essentially a non-linear optimisation task. The simplest optimisation technique makes use of the Jacobian of the objective function to determine the search direction. A 'learning' rate term which influences the rate of weight adjustment [RUMELHART 87] is used as the basis for the back-error propagation algorithm, which is a distributed gradient descent technique. In order to train the network with this approach, a representative process input data set is presented to the network. At each time instance, the input set is propagated through the network to give a prediction of output. The error in prediction is then used to update the weights based upon gradient information, in order to drive the cost function to a minimum. The error in prediction is thus in a sense back-propagated through the network to update the weights. Such an approach to training is termed supervised learning since at any time instant both input and output data is available. The search direction can be generalised by including a term which represents the 'learning' rate and influences the rate of weight adjustments [RUMELHART 87]. This formed the basis for their *back-error propagation* algorithm, which is a distributed gradient descent technique. The popularity of this learning technique can be judged by the commonly adopted, somewhat misleading, description of feedforward networks as

Backpropagation networks. There are, however, a number of problems with the backpropagation approach. In common with other descent algorithms, difficulty arises when the search approaches the minima. If the surface is relatively flat in this region then the search becomes inefficient. Therefore, in most neural network applications, a 'momentum' term is added. The current change in weight is then forced to be dependent upon the previous weight change. Although this modification does yield improved performances, gradient techniques can require significant convergence times in large dimensioned problems. Furthermore, in adopting a down-hill search technique, the question arises as to whether the minimum is local or global. Although it could be argued that a momentum term may take the solution over a local minima, global optimality is not assured.

A more appealing method is that of *Conjugate Gradients* [LEONARD 90]. Although quasi-Newton methods are usually more rapidly convergent and more robust than conjugate gradient methods, they require significantly more storage. An advantage of the conjugate gradient method is that it relinquishes the need for second derivatives of the objective function whilst retaining convergence properties of second order techniques. A conjugate gradient methodology is thus a well established contender for problems with a large number of variables such as the training of an artificial neural network. The basic philosophy is to generate a conjugate direction as a linear combination of the current steepest descent direction and the previous search direction. With this technique minimisation is initiated as with steepest descent. After this iteration a new direction of search is required. This direction is chosen so that it is conjugate with the initial search direction. Minimisation then proceeds in the newly defined direction. It should be noted, however, that when this technique is applied to non quadratic functions, the exact minimum will not be found in a finite number of steps. Practical experience suggests that resetting the algorithm to the steepest descent direction every n iterations (where n is the number of network weights) is superior to the repeated use of the conjugate gradient method.

An alternative approach [BREMERMANN 89] attempts to avoid the solution becoming locked into a local minima. Postulating that weight adjustments occur in a random manner and that weight changes follow a multivariate zero mean Gaussian distribution, the algorithm adjusts weights by adding Gaussian distributed random values to old weights. The new weights are accepted if the resulting prediction error is smaller than that from the previous set of weights. This procedure is repeated until the reduction in error is negligible. Parallels have been drawn to bacterial 'chemotaxis', and the procedure is referred to as the *chemotaxis* algorithm. During minimisation the allowable variance of the increments can be adjusted to assist network convergence and aid in the avoidance of local minima. The algorithm is of the graded learning type. Network outputs do not need to be available at every time instant, merely a measure of quality of fit needs to be given periodically. Applications of chemotaxis based network optimisation have been shown to be very encouraging [MONTAGUE 91], [MONTAGUE 92].

Radial Basis Function Networks

Recently there has been an increasing interest in Radial Basis Functions (RBF) within the engineering community as a traditional, and powerful, technique for interpolation in multidimensional space [MICCHELLI 86], [POWELL 85], [POWELL 87]. A generalised form of the RBF has found wide application in areas such as image processing, signal processing, control engineering, etc. A RBF is a function which has built into a distance criterion with respect to a centre. Functions like this can be used very effectively for interpolation and for smoothing of data. A recent application of radial basis functions is in the area of neural networks where they can be used as a replacement for the sigmoidal transfer function. Like most feedforward networks, RBF networks have three layers, namely an input layer, the hidden layer with the RBF nonliterary and a linear output layer (figure 2). The function of the input layer of a RBF network is to distribute all inputs unaltered to each of the hidden layer nodes. The weights on the links between the input layer and the hidden layer are all set to unity and do not change during training. A radial basis function expansion with n_i-inputs, n_H hidden layer nodes and a scalar output implements a mapping according to:

$$f(x) = b_0 + \sum_{i=1}^{n_H} w_i f(|| x - c_i ||)$$

where

w_i are the parameters or weights of c_i,
c_i are the RBF centres and n_H is the number of centres.
b_0 are the bias parameters

The functional form $f(.)$ is pre-selected with the centres c_i being some fixed points in n-dimensional space appropriately sampling the input domain. For a given set of centres, the second, or hidden, layer performs a fixed nonlinear transformation which maps the input space onto a new space in a similar manner to that in the sigmoidal feedforward network. Each term $f(|| x - c_i ||)$ forms the activation function in a unit of the hidden layer. The output layer then implements a linear combination on this new space. With the RBF centres regarded as adjustable parameters a network structure in feedforward form results.

The most popular choice for $f(.)$ is the Gaussian form. The Gaussian RBF also has some nice mathematical features that enable generation of reliability intervals and validity indices for each calculated estimate [LEONARD 91]. The output generated by the hidden layer O_H is given by:

$$O_H = \exp\left[-\sum_{i=1}^{n_H} (x_i - c_{H,i})^2 / s_i^2 \right]$$

where each Gaussian is characterised by the function centre, $c_{H,i}$ and the function width, s_i. The RBF network is solved [HOFLAND 92] by first computing, for each node of the hidden layer, the distance between the current input to the network and the centre of the RBF using the relationship:

$$D(\underline{x}, \underline{c}) = (\underline{x} - \underline{c})^T (\underline{x} - \underline{c})$$

where $\underline{c}$ is the vector of coordinates of the RBF centre. The result is then passed through the nonlinearity $f(.)$. The output generated by the hidden layer is fed into the output layer which then generates a weighted sum:

$$O_j = \sum_{i=1}^{n_H+1} w_{ij} O_{Hi}$$

where w_{ij} is the weight on the link between output node j and hidden layer node i, O_{Hi} is the output generated by hidden node i and n_H is the number of nodes in the hidden layer. The (n_H+1)th. node of the hidden layer provides a constant unity output which allows for the implementation of bias on the output layer.

The weights on the connections between the hidden layer and the output layer are calculated by solving the equation:

$$AW = B$$

The matrix A is a $[n \times (n_H+1)]$ matrix which stores the results of the n_H hidden layer nodes for each of the training examples presented to the network plus the constant value node, and where n is the number of examples in the training set. The second layer weight matrix W is a $[(n_H+1) \times m]$ matrix where m is the number of output nodes. The matrix B is a $n \times m$ matrix which stores the n training outputs of the m output nodes, associated with each of the n training inputs to the network.

Since the system of equations is overdetermined, a Singular Value Decomposition (SVD) based generalised least squares minimisation procedure is used. RBF networks implemented in this way do not suffer from the problem of becoming locked into local minima as do the backpropagation networks. There are two parameters in the RBF networks that determine the modelling capability of such a network, namely n_H and P, where n_H is the number of RBFs used in the hidden layer, and P is the parameter used to determine the spread s_j of each RBF (cf. placement of RBF centres). To determine the spread the P-nearest neighbour method is used. This method selects for a particular cluster C_i, a set of P clusters whose cluster centres are nearest to cluster C_i. The spread s_j is calculated using the relationship:

$$s_j = 1/P * [\sum_{i=1}^{P} D(\underline{c}_i, \underline{c}_j)]^{0.5}$$

where $\underline{c}_i$ and $\underline{c}_j$ are the cluster centre coordinates. The values of n_H and P are determined iteratively as part of the training procedure.

The predictive error is calculated using the jack-knifing method [HOFLAND 92]. This method can generate an unbiased estimate of the predictive error without making any assumption about the distribution function of the error. The method essentially splits the training set in two partitions, namely a partition that the network will actually be trained upon and a partition that is used to test the network. The error generated using the test partition is stored for later use. After the newly trained network has been tested, the original training set is partitioned differently and the whole process is repeated. This continues until all exemplars in the original unpartitioned training set have been used for testing exactly once. Following this, the stored errors are averaged to form the estimate for the predictive error for a network with the chosen values of n_H and P. According to the theory of jack-knifing the best results are actually obtained when the cardinality of the test partition is one.

A number of different methods of placement of the RBF centres have been reported [HUNT 91]. One method is to position the RBF centres evenly spaced out along all dimensions of the input space which is usually a hypercube with coordinates in the interval [-1, 1]. Although this method is simple it has a few drawbacks, especially when the dimension n becomes large since the number of RBFs will increase exponentially. Another approach actually clusters the network inputs in the training set and positions the RBF centres at the centres of each of the clusters. For this purpose k-Means Clustering is used [MACQUEEN 67]. This method does not suffer from the same problems as the first one since it will position RBF centres in those parts of the input space where the data points actually are.

Because of the way in which the distance measure is incorporated into an RBF network, it is impossible for RBF networks to selectively ignore certain inputs. A sigmoidal feedforward network, however, can selectively ignore inputs. RBF networks do not have this capability and therefore care has to be taken with the data presented to the network. It is extremely important to only use inputs that have a fairly strong, not necessarily linear, correlation with the outputs. Totally uncorrelated data in either the linear or the non-linear sense should not be presented to the netwok.

Dynamic Neural Networks

The basic feedforward network performs a non-linear transformation of input data in order to approximate output data. This results in a *static* network structure. In some situations such a steady state model may be appropriate. However, a significant number of projects require a description of system dynamics. The most straightforward way to extend this essentially steady state mapping to the dynamic domain is to adopt an approach similar to that taken in linear ARMA (Auto-Regressive Moving Average) modelling. Here a time series of past process inputs (u) and outputs (y) are used to predict the present process outputs. Important process characteristics such as system delays can be accommodated for by utilising only those process inputs beyond the dead time. Additionally any uncertainty in delay can be taken account of by using an extended time history of process inputs. Inevitably a significant number of network inputs result.

The use of network models as predictors may be problematical if care is not taken in training the network properly. If the models are used to predict more than one-step-ahead in time, and have only been trained for such a task, then the ARMA time series approach is not appropriate. The one-step-ahead ARMA network model does not capture the process dynamics. Essentially, the autoregressive nature of this form of network results in the need to predict $y(t+n)$ from the estimates of $y(t+n-1)$. Errors in the estimate of y thus accumulate as the prediction horizon increases. The problem of the ARMA network approach can, however, be overcome by minimising the network prediction error over time. That is the network training minimises the estimate of $y(t)$ and all other output predictions up to a specified prediction horizon. This approach is called 'backpropagation-in-time' [BHAT 90]' [SU 91]. Recent studies have shown that the incorporation of dynamics into the network in this manner is highly beneficial in many real process application studies [MONTAGUE 91]' [MONTAGUE 92]. Although perhaps the most concise network representation of a dynamic system is obtained by using network inputs comprised of past input and output data, the requirement to model processes over a wide dynamic range (models containing both very large

249

and very small time constants) can result in large network structures leading to network training and convergence problems.

An alternative philosophy is to modify the neuron processing and interconnections in order to incorporate dynamics inherently within the network. Since a dynamic network model is not autoregressive the prediction problem does not arise. In addition to the sigmoidal processing of nodes, the neurons (or transmission between neurons) can be given dynamic characteristics. For example, dynamic processing of the form:

$$\frac{N(s)}{D(s)} \exp(-st_d)$$

where N(s) and D(s) are polynomial functions in the Laplace operator s, and the time delay is represented by a Pade approximation [MONTAGUE 91], [MONTAGUE 92]. Parallels with the biological neuron might be considered to enhance the basic feedforward network. It has been shown [TERZUOLO 68] that the transfer of information along a synapse (the interconnections) had associated dynamics which could be modelled by a first order transfer function. Furthermore, it has also been demonstrated [WATANABE 69] that a time delay was associated with the transmission. Other workers have have modelled the temporal properties of synapses and dentrites by Finite Impulse Response filters. Thus the model of the function of the individual neuron might be extended account for dynamics and dead-time. Whilst this might be biologically motivated, no claims are made as to its biological plausibility. However, from a engineering point of view it does acknowledge the importance of time delays in the modelling of dynamic processes.

Assessment of Model Validity

A number of model validity tests for non-linear model identification procedures have been developed. For example, the statistical chi-squared test [LEONTARITIS 87], the Final Prediction Error Criterion [AKAIKE 74], the Information Theoretic Criterion [AKAIKE 74] and the Predicted Squared Error criterion [BARRON 84] (PSE). The PSE criteria, although originally developed for linear systems, can be applied to feedforward nets providing that they can be approximated by a linear model.

Final Prediction Error (FPE) = $(E/2N) (N + N_w)/(N - N_w)$

Information Theoretic Criterion (AIC) = $\ln(E/2N) + 2 N_w/N$

Predicted Squared Error (PSE) = $E/2N + 2^*(\sigma^2)N_w/N$

where σ^2 is the prior estimate of the true error variance and is independent of the model being considered.

It can be seen that these tests make use of functions that strike a balance between the accuracy of model fit (average squared error over N data points, E/2N) and the number of adjustable parameters or weights used (N_w). Because of this, minimisation of these test functions results in networks (models) that are neither under nor over dimensionalised. A procedure involving 'train - test - validate' experiments is used with different network dimensions to minimise a selected validation function.

NARMAX modelling approaches have used a number of model validation tests based on the autocorrelation of the residuals [BILLINGS 83], [BILLINGS 86]. These correlation tests should also help with neural network model validation:

$$\Psi_{\varepsilon\varepsilon}(\tau) \quad = \quad E\{\ \varepsilon(t\text{-}\tau)\ \varepsilon(t)\ \} \qquad\qquad = \delta(t) \qquad \forall\ \tau$$

$$\Psi_{u\varepsilon}(\tau) \quad = \quad E\{\ u(t\text{-}\tau)\ \varepsilon(t)\ \} \qquad\qquad = 0 \qquad \forall\ \tau$$

$$\Psi_{u^{2'}\varepsilon}(\tau) \quad = \quad E\{\ [u^2(t\text{-}\tau)\text{-}u^{\text{-}2}(t)]\ \varepsilon(t)\ \} \qquad = 0 \qquad \forall\ \tau$$

$$\Psi_{u^{2'}\varepsilon^2}(\tau) \quad = \quad E\{\ [u^2(t\text{-}\tau)\text{-}u^{\text{-}2}(t)]\ \varepsilon^2(t)\ \} \qquad = 0 \qquad \forall\ \tau$$

$$\Psi_{\varepsilon(\varepsilon u)}(\tau) \quad = \quad E\{\ \varepsilon(t)\ [\varepsilon(t\text{-}1\text{-}\tau)\ u(t\text{-}1\text{-}\tau)]\ \} \qquad = 0 \qquad \tau \geq 0$$

Normalisation to provide all the tests with a range of ±1 and approximate 95% confidence bands at $1.96/\sqrt{N}$ allow the tests to be independent of the excitation signal amplitudes and hence easier to interpret. Validation of the identified network model can also be checked by observing its predictive qualities for both one-step-ahead and multi-step-ahead predictions.

Multivariate Statistical Methods & Neural Networks

The effective treatment of noisy and highly correlated and collinear data can be achieved using linear statistical methods such as Principle Component Analysis (PCA) and Projection to Latent structures, or Partial Least Squares, (PLS). The aim is to develop models relating process input/output measurements to the prediction of future variables, in this presentation such variables as biomass or penicillin titre, which may be very difficult to measure continuously. PCA and PLS project measurable data onto low dimensional subspaces which contain all the relevant information about the process. Current interest in the two approaches has grown quite rapidly following initial work in socio economics and chemometrics [WOLD 82], [WOLD 87].

Principal Component Analysis (PCA) and Projection to Latent structures, or Partial Least Squares, (PLS) are multivariable statistical techniques which consider all the noisy and highly correlated measurements made on a process, but project the information down onto a low dimensional subspaces which contain all the relevant information about the process. The development of PLS is more recent and current interest has grown quite rapidly following initial work in socio economics and chemometrics. In the application of PCA and PLS, the measured variables (eg. temperature, pressure, flowrate, pH, speed, etc) and the product quality or property data (eg. molecular weight, stickiness, titre, laboratory assays, etc.) are respectively amalgamated into an 'input' matrix [X] and an 'output' matrix [Y].

The methods address the problems of:

a) being able to deal with collinear data of high dimension in both the independent (X) and dependent (Y) variables.
b) being able to substantially reduce the dimension of the problem.
c) being able to provide good predictions of (Y) when both process (X) and quality (Y) variables are present.

Principal Component Analysis (PCA)

PCA is a procedure used to explain the variance in a single data matrix (X) of dimension (n*k). PCA calculates a vector, called the first principal component, which is a linear combination of the k measure variables which explains the greatest amount of variability ($t_1 = Xp_1$). In the k-dimensional variable space the loading vector p_1 defines the direction of greatest variability, and the score vector t_1 represents the projection of each object (observation vector) onto p_1. Alternatively p_1 and t_1 are the first eigenvectors of X^TX and XX^T respectively. The second principal component is orthogonal to the first principal component and explains the greatest amount of the remaining variability. It is obtained by fitting a least squares line through the residuals left after fitting the first principal component. That is, $t_2 = E_1 p_2$ where $E_1 = X - t_1 p_1^T$ is the residual matrix left after removing the predictions of the first component.

This approach is continued until k principal components are obtained. Thus the axes of X have been rotated to a new orthogonal basis. For large data sets it is often found that the first A principal components, where A<<k, explain most of the variation in the data matrix X. By stopping at this point and summarising the information in X using the new variables (t_i's), defined as those linear combinations of the x's ($t_i = Xp_i$, i=1,2,...A), given by the first A principal components, the dimensionality of the data space has been reduced from k to A. In geometric terms this is equivalent to approximating the k dimensional observation space by the projections of the observations onto a much smaller A dimensional hyperspace.

In PCA it is very important that the data is scaled correctly since the variance contribution of a particular x to the total variation in X is dependent upon its units of measurement. In general each variable should be scaled relative to the others in terms of its relative importance. Scaling is usually to a common unit variance and care is taken not to scale up the variance of variables that are almost constant so as not to disrupt the natural relationships amongst variables of the same type.

Partial Least Squares (PLS)

Often one group of variables (Y) are of greater importance, eg product quality variables, and should be included in the monitoring problem. Unfortunately these variables are usually measured much less frequently than the normal process variables. A major problem is to use the information contained in the process variables (X) to predict, monitor and detect changes in the output variables (Y).

The PLS algorithm does not attempt to calculate a regression relationship directly and thus avoids singularity problems associated with the inversion of X^TX. PLS builds the relationship sequentially, one dimension at a time. Each PLS dimension defines two new latent variables, t_i in X and in u_i in Y, such that the correlation between t_i and u_i is maximised.

The final PLS representations of the X and Y spaces are given by:

$$X = t\,p' + E$$

$$Y = u\,q' + F$$

These two relationships can be converted into an equivalent regression relationship between the original X and Y matrices, that is $Y = X\beta$. They also formulate a partial outer model. An inner model $u = f(t)$ is the used to relate the score vectors. For linear PLS this becomes $u = bt + h$.

If PLS is used on a data set of uncorrelated variables or if the number of transformed variables equals the original number of variables, the results obtained are equivalent to multiple linear regression. The advantages of PLS are most evident in the computations and predictions in systems when the number of independent underlying latent variables in the X and Y spaces is much smaller than the dimension of these matrices.

For nonlinear PLS approaches, $f(t)$ takes on a simple nonlinear form, $u = N(t) + h$. Typical relationships have include second or third order polynomial forms [WOLD 89].

Real and apparent nonlinear responses can be attributed to a number of different sources - instrumental, physical and chemical. Whilst chemical equilibria are an important source of variability in measurements, it is well known that multivariate calibration techniques can model these effects, at least in a least squares sense under ideal conditions. There are compelling

253

reasons why multivariate nonlinear techniques and mathematical methods for reducing the effect of nonlinear responses in linear multivariate analysis should be developed. Traditional methods for reducing or avoiding nonlinear responses will not always work. Furthermore, some sources of nonlinearity within the process cannot be reduced or eliminated. It practice it would appear that in low noise situations linear approaches such as Principal Component Regression (PCR), and especially PLS, can give good approximations given the incorporation of additional factors in the model to help achieve local linearisation of the nonlinearities. Second and higher order PCR techniques and Neural Networks provide improved routes to better approximations. A few algorithms have more recently appeared describing nonlinear statistical approaches - Projection Pursuit Regression (PPR) [FRIEDMAN 91a], Multivariate Adaptive Regression Splines (MARS) [FRIEDMAN 91b], and nonlinear PCR and neural networks [GEMPERLINE 91].

Autoassociative Neural Networks

Autoassociative networks are feedforward networks whose inputs and outputs are identical, with network training aimed at approximating the resulting identity mapping between network input and output. A typical network topology is shown in figure 3 indicating the key feature of autoassociative networks - the data compression or bottle-neck inner layer. It is this network bottle-neck layer that provides the topology with the very powerful properties of feature extraction. Both the first and final hidden layers have dimensions greater than (or equal to) the input/output layer and significantly greater dimension than the middle feature layer. Following convergence the network bottle-neck provides information which describe significant features, or signatures, of the data.

The concept of using an artificial neural network to extract important features has been discussed by a number of researchers [BALDI 89], [FOLDIAK 89], [HINTON 85], [KOHONEN 84], [KRAMER 91], [OJA 82]. Of particular interest is the ability of the network topology to provide a transformation which has been referred to as nonlinear principal component analysis (NLPCA) [KRAMER 91]. It is arguable, however, that the bottle-neck layer actually provides principal component information purely by analogy to the linear case PCA case. This important point is not addressed here, rather use is made of the data compression layer to provide nonlinear data feature information in the form of 'fault or malfunction signatures'. The first hidden layer provides a mapping into the lower dimensioned feature space. The final hidden layer provides a de-mapping to reconstruct the network input data at the network output. The application of this technique to an industrial polymerisation process and to a bioprocess has recently been described [PEEL 93].

Neural Networks in State Estimation

In many process control situations, due to sampling limitations, the infrequent measurement of some process outputs prevents the early detection of load disturbances. This can result in large deviations from setpoints and long disturbance recovery times. Often, these adverse effects cannot be acceptably overcome even by the use of existing advanced control algorithms and can lead to unsatisfactory system performance. The problem of controlling infrequently measured process outputs has long been studied and publications in this area date back to the early 1970s.

A approach is to use the information provided by other easily measurable variables to provide an estimate of the controlled output. The estimated outputs can then be used for overall plant control. Control schemes based on the feedback of output estimates are often termed 'inferential control' schemes. An ideal situation arises when the plant states are completely observable from the secondary outputs. Kalman filtering techniques can then be employed to estimate plant states using secondary output measurements, and hence estimates of the controlled output computed from its relationship with the states. Control of the plant is achieved by feed-back of either the state estimates or the output estimates to appropriate controllers. The literature on the above methods is extensive, and previous publications are examples of the more recent contributions describing the application issues of these techniques. However, the application of Kalman filtering techniques is limited to those processes that are completely observable from the secondary outputs. For most plants, such a set of secondary outputs can be difficult to determine and, in some cases, may not even exist.

Alternatively, it is possible to make use of both measurements of the controlled output and secondary output within an adaptive inferential control framework. In such a scheme, the infrequent measurements of the controlled output are used only for parameter estimation while output estimation, and hence plant control, is achieved using a secondary output at its faster sampling rate. Estimators based on such a scheme has been developed by Morris and colleagues and are termed software-sensors or soft-sensors [THAM 91]. There are a number of industrial situations where infrequent sampling of the 'controlled' process output can present potential operability problems. Common examples are in product composition control of distillation columns and chemical reactors. In these cases, the sampling delay is a direct result of the long cycle time of on-line composition analysers. Due to the potential problems associated with infrequent sampling, the control of product quality is commonly achieved by regulating an *a priori* chosen process variable, correlated with the product quality measure, at a setpoint. Such applications arise, for example, in polymerisation reactors, in which measurements of reactor feed rate and one or more reactants can be used to infer polymer properties, or in bioreactors where measurements of feed flow, pH, temperature and off-gas CO_2 can be used to infer biomass and product titre, that would otherwise only be available by relatively slow on-line or off-line analysis.

An Industrial Continuous Polymerisation Reactor

In this application a continuous polymerisation reactor is required to meet operating schedules demanding significant polymer grade changes to meet differing market demands. The melt flow index (MFI) of the product was estimated using measurements of reactor feed rate, reactor coolant flow rate and hydrogen concentration above the reaction mass. The secondary variables were measured at 10 min intervals, while measures of MFI were obtained from laboratory analyses every 4 hours. Figure 4a shows the training (full line) and test (dots) data used for model development where the directional dynamics of the process can easily be observed. Figure 4b shows the results of a process run which included a major change in polymer grade at around 100 hours. The figure compares the laboratory assayed melt flow index (step-like response) and its estimates. Since only two laboratory samples per work-shift were available the predictions from the neural network model led the "true" laboratory measurements by several hours. The tighter control that is now possible on the plant has led to the halving of the off-specification production during grade changes. In addition, the study highlights the fallibility of some laboratory assays. It can be seen that on several occasions when the "true" laboratory assay has shown a discrepancy with the prediction, which has given operators the confidence to use the network prediction and have laboratory assays re-checked.

Since continuous processes tend to operate at essentially steady state, it could be argued that the estimation or inferential measurement problem was roughly linear. In this case a fixed model, or adaptive, linear time series representation would be quite adequate. The same cannot be said of batch reactions. Here the nonlinearities associated with the different phases of reactor operation cause the performance of adaptive mechanisms to significantly degrade. This is where neural network based process representations (models) provide a method of capturing the nonlinear dynamics in a relatively straightforward manner. Both simulation and industrial applications of neural network modelling have demonstrated their significant potential in process modelling and control [WILLIS 91], especially in bioprocessing [LANT 90], [DIMASSIMO 91], [DIMASSIMO 92].

An Industrial Penicillin Fermentation

Here the industrial fermentation for the production of penicillin G by *penicillium chrysogenum* is achieved in 100,000 and 200,000 litre fermenters operating in fed-batch form. The complexity of the fed batch fermentation presents a good example to highlight the possible benefits of applying artificial neural network biomass estimation. Due to industrial confidentiality a thorough description of the fermentation process is not permissible except to emphasise that the fermenter biomass levels and rate of biomass change need to be regulated to achieve good performance and desired penicillium penicillin titre. Numerical annotation of the industrial data plots is also restricted for reasons of commercial confidentiality.

The only method available to determine biomass is by off-line laboratory analysis. The delay introduced by this sampling procedure and the frequency at which samples can be taken reduce the effectiveness of control. A complex non-linear relationship is known to exist between on-line measured variables, such as off-gas concentrations, and biomass levels in the fermenter. The objective was to model this relationship using the neural network modelling procedure. A series of experimental trials were performed and the results were utilised to provide training data linking on-line measurements to laboratory assays of biomass levels. The present operating regime of off-line analysis to determine fermenter condition results in a conservative feeding strategy. To increase the frequency of information routinely available from the fermentation would require a move from off-line analysis to on-line measurement or estimation. The objective being to utilise more frequent process knowledge provided by the software sensor within a feedback control scheme in order to directly improve fermentation operation.

The adoption of an artificial neural network nonlinear representation requires the specification of appropriate process measurements as network inputs in order to predict the output biomass concentration. The ultimate objective is to maximise penicillin production. A two staged approach has been adopted to achieve this: firstly the aim was to control specific growth rate to a pre-specified trajectory and secondly to determine the trajectory of specific growth rate which maximises penicillin production. Biomass concentration in the fermenter is a key variable in the overall control philosophy. A neural network was constructed to provide inferences of biomass using on-line measurements of oxygen uptake rate (OUR), carbon dioxide evolution rate (CER) and substrate feed were used as inputs. Additionally, since the characteristics of the fermentation are also a function of time, fermentation age was also considered an important input. Since there are in this case two primary substrate feeds, utilised in a varying ratio, it was useful to separate the feed in its two components. Both sigmoidal feedforward networks and radial basis function networks will be discussed and evaluated. The network topology was specified as four inputs, two hidden layers and one output. The network was then trained on selected fermentation data which spanned the normal operating conditions. It was found that the most appropriate topology was six nodes per layer in the two hidden layers. The performance of the trained neural network estimator, when applied to predict the behaviour of two batches operating at different biomass levels, can be seen in figure 5. A low and high sugar feedrate over the entire course of the batch (compared with standard operation) was employed in the fermentations. Here, the estimates of biomass have been obtained solely with the on-line information. Network estimates are compared against off-line biomass measurements. Although the biomass assay results are corrupted with a high level of noise, especially towards the end of the fermentation, it can be seen that the network estimates are representative of the underlying process behaviour at the two different operating conditions.

Figure 6 highlights a complex series of behaviours resulting from process problems. Acceptable estimator performance is seen to be achieved until 60 hours into the fermentation. At this time the estimates were observed to diverge from the laboratory assays. Data validation tests including the calculation of changes greater than three standard deviations, changes in mean and median, rates of deviation, etc., can be used to indicate bad data, process disturbances or potential plant or process problems. Together with data reconciliation using other measurements, operational experience, etc., the process operators can be guided to check possible malfunction locations. In this case the problems was the CO_2 analyser calibration. Bioprocess supervision using a real-time knowledge based system, alongside the algorithmic methods mentioned, can provide for a major step forward in bioprocess supervision [AYNSLEY 93]. An uncharacteristic drift in calibration had resulted in significant error and re-calibration can be seen to restore estimator performance. Errors in calibration, leading to offsets could, however, be accommodated to a certain extent by taking into account off-line laboratory assay information. Towards the end of the fermentation (140 hours onwards) the estimates can be seen to diverge once again, here off-line sample analysis confirmed the presence of a contaminant. Upon contamination the process characteristics change significantly and therefore the network would not be expected to provide reasonable estimates.

To date plant results have been particularly encouraging, with the poor performance observed in some batches being attributed to problems experienced with the process. This type of behaviour hints at a potential for fault detection. In particular although contamination occurs relatively infrequently, early detection could yield significant savings in feed and enhance penicillin recovery. It is here that the use of knowledge based systems can also play an important role [AYNSLEY 93].

For comparison with these results a radial basis function network approach was chosen to illustrate the use, of confidence intervals in the prediction of biomass. The network was configured with four inputs, one hidden layer with 8 clusters and one output node and trained on the two different sets of fermentation data. Figures 7 and 8 show the quality of the fits obtained with 95% confidence limits calculated as part of the network algorithm [LEONARD 91]. The effects of laboratory biomass assay variability (data noise) can be clearly observed, especially towards the end of the fermentation period. In spite of this it can be seen that the network estimate of biomass appears to be representative of the underlying process behaviour. This is confirmed by the plots of the confidence limits of the estimation.

In order to determine the quality of the neural network model, estimator network performance was assessed on new 'unseen' data. The performance of the estimator can be observed in figure 9. Here, the estimates of biomass have been obtained solely with the fermenter on-line measurements as network inputs with the network estimates being compared against interpolated off-line biomass measurements. Again, although the biomass assay results are corrupted with a high level of noise, it can be seen that the neural network estimate is very acceptable. Indeed the bioprocess operators were quite willing to accept the network prediction. However, inspection of the confidence limits at around time 150 hours shows a marked non-confidence in the network prediction.

Re-appraisal of the training data model fits shows that around this time into the fermentation data set 1 (figure 7) shows a rising trend in biomass whilst data set 2 (figure 8) shows a falling trend. In terms of numerical assessment of the 'unseen' data it is quite understandable that the confidence predicted by the network algorithm takes this discrepancy in biomass trends which results in a very low confidence in the prediction. This analysis of the data and network predictions emphasises the need to understand the process being modelled and not to treat the problem purely as a data fitting exercise. Other useful bio-representations and relationships might also be developed. For example, in the fermentation process discussed above, biomass estimates which are provided by gas data and feed data fail when the CO_2 analyser goes off-line. Another neural network model fitted to alkali addition rate data is able to provide estimates of biomass concentration during such periods. This would result in a significant improvement in the information available to process operators. Such sensor redundancy inevitably leads to more robust fermentation supervision and control policies.

Neural Networks in Feedback Process Control

Inferential Control

Whilst inferential estimation schemes, operating in 'open-loop', can be used to assist process operators with the availability of fast and accurate product quality estimates, the possibility of closed loop inferential control becomes very appealing. Such a cotrol strategy could be implemented manually, with conventional regulators or adaptively. Here the inferred estimates of the controlled output are used directly for feedback control. The effective elimination of a time delay caused by the use of an on-line analyser or the need to perform off-line analysis affords the opportunity of tight product control, even through the use of standard industrial controllers. Consequently reductions in product variability caused by process disturbances, and corresponding reductions in off-specification product, can be achieved. Whilst the use of a secondary variables is, strictly speaking, not necessary, given a suitable choice of secondary variable the resulting adaptive inferential control strategy will inherently provide for feedforward regulation. This is because disturbance effects will manifest themselves first in secondary variable responses and hence will be reflected in the primary output estimate.

A potential problem with using neural network model-base estimators is that network models might have been identified using data collected from the plant which may have some of its control loops still closed. The resulting model will then have been identified with correlated data and will not be representative of the underlying process behaviour. When such a model is used within a feedback control loop (manual or automatic) it will be subject to new process data that is further correlated and the model predictions will degrade. In this case it is important to identify a new network using the loop data now available and when this new network predictions are deemed better than those of the previous network model it should replace the 'old' model.

Biomass Inferential Control

Whilst inferential estimation schemes, operating in 'open-loop', can be used to assist process operators with the availability of fast and accurate product quality estimates, the possibility of closed loop inferential control becomes very appealing. Such a control strategy could be implemented manually, with conventional regulators, or even adaptively. Here the inferred estimates of the controlled output are used directly for feedback control. The effective elimination of a time delay caused by the use of an on-line analyser or the need to perform off-line analysis affords the opportunity of tight product control, even through the use of standard industrial controllers. Consequently reductions in product variability caused by process disturbances, and corresponding reductions in off-specification product, can be achieved.

The application of neural network based inferential PI(D) control to an industrial fermentation process has been achieved. Figure 10 shows the performance resulting from biomass being controlled to a set point profile predetermined by the fermentation technologists. A sigmoidal neural network was used in this study with four bioprocess inputs, one hidden layer with five hidden nodes and one output node predicting biomass. Additional biomass assays are also shown to indicate the capability of the network estimator. The tuning of the controller here caused a number of problems such as operation in different fermentation phases (severe nonlinearities), quite severe disturbances in gas flow rate and growth dynamics. Broadly speaking, the impact that these problems have on the fixed gain controller can be substantially reduced with an auto-tuning approach. These studies, together with other studies of estimation and control of fermentation processes using neural networks, has demonstrated the significant potential of the application [DIMASSIMO 91], [DIMASSIMO 92].

Auto-tuning Feedback Control

Well over 90% of the controllers/regulators on existing process control loops are of the Proportional-Integral-Derivative (PID) type. Since the number of PI(D) regulators found in the process industries is so significant, the tuning of such controllers becomes a very important issue. Indeed the importance attached to PI(D) controller tuning and the use of automatic techniques of achieving this is evidenced by the number of publications addressing the subject. It is not the place of this paper to review the multitude of methods available, especially since such review articles are available [ASTROM 93]. Recent studies, however, have addressed the use of neural networks to provide non linear models for PI(D) controller tuning [WILLIS 93]. One approach taken is to adopt a controller modelling and supervision philosophy shown schematically in Figure 11 where the use of a non linear model is shown to be able to provide for better controller tuning than that possible using other linear approaches.

Neural Network Model Based Process Control

Given a dynamic neural network model of the process it can be employed in a number of ways within a model based control strategy [HUNT 92], [THIBAULT 91], [PSALTIS 88], [NARENDRA 90]. One very important dynamic modelling approach is that of inverse modelling. Inverse modelling plays a key role in process control. The simplest approach is direct inverse modelling or generalised inverse modelling [PSALTIS 88].

Here, a synthetic training signal is introduced into the system and the system output used as an input to a neural network placed in the feedforward part of the control loop. The network is then trained on the error between the system input training signal and the system output. Such a structure forces the network to represent the plant inverse. There are, however, a number of drawbacks with this approach relating to persistencyexcitation and the possibility that an incorrect, or even unstable, inverse might be modelled [HUNT 92], [PSALTIS 88].

One approach which aims to overcome the problems of the direct inverse method is known as specialised inverse learning [PSALTIS 88]. Here the neural network inverse model in the control loop forward path is trained on the error between the plant output and a training signal replacing the control loop reference set point. A trained feedforward model (trained using standard feedforward methods) is also included in the inverse network training structure and located in parallel with the plant. If the plant signals are noisy the network can be trained on the error between the training signal and the feedforward model output. The overall training procedure effectively learns the identity mapping across the inverse model and the feedforward model.

Another important approach is that of model reference control. This requires that the plant output follows a stable reference model. A nonlinear approach using neural networks has been proposed [NARENDRA 90], which is related to the inverse model training approaches briefly reviewed above. The network training will force the controller to be appear to be a de-tuned inverse model controller, the de-tuning resulting from the dynamics of the reference model.

Internal model control (IMC) [GARCIA 82] provides a control loop design philosophy that has wide ranging implications in the area of process control. In the IMC approach the roles of the inverse model and the feedforward model have been analysed and discussed at length by Morari and colleagues (eg. see [GARCIA 82]). Here feedforward and inverse models are used directly within the feedback loop and can be linear or nonlinear [ECONOMOU 86]. In internal model control the plant (feedforward) model is placed in parallel with the plant, with the plant control input as the feedforward network input. The difference between the real plant output and the plant model output is used for feedback control. The network in the forward path of the control loop, and in series with the plant, is related to the inverse plant model [HUNT 91]. This form of nonlinear control, like its linear counterpart, is only applicable to open loop stable systems.

Although it is not the remit of this paper to survey in detail all potentially applicable neural network based controllers, since other surveys are available [HUNT 92], [THIBAULT 91], it is useful to look in a little more detail at one approach that is receiving increasing industrial attention - model based predictive control.

Neural Network Model Based Predictive Control

Since the early 1970's numerous model based predictive control algorithms have been proposed, with the prime development objective being to increase performance and robustness of process regulation. A strategically very useful control algorithm is one which minimises future output deviations from the desired setpoint, whilst taking suitable account of the control sequence necessary to achieve this objective. This concept is not new and is common to most predictive control algorithms [PETERKA 84]. However, the attraction of using the neural network instead of other model forms, such as autoregressive moving average and nonlinear autoregressive moving average representations within the control strategy, is the ability to effectively represent complex nonlinear systems.

Good examples of predictive control are Model Algorithmic Control, Dynamic Matrix Control and Generalised Predictive Control [RICKER 91]. Algorithms such as Dynamic Matrix Control have been found to provide major cost benefits on many industrial systems with pay-backs achieved in relatively short time scales. In almost all such control schemes a linear description of the process is assumed. If the system dynamics are relatively linear around the operating region, then the use of a linear model based control algorithm may lead to acceptable performance. However, in situations where the process is highly non-linear, the linearity assumption could well be detrimental to control system robustness. In this situation a common approach has been to adopt an adaptive control policy. Although the techniques for on-line adaptation are fairly standard, in a 'real' process environment the demands placed upon adaptive estimation schemes by everyday process occurrences can be extremely severe. Jacketing procedures are used to provide algorithm robustness. Whilst effective control system jacketing is essential, the consequences of failure in an adaptive scheme has resulted in their industrial application being far from common. In many situations a fixed linear model is used even if the system is known to be non-linear. As a result performance may be sacrificed in order to maintain robustness in the face of model / process mismatch.

The predictive control algorithm is centred around an iterative solution of the following cost function:

$$J = \{ \sum_{i=1}^{N_L} \sum_{n=N_{1,i}}^{N_{2,i}} [w_i(t+n)-y_i(t+n)]^2 + \sum_{i=0}^{N_{u,i}} [q_i \, \Delta u_i(t+i)]^2 \}$$

263

where $y_i(t)$, $u_i(t)$ and $w_i(t)$ are the controlled output, manipulated input and set-point sequences of control loop 'i' respectively. $N_{1,i}$ and $N_{2,i}$ are the minimum and maximum output prediction horizons. $N_{u,i}$ is the control horizon and q_i is a weighting which penalises excessive changes in the manipulated input of loop 'i'. N_L is the number of individual control loops.

In the performance function, the terms $y_i(t+n)$, $n = N_1, ..., N_2$, represent a 'sequence' of future process output values for each respective loop, $i=1, N_L$, which are unknown. Thus during minimisation, the sequence of process outputs is replaced by their respective n-step-ahead predictions. With the ability to predict the future outputs $y(t+n|t)$, together with known future set-points, the future controls which will minimise the performance function can be determined. In common with most predictive control strategies, beyond the control horizon, N_u, it is assumed that the control action remains constant. The optimisation algorithm therefore searches for N_u control values in order to minimise the performance function. Since neural network model is nonlinear, an analytical solution of the cost function is not possible and a nonlinear programming problems needs to be solved at every time step

There are a number of methods [FLETCHER 80] available to minimise the cost function, J. Most algorithms employ some form of search technique to scan the feasible space of the objective function until and extremum point is located. The search is generally guided by calculations on the objective function and/or the derivatives of this function. The various procedures available may be broadly classified as either 'gradient based' or 'gradient free'. It is important to note that care needs to be taken in the choice of an appropriate optimisation routine in that a quadratic minimisation of J will only provide a single 'global' solution. The requirement, however, is for a truely global solution where multiple solutions are possible [MARQUARDT 63]. A gradient free method is adopted which enables the extremum of a function to be located using a sequential unidirectional search procedure. Starting from an initial point, the search proceeds according to a set of conjugate directions generated by the algorithm until the extremum is found.

It is also required that set-points are tracked with zero errors. Since control is based upon the prediction of future outputs obtained from a model of the process, offsets may occur due to disturbances and plant-model mismatch. Following the concept of internal model control, any discrepancy between model and process responses can be estimated at each sample instant as:

$$d_i(t) = y_i(t) - y_i(t|t-1)$$

where $y_i(t)$ is the actual process output of loop 'i' at time t and $y_i(t|t-1)$ is the corresponding process model output. In circumstances where the process noise is significant then the estimate of prediction offset can be filtered in order to reduce the effect of noise and enhance stability. This estimate is then used to 'correct' the predictions obtained from the model:

$$y_{ic}(t+n \mid t) = y_i(t+n \mid t) + d_i{}^f(t) \qquad\qquad n = N_1, \ldots N_2$$

where the subscript 'c' denotes a corrected value and f refers to a filtered value. It is the sequence of corrected predictions that are then used in the cost function minimisation.

Since processes are normally affected by physical or operational limitations, any advanced process control algorithm should have the ability to account for such constraints [WILKINSON 90]. Failure to do so can result in poor closed loop performances, and possibly even system instability. Although it is usual for the control signals, say, to be rate and magnitude limited, such constraints have been commonly implemented by limiting or clipping the changes and the magnitude of the control signal: a technique which has worked satisfactorily within the framework of conventional control policies. With long range predictive control algorithms however, a sequence of control moves is usually calculated, ie. NU>1. Although the control may be implemented in a receding horizon manner, and the implemented control signal may have been 'clipped', other moves in the sequence may still lie outside the acceptable bounds. Indeed, control signal clipping could, in some cases, lead to an unstable closed loop. In the case of multivariable systems, simple clipping of calculated control moves can also cause degeneracy in the controller decoupling properties. To maintain integrity of control and to fulfil overall operational requirements, it is therefore necessary to explicitly consider all process constraints in the formulation of the control cost function. In fact limits on process outputs can only be implemented when they are explicitly taken into account in the cost function minimisation. In a similar manner to output constraints, constraints can be imposed on auxiliary outputs (ie. those outputs that are important but not controlled).

Application to a Binary Distillation Column

The process used to demonstrate the performance of the controller is a comprehensive nonlinear simulation of a 10 stage pilot plant distillation column which is installed at the University of Alberta, Canada. Figure 12 shows a schematic diagram of the plant. The column is modelled by a comprehensive set of dynamic heat and mass balance relationships. Both the column and the nonlinear model have been used by many investigators to study different advanced control schemes [THAM 91a], [THAM 91b]. The multi-input multi-output objective is to control the top and bottom product compositions using steam flowrate to the reboiler and reflux flowrate as the manipulated inputs. The column feed is a 50/50 wt%

methanol water mixture which is separated into top and bottom product compositions of 95 wt.% and 5 wt.% respectively.

In the first study the relationships between the bottom product composition and steam flow rate were modelled taking into account the effects of reflux flow rate changes (interaction term) and feed flow rate (disturbance term). A first order plus time delay linear transfer function model was developed by purturbing the plant about its steady state operating point. A one-step-ahead predictive neural network model, incorporating filter elements, (4-4-4-1) was then developed. Although a number of approaches are now becoming available for hidden unit specification [WANG 93], in this case network the number of hidden nodes was increased until the fit on both training and test data satisfied the network validation tests. Network training used the chemotaxis algorithm. The output prediction horizons were fixed at $N_1 = 1$ and $N_2 = 7$, and the control horizons set to $N_u = 1$. Analysis of the trained network revealed that although one-step-ahead predictions of the process output were quite good, its performance as an n-step-ahead predictor was, as expected, very poor [MONTAGUE 91], [MONTAGUE 92]. Re-training of the network as a dynamic n-step-ahead predictor resulted in a significantly improved performance.

The MISO control of bottom composition was first investigated. Here the performance of the n-step-ahead nonlinear predictive controller was compared to both a fixed linear predictive control strategy and a conventional PI controller. The control objective here was to regulate the bottom product composition to a series of feed flow rate disturbances - a 15% step up from steady state, a return back to steady state, a 15% step down from steady state, followed by a final return back to steady state flow rate. Figure 13 illustrates the improved performance achieved by using a nonlinear process representation. Although the PI controller was not 'optimally' tuned (IAE = 78.1), the main reason for this large value of IAE is the lack of any feedforward action. Now, using the linear model within the predictive control scheme, shows a significant improvement in control performance (IAE = 62.1). Use of the neural network model, however, shows a further improvent with an IAE of 5.7. Although feedforward terms were naturally included in the predictive control studies, it can be seen that the use of a nonlinear model does provide a major reduction in output deviations from set point.

Next the full multi-input multi-output predictive controller was studied. The first step in the implementation procedure was the determination of the MIMO process model. The neural network model used had 8 inputs, 2 hidden layers with 6 neurons in each layer and two outputs (8-6-6-2). The inputs to the network were time histories of steam flowrate and reflux flow rate at times 't-1', 't-2', 't-3' etc. Whilst first order neuron output filters incorporated some dynamic information into the network, the 'time history' of inputs served to accommodate the delays inherent in the system. In addition both top and bottom product composition measurements at time 't-1' were included. The outputs from the network were estimates of top and bottom product composition at time 't'. As with linear long range predictive control strategies the output horizon was set to predict a significant proportion of plant rise time. In this case the values used for both loops were $N_1 = 1$ and $N_2 = 6$, while both the control horizons were chosen to be $N_u = 1$. A small weighting was also used to reduce excessive control activity. Additionally, control action constraints were specified for both magnitude and rate limits.

The MIMO set point response characteristics of the bottom composition loop and the resulting disturbance rejection properties of the top composition loop were investigated. The predictive controller, based upon a n-step ahead neural network model, was compared to a conventional Proportional plus Integral (PI) controller. Previous studies [MONTAGUE 91], [MONTAGUE 92] have already addressed the comparable performance of linear model and neural net based predictive control algortithms. The control objective here was to follow a series of setpoint changes in the bottom loop whilst simultaneously maintaining top product composition at 95 wt% methanol. The integral of the absolute set-point tracking error (IAE) was used to quantify the performance characteristics of the two controllers.

Figures 14 and 15 show the controlled performances achieved. Non linear predictive regulation of the top composition to the sequence of top product set point changes (IAE = 3.0) is seen to be considerably improved over that achieved using PI regulation (IAE = 11.25). For the bottom composition controller the predictive controller provided an IAE of 22.24 compared to that obtained with conventional PI control (IAE 31.58). This improvement can also be compared with the performance of a 1-step ahead network prediction [MONTAGUE 91] (plots not shown) where an 18% improvement in top composition performance was observed when using the n-step ahead predictive network model, with a 22% improvement in the bottom loop regulation when using the n-step ahead network predictor. The differing control profiles shown in Figures 14 and 15 are characteristic of the respective control types. An important aspect of these results is the decoupling capabilities of the multivariable controller. Here the ability of the network model to predict the effect of bottom loop disturbances on the output of the top loop enables more effective regulation of the top product composition. Experience in such applications suggest that in addition to the improved performance, increased control system robustness results from reduced process/model mismatch.

Application to a C3 Splitter

In this study, the performance of four controllers were compared by application to the control of column pressure which is a critical variable for high performance process operation. The important variables, identified through analysis of column operating data, made up from 27 measured variables, by multivariate statistical methods, were column feed, cooling water temperature, column vent, bottoms level and reboiler energy. A dynamic artificial neural network topology, incorporating filters as discussed earlier, was used with a single hidden layer. The network was trained using the Levenberg-Marquardt nonlinear algorithm. Comparison of a linearised model with the neural network model responses (not shown) demonstrated the inadequacy of adopting a linear approach. The neural network model provides an excellent representation with an average error of Â1.5% on the test data. Figures 16 and 17 compare the performance of a linear model based predictive controller with that using a neural network nonlinear model based algorithm. Clearly the improvement in control performance observed with the neural network model emphasises the results obtained from the binary distillation column simulation study shown in figure 13. Next, four controllers were compared - a de-tuned PI controller, a high-gain PI controller, and the linear model-based predictive controller and neural network based predictive controller. The two PI controllers were tuned using a robust controller design package. Figure 18 compares the control performance of the four controllers. It was observed that the performance acieved by the high-gain controller required excessive levels of reboiler activity which would be unacceptable in long term operation. The neural network model predictive controller is seen to out-perform all the other controllers, reducing the standard deviation about set point of the linear controller by 50% and the high-gain controller by 25%. The results demonstrate that any form of model predictive control is only suitable if a reliable and representative dynamic model can be developed from the process data. The linear model had an error of Â10% on the test data set but was out-performed by the robust PI regulator. The neural network model had an error of Â0.5% and outperformed the conventional PI regulator by 25%. It is an appreciation of this balance that needs to be addressed by those involved in model predictive control and the acceptance that neural networks can be succesfully used to model, with similar effort to that require in developing linear models, nonlinear process dynamics.

Process Monitoring and Fault Detection

An increasingly strategically important area of process control is that associated with Statistical Process Control (SPC). Here the aim is to monitor process operation and performance in order to be able to detect the occurrence of off-specification production, important process disturbances and process malfunctions (faults). The early detection of process malfunctions, followed by the location of their causes, can lead to significant improvements in process operation.

SPC requires the efficient handling of large amounts of monitored plant data which often is subject to measurement errors, is ill conditioned, highly correlated and collinear. The effective treatment of such data is possible using linear statistical methods such as PCA and PLS to develop models relating process input/output measurements to the prediction of future 'quality' variables which can be very difficult to measure continuously. Unfortunately, though, many chemical processes are highly nonlinear and time varying and nonlinear modelling procedures are more appropriate.

A comprehensive steady state model of an industrial continuous polymerisation reactor [KIPARISSIDES 86], [MACGREGOR 91a] is used for the study. Thirteen process variables (reactor temperatures, solvent flowrates and initiator concentrations) are frequently monitored and five polymer quality variables (molecular weight properties, cumulative conversion and branching properties) are available from infrequent off-line measurements. It is interesting to note that the application of standard SPC charting methods is not particularly useful in highly dimensioned processes. For example with the polymer reactor studied here SPC charts would be required for all the 13 on-line measured variables as well as the 5 polymer quality variables.

Monte Carlo simulations provide two nominal operating regions for the process, which have been validated against industrial data. In practice such nominal data sets would be available from process data bases of monitored plant variables. Further simulations involving the introduction of systematic process faults (reactor fouling and efficiency changes) provide 'faulty' data sets representing the effects of important process malfunctions. The 'on-line observed variables' are consolidated in the 'input' matrix X, whilst the polymer property variables are consolidated in the 'output' matrix Y.

The use of multivariate statistical methods in process monitoring and physical property inferencing has been addressed in recent years [MACGREGOR 91b], [KRESTA 91]. The approach adopted here is to take advantage of the dimensionality reduction imposed by the 'bottleneck' network topology which explains the intrinsic features of the process data. Changes in process operation, due say to process malfunctions (faults), are revealed by a movement in the observed features away from the nominal, or reference, region. These attributes provide fault signatures to aid process monitoring.

The linear PCA results in Figure 19 are presented as three dimensional plots of the orthogonal latent vector space $\{t_1\text{-}t_2\text{-}t_3\}$ (ie a plot of the scores), whilst the autoassociative network feature analysis results in Figure 20 are presented as three dimensional plots of the feature vector space $\{f_1\text{-}f_2\text{-}f_3\}$ which are the outputs of the bottleneck layer. On all the plots the scatter points are denoted by:

- •   Reactor Nominal Steady State Operating Regions

- ×   Reactor Fouling Fault

- +   Initiator Impurity Fault

- []   Initial Solvent Flowrate Fault

Linear PCA on the observed data set [X] results in the plot shown in figure 19. In this case studied most of the process variability tends to be highlighted by the first three latent vectors (t_1, t_2 and t_3). It can be seen that the data monitored under the two nominal operating conditions clusters in two separate regions. On the occurrence of non-nominal operation, for example when the process is subject to faulty operation / malfunctions, the score vectors tend to move away from the nominal regions indication that the process is no longer operating within its expected region of operation. This is indicated on the plot for the process malfunctions - reactor fouling, initiator impurity and initial solvent flow rate changes. Fouling effects are indicated by the plots moving off in a North-Easterly direction for both the nominal operating regions, whilst initiator impurity changes are indicated by the plots moving off in North-Westerly directions. Initial solvent flow changes appear to be indicated by a plot in a South-Easterly direction from the first nominal data region and North-Westerly from the second nominal data region. It is pointed out, however, that rotation of the axes reveals that the fault plots are also out from the page. Reconciliation of these results with other process information should lead to confirmation of the location of the source of the malfunction.

The equivalent 'three feature' neural network approach is demonstrated by the plots in figure 20. A (14-14-3-14-14) topology was adopted to interpret the input data matrix [X]. The reactor fouling malfunctions are clearly distinguished by the vectors moving in an approximate North-Easterly direction from both nominal operating regions. The initiator impurity problems are indicated by a quite distinctive North-Westerly movements, whilst the solvent flow rate malfunction results in the feature vectors appearing to move out of the plot. Again, rotation of the axes can provide a more distinctive view of the three dimensional information. Although much more work needs to be done with feature analysis there is evidence [PEEL 93] that nonlinear feature detection can provide improved isolation of fault signatures.

Safety Critical Aspects of Neural Networks

The use of neural network modelling procedures is not a straight forward case of "throwing" data at a network and using the subsequently generated model within some form of information-giving or control loop. Prior to any neural network application the problem must be well defined - "a problem well defined is a problem half solved".

The process testing procedures are of paramount importance - the use of non-appropriate testing signals will certainly lead to non-representative network models. The use of non-representative models will lead to poor predictive properties. Test signals need to be of the correct spectral density and span the whole operating space. Persistent excitation is not sufficient. It is also imperative that the data generated is well understood and validated before ever applying to network training. Lack of understanding here can easily lead to disaster. In addition, a good understanding of the problem and the data can lead to the extraction of known nonlinear artifacts allowing the network training to be achieved more easily and reliably.

An additional potential problem with using neural network model-base estimators is that network models might have been identified using data collected from the plant which may have some of its control loops still closed. The resulting model will then have been identified with correlated data and will not be representative of the underlying process behaviour. When such a model is used within a feedback control loop (manual or automatic) it will be subject to new process data that is further correlated and the model predictions will degrade. In this case it is important to identify a new network using the loop data now available and when this new network predictions are deemed better than those of the previous network model it should replace the 'old' model.

Nonlinear systems have many minima - which is the "correct" minimum to use; has the "right" minimum been achieved? This is especially a problem with recurrent networks used for dynamic modelling purposes. How sensitive is the minimum chosen to model-process mismatch, process parameter variations, etc. The validation tests set out in this paper may help in identifying "best" minima. Can some types of recurrent network exhibit "chaotic" behaviour given certain data patterns?

Given a dynamic neural network model of the process it can be employed in a number of ways within a model based control strategy. Internal Model Control requires the development of an inverse model. Although there have been methods proposed to generate inverse models, with nonlinear systems there is always the possibility that an unstable model might be generated in complex systems. This would directly lead to an unstable control policy. Again, the use of appropriate plant excitation signals is very important here.

The stability of the models developed in the control of time varying systems needs study, as does the reliable, stable use of on-line adaptive controllers based upon neural network models. would these be allowed in transportation systems for example ?

Jacketing procedures, similar to those used in on-line linear-model-based controllers, will need to be used for safe application. Again, as in adaptive control, neural network based control schemes will be implemented ahead of stability and convergence proofs, which will follow in time.

Neural network based condition monitoring will open up complex nonlinear processes to reliable early warning of potential malfunctions. Here the synergy between multivariate statistical approaches and the nonlinear modelling capabilities of neural networks can be exploited to improve the safety of potentially dangerous plants. This will clearly be an expanding area of study and application.

The availability and perceived ease of use of neural network software can be very comforting - the feel-good factor. Models are developed for specific well founded reasons and in specific regions. We should not be tempted to extrapolate beyond these original reasons or operational regions. Care needs to be exercised in not massaging new data to fit the model. Throw discredited models away - **do not fall in love with your model !**

Neural networks are neither a panacea nor pragmatic solution, they are another modelling and classification technique closely related to well known statistical and nonlinear approximation approaches. They should be treated as such and not used for what sometimes passes as their placebo effect.

Concluding Remarks

Artificial neural networks provide an exciting opportunity for the process engineer. It is possible to rapidly develop models of complex operations that would normally take many man months to model using conventional structured techniques. It is essential to understand, however, that neural network modelling is no replacement for a good understanding of process behaviour.

Neural networks have shown good potential as 'software-sensors'. The additional benefits obtained from accurate output prediction, using secondary variables, allows for the effects of load disturbance compensation in a feedforward sense. It was also demonstrated that significant improvements in process regulation could be achieved when a network model was used directly for control purposes.

Multivariate projection approaches of PCA and PLS have much to offer process monitoring and supervision. However, there may be processes in which the nonlinearities cause the linear approaches to not completely interpret the nonlinear correlations. here, the autoassociative neural network topology could provide a method of feature extraction which can be interpreted in terms of process fault signatures. This property can be related to that provided by nonlinear principal component analysis. The neural network feature detection approach, however, has the ability to interpret nonlinear process information and thus to hopefully provide the potential for clearer distinctions between different process faults. Time dependent neural network feature detection may well provide a solution to batch monitoring problems as well as making a valuable addition to existing multivariate statistical process monitoring and control procedures.

The paper has also reviewed the dynamic modelling capabilities of artificial neural networks showing that care needs to be exercised. This was highlighted by considering the consequences of using auto-regressive structures for dynamic modelling. A low error in single step ahead predictions can be deceptive as an indicator of 'quality' of the process model. Given that a representative dynamic model can be obtained then it is relatively straightforward to incorporate the artificial neural network within an industrially acceptable multivariable control framework. As the control solution is iterative in nature the approach may only be suitable for systems which are not time critical. It is believed that the artificial neural network approach to generic system modelling, if necessary coupled with inferential and predictive control approaches, could result in improved process supervision and control performance without needing to undertake major model development programmes.

It is stressed, though, that the field is still very much in its infancy and many questions still have to be answered. Determining the 'best' network topology is one example. What is best for one problem may well not be best for another. Currently, rather *ad hoc* procedures tend to be used. This is not good and this arbitrary facet of an otherwise promising philosophy is a potential area of active research. A formalised technique for choosing the appropriate network topology is desirable. More analysis of network structure is needed to explore the relationship between the number of neurons required, to totally characterise the important process information, and the dimensionally finite topological space. There is also no established methodology for determining the robustness and stability of the networks. This is perhaps the one of the most important issues that has to be addressed before their full potential might be realised.

Ultimately, problems tend to escalate rapidly with complex systems and complex data. It is here that significant advantages might be gained by incorporating principal component analysis and partial least squares approaches to allow for major reductions in dimensionality to an underlying latent data set. Integration of these methods with neural network representations should provide for their extension to accommodate nonlinear correlated data handling.

Given the resources and effort that are currently being infused into both academic and industrial research in neural network technology, well within the decade neural networks will have been firmly established as a valuable tool. In the short term they will indeed provide some interesting, useful and pragmatic solutions for solving engineering problems. A panacea, however, they will not be.

Acknowledgements

The support of the Department of Chemical and Process Engineering, University of Newcastle and Company members of the International Neural Network Club is gratefully acknowledged. Special thanks are due to Professor Costas Kiparissides of the University of Thessaloniki for providing the polymer model data. The many enlightening discussions with colleagues in the research group are also gratefully acknowledged.

References

[LYUNG 83] Ljung, L. and Soderstrom, T., "Theory and Practice of Recursive Identification", MIT Press, 1983.

[BILLINGS 83] Billings, S.A., and Voon, W.S.F., "Structure Detection and Model Validity Tests in the Identification of Nonlinear Systems", Proc. IEE Pt. D, 130, oo 193-199, 1983.

[BILLINGS 86] Billings, S.A., and Voon, W.S.F., "Correlation Based Model Validity Tests for Nonlinear Models", Int. J. Control, 44, No.1, pp235-244, 1986.

[BILLINGS 91] Billings, S.A., Jamaluddin, H.B., and Chen, S., "Properties of Neural Networks with Applications to modelling Nonlinear Dynamical Systems", Int. J. Control, 1991.

[CHEN 90a] Chen, S., Billings, S.A., and Grant, P.M., "Nonlinear System Identification Using Neural networks", Int. J. Control, Vol 51, No. 6, pp 1191-1214, 1990.

[CHEN 90b] Chen, S., Billings, S.A., Cowan, C.F.N., and Grant, P.M., "Practical Identification of NARMAX Models Using Radial Basis Functions", Int. J. Systems Science, 21, pp 2513-2539, 1990.

[BHAT 89]   Bhat N., Minderman, P. and McAvoy, T.J., "Use of neural nets for Modelling of Chemical Process Systems", Preprints IFAC Symp. Dycord+89, Maastricht, The Netherlands, Aug. 21-23, pp 147-6, 1989.

[IEE 88-90]   IEEE, Special Issues on Neural Networks, Control Systems Magazines - nos. 8, 9, 10, 1988, 1989, 1990.

[HUNT 92]   Hunt, K.J., Sbarbaro, D., Zbikowski, R., and Gawthrop, P.J., "Neural Networks for Control Systems - A Survey", Automatica, vol 28, No. 6, pp 1083-1112, 1992.

[THIBAULT 91]   Thibault, J., and Grandjean, B.P.A., "Neural Networks in Process Control - A Survey", IFAC Conference on Advanced Control of Chemical Processes", Toulouse France, pp 251-260, 1991.

[CYBENCO 89]   Cybenco, G., "Approximations by Superpositions of a Sigmoidal Function", Math. Cont. Signal & Systems, 2, pp 303-314, 1989.

[WANG 92]   Wang, Z., Tham, M.T., and Morris, A.J., "Multilayer Neural Networks: approximated canonical decompoisition of nonlinearity", Int. J Control, 56, pp 665-672, 1992.

[WANG 93]   Wang, Z., Di Massimo, C., Montague, G.A. and Morris, A.J., "A Procedure for Determining the Topology of Feedforward Neural Networks", Neural Networks, vol 7, 2, pp 291-300, 1993.

[RUMELHART 87]   Rumelhart, D.E. and McClelland, J.L., "Parallel Distributed Processing: Explorations in the Microstructure of Cognition", Vol.1, Chp. 8, MIT Press, Cambridge, 1987.

[HOLDEN 76]   Holden A.V., "Models of the stochastic activity of neurones", Lecture Notes in Biomathematics, 12, Springer-Verlag, 1976.

[LEONARD 90]   Leonard, J. A. and Kramer, M.A., "Improvement of the back-propagation algorithm for training neural networks." Computers and Chem. Engng., 14, pp 337-341, 1990.

[BREMERMANN 89]   Bremermann, H.J. and Anderson, R.W., "An alternative to Back-Propagation: a simple rule for synaptic modification for neural net training and memory", Internal Report, Dept. of Maths, Uni. of California, Berkeley, 1989.

[MONTAGUE 91]   Montague G.A., Willis M.J., Morris A.J. and Tham M.T., "Artificial Neural Network based multivariable Predictive Control". ANN'91, Bournemouth, England, 1991.

[MONTAGUE 92]   Montague G.A., Tham, M.T., Willis, M.J., and Morris, A.J., "Predictive Control of Distillation Columns Using Dynamic Neural Networks", 3rd IFAC Symposium DYCORD+'92, Maryland USA, April, pp 231-236, 1992.

[MICCHELLI 86]   Micchelli, C.A., "Interpolation of scattered data: distance matrices and conditionally positive definite functions", Constructive Approximation, 2, pp 11-22, 1986.

[POWELL 85]   Powell, M.J.D., "Radial Basis Functions for Multivariable interpolation : a review", Proceedings of the IMA Conference on Algorithms for the Approximation of Functions and Data, RMCS Shrivenham, 1985.

[POWELL 87]   Powell, M.J.D., "Radial basis function approximations to polynomials", 12th Biennial Numerical Analysis Conf., Dundee, pp 223-241, 1987.

[LEONARD 91]   Leonard, J.A., and Kramer, M.A., and Ungar, L.H., "A Neural Network Architecture that Computes its Own Reliability", Private communication, 1991.

[HOFLAND 92]   Hofland, A., Montague, G.A., and Morris, A.J., "Radial Basis Function Networks Applied to Process Control", Proc. American Control Conference, Chicago, pp 480-484, 1992.

[HUNT 91]   Hunt, K.J., Sbarbaro, D., "Neural Networks for Internal Model Control". IEE Proceedings-D, Vol. 138, No. 5, pp 431-438, 1991.

[MACQUEEN 67]   MacQueen, J., "Some Methods for Classification and Analysis of Multivariate Observations". Proceedings of the Fifth Berkeley Symposium on Mathematical Statistics and Probability, U.C. Berkeley Press, CA., 1967.

[BHAT 90]   Bhat N., and McAvoy, T.J., "Use of neural nets for Dynamic Modelling and Control of Process Systems", Computers Chem. Eng., vol 14, No 4/5, pp 573-583, 1990.

[SU 91]   Su, H.T., and McAvoy, T.J., "Identification of chemical processes using neural networks"", Proc. American Control Conference, Boston USA, pp 2314-2319, 1991.

[TERZUOLO 68]   Terzuolo C.A. and Bayly E.J., "Data transmission between neurons", Kybernetik, 5, pp 83-84, 1968.

[WATANABE 69]   Watanabe, Y., "Statistical measurement of signal Transmission in the central nervous system of the crayfish", Kybernetik, 6, pp 124-130, 1969.

[LEONTARITIS 87] Leontaritis, I.J., and Billings, S.A., "Model selection and validation methods for non-linear systems", Int. Journal of Control, 45, pp 311-341, 1987.

[AKAIKE 74]   Akaike, H., "A New Look at the Statistical Model Identification", IEE Trans. Auto. Cont., AC-19, 6, pp 716-723, 1974.

[BARRON 84]   Barron, A.R., "Predicted squared error: a criterion for automatic model selection", Self Organising Methods, S.J. Farlow (Ed.), pp 87-103, 1984.

[WOLD 82]   Wold, H., "Soft Modelling, The Basic Design and Some Extensions", Systems Under Indirect Observation, Joreskog and Wold (Eds.), 1982.

[WOLD 87] Wold, S., "PLS Modelling with Latent Variables in Two or More Dimensions", Frankfurt PLS Meeting, 1987.

[WOLD 89]   Wold, S., Kettaneh-Wold, N., and Skagerberg, B., "Nonlinear PLS Modelling", Chemometrics and Intelligent Laboratory Systems, Elsevier Science, Amsterdam, pp 53-65,1989.

[FRIEDMAN 91a]   Friedman, J. H. and W. Stuetzle, "Projection pursuit regression", Journal of the American Statistical Association, Vol.76, pp 817-823, 1991.

[FRIEDMAN 91b] Friedman, J. H., "Multivariate adaptive regression splines", The Annals of Statistics, Vol.19, pp 111-141, 1991.

[GEMPERLINE 91] Gemperline, P. J., J. R. Long, and V. G. Gregoriou, "Nonlinear multivariate calibration using principal components regression and artificial neural networks", Anal. Chem., Vol.63, pp 2313-2323, 1991.

[BALDI 89] Baldi, P., and Hornik, K., "Neural Networks and Principal Component Analysis: Learning from Examples without Local Minima", Neural Networks, 2, pp 53-58, 1989.

[FOLDIAK 89] Foldiak, P., "Adaptive Network for Optimal Linear Feature Extraction", Int, Joint Conf. on Neural Networks, Washington, vol I, pp 401-405, 1989.

[HINTON 85] Hinton, G.E., "Learning Distributed Representations of Concepts", Proc. 8th. Ann. Conf. of the Cognitive Sci. Soc., pp 1-12, 1985.

[KOHONEN 84] Kohonen, T., "Self Organisation and Associative Memory", Springer-Verlag, Berlin, 1984.

[KRAMER 91] Kramer, M.A., "Nonlinear Principal Component Analysis Using Autoassociative Neural Networks", AIChE Journal, Vol 37, No. 2, pp 233-2433, 1991.

[OJA 82] Oja, E., "A Simplified Neuron Model as a Principal Component Analyser", J. Math. Biology, 15, pp 267-273, 1982.

[PEEL 93] Peel, C., Morris, A.J., and Kiparissdes, C., "Process Condition Monitoring and Neural Network Feature Detection", Proc. IFAC World Congress, vol IV, Sydney, Australia, pp 65-68, 1993.

[THAM 91] Tham. M.T., Morris, A.J., Montague, G.A., and Lant, P.A., "Soft Sensors for Process Estimation and Inferential Control", J Proc. Control, vol 1, pp 3-14, 1991.

[WILLIS 91] Willis, M.J., Di Massimo, C., Montague, G.A., Tham, M.T. and Morris, A.J., "Artificial neural networks in process engineering". Proc. IEE, Pt D., 138, No 3, pp 256-266, 1991.

[LANT 90] Lant, P., Willis, M.J., Montague, G.A., Tham, M.T. and Morris,A.J., "A Comparison of Adaptive Estimation with Neural Network Based Techniques for Bioprocess Application", Proceedings of the American Control Conference, San Diego, USA, pp 2173-2178, 1990.

[DIMASSIMO 91] Di Massimo, C., Willis, M.J., Montague, G.A., Tham, M.T. and Morris, A.J., Bioprocess model building using Artificial Neural Networks. Bioprocess Engineering, 7, 77-82, 1991.

[DIMASSIMO 92] Di Massimo, C., Montague, G.A., Willis, M.J., Tham, M.T., and Morris, A.J., "Towards improved penicillin fermentation via artificial neural networks", Computers and Chemical Engineering, vol 16, No. 4, pp 283-291, 1992.

[AYNSLEY 93] Aynsley, M., Hofland, A., Morris, A.J., Montague, G.A. and Di Massimo, C., "Artificial Intelligence and the Supervision of Bioprocesses (Real-Time Knowledge-Based Systems and Neural Networks), Advances in Biochemical Engineering Biotechnology, vol 48, pp1-27, 1993.

[ASTROM 93] Astrom, K.J., and McAvoy, T.J., "Intelligent Control", J. Proc. Cont., vol 2, No 3, pp 115-127, 1993.

[WILLIS 93]   Willis, M.J., and Montague, G.A., "Auto-tuning PI(D) Controllers with Artificial Neural Networks", Proc. IFAC World Congress, Sydney Australia, vol 4, pp 61-64, 1993.

[PSALTIS 88] Psaltis, D., Sideris, A. and Yamamura, A.A., "A multilayered neural network controller", IEEE Control Systems Magazine, April, pp 17-21, 1988.

[NARENDRA 90] Narendra, K.S. and Parthasarathy, K., "Identification and Control of Dynamical systems using Neural Networks, IEEE Trans Neural Networks, 1, 1, pp 4-27, 1990.

[GARCIA 82] Garcia, C.E., and Morari, M., "Internal Model Control - 1. A unifying review and some new results", Ind. Eng. Chem. Process Des. Dev., 21, pp 308-323, 1982.

[ECONOMOU 86] Economou, C.G., Morari, M., and Palsson, B.O., "Internal Model Control - 5. Extension to nonlinear systems", Ind. Eng. Chem. Process Des., 25, pp 403-411, 1986.

[PETERKA 84]   Peterka, V., "Predictor based self-tuning control". Automatica, 20, pp 39-50, 1984.

[RICKER 91]   Ricker, N.L., "Model predictive control: state of the art". Preprints CPC IV, South Padre Island, Texas, 1991.

[FLETCHER 80] Fletcher, R. "Practical methods of optimization", Volume 1. John Wiley, 1980.

[MARQUARDT 63]   Marquardt, D., "An algorithm for Least Squares Estimation of Nonlinear Parameters", SIAM J. Appl. Math., 11, pp 431, 1962.

[WILKINSON 90]   Wilkinson, D.J., Morris, A.J., and Tham, M.T., "Multivariable Constrained Generalised Predictive Control (A comparison with QDMC)", Proc. American Control Conference, Boston USA, pp 1620-1625, 1990.

[THAM 91a] Tham, M.T., Vagi, F., Morris, A.J. and Wood, R.K., "On-Line Multivariable Adapative Control of a Binary Distillation Column". Canadian Journal of Chemical Engineering, 69, pp 997-1009, 1991.

[THAM 91b]   Tham, M.T., Vagi, F., Morris, A.J. and Wood, R.K., "Multivariable and Multirate Self-tuning Control: a distillation column case study". IEE Proceedings-D, 138, 1, 9-24, 1991.

[KIPARISSIDES 86] Kiparissides, C. and Mavridis, H., "Mathematical Modelling and Sensitivity Analysis of High Pressure Polyethylene Reactors", in Chemical Reactor Design and Technology, NATO ACI Series, H.I. de Lasa, ed., Martinus Nijhoff, Boston, 1986.

[MACGREGOR 91a] MacGregor, J.F., Skagerberg, B., and Kiparissides, C., "Multivariate Statistical Process Control and Property Inference Applied to Low Density Polyethylene Reactors", IFAC Conference ADCHEM'91, Toulouse, France, pp 131-135, 1991.

[MACGREGOR 91b]   MacGregor, J.F., Marlin, T.E., Kresta, J., and Skagerberg, B., "Multivariate Statistical Methods in Process Analysis and

Control", Proceedings CPC-IV Conference, South Padre Island, Texas, pp 18-22, 1991.

[KRESTA 91]   Kresta, J., MacGregor, J.F., and Marlin, T.E., "Multivariate Statistical Monitoring of Process Operating Performance", Canadian Journal Chemical Engineering, 68, No. 1, 1991.

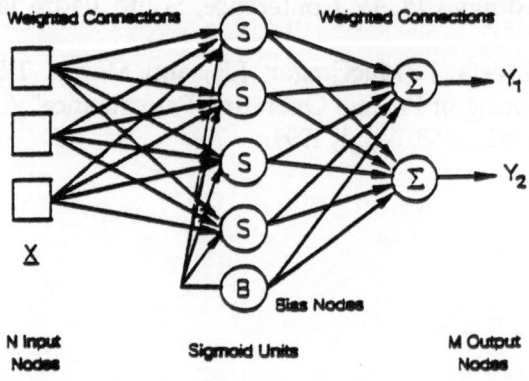

Figure 1. Sigmoidal Feedforward Artificial Neural Network

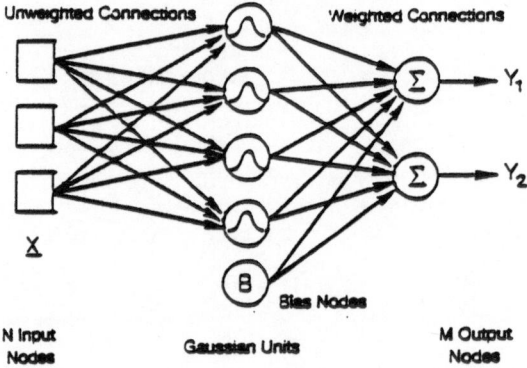

Figure 2. Radial Basis Function Artificial Neural Network

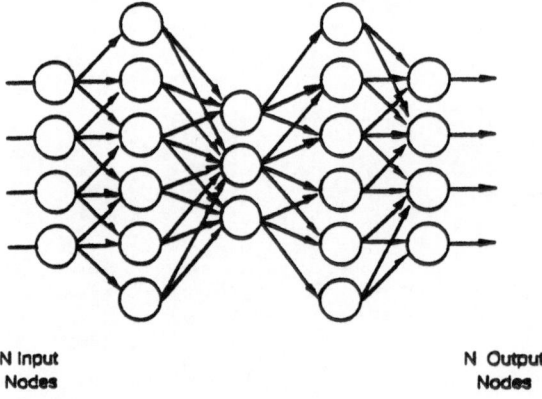

Figure 3. Autoassociative Artificial Neural Network

280

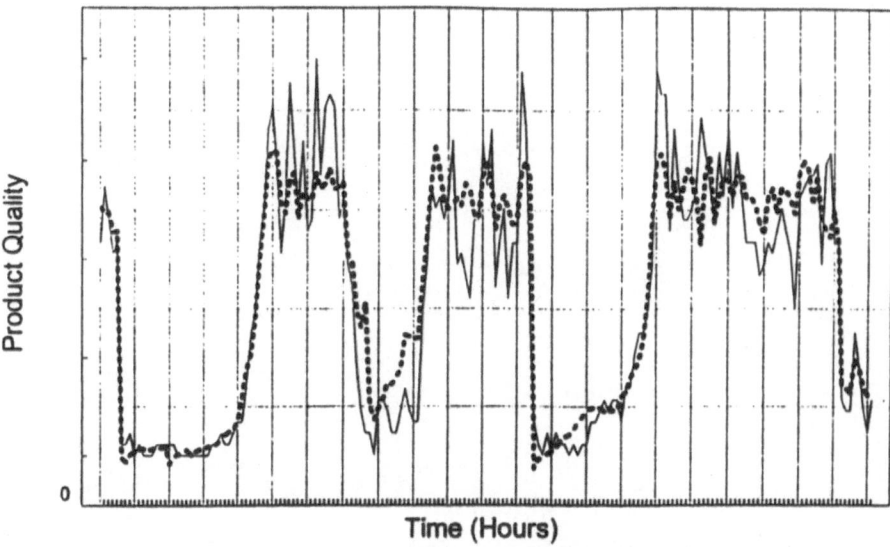

Figure 4a. Continuous Polymerisation: Estimation of Polymer Properties
(Training and Testing Data)

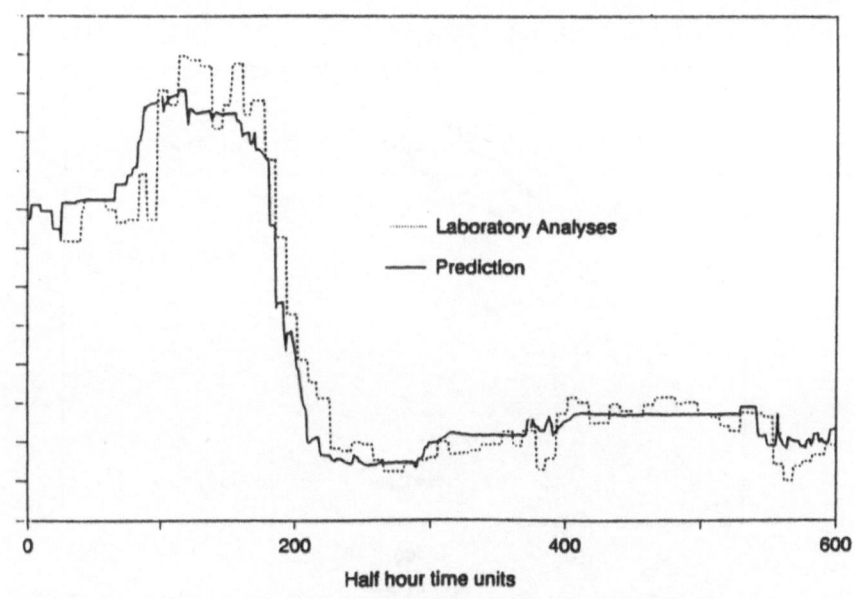

Figure 4b. Continuous Polymerisation: On-line Estimation of Polymer Properties

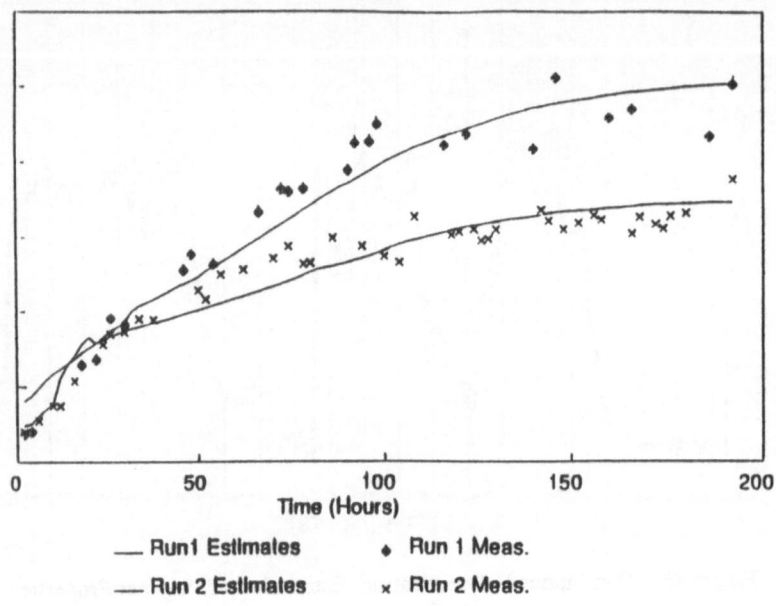

Figure 5. Typical Fed Batch Fermentations: Estimation of Biomass

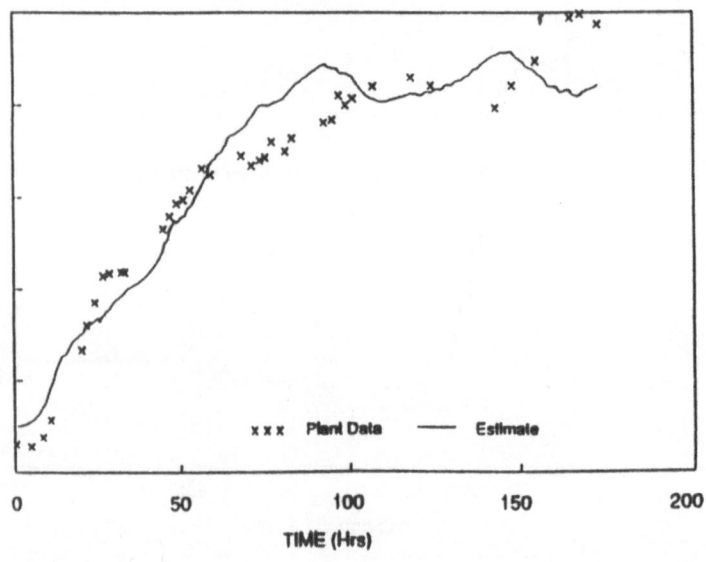

Figure 6. A Complex Fed Batch Fermentation: Estimation of Biomass

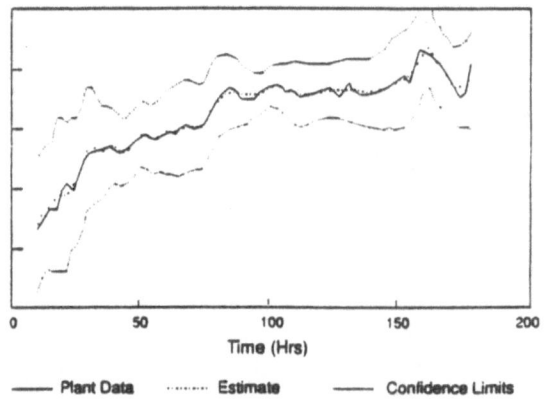

Plant Data ······· Estimate ——— Confidence Limits

Figure 7. Fed Batch Fermentation: RBF Estimation of Biomass
(with 95% prediction confidence intervals)

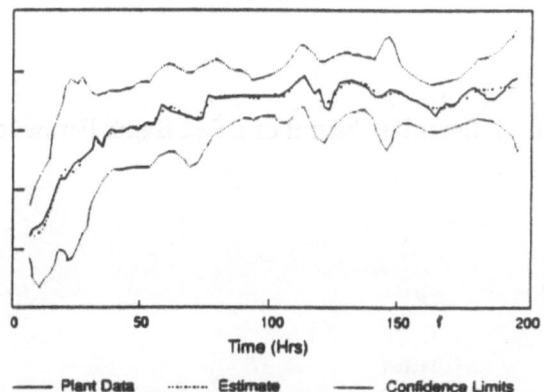

Plant Data ······· Estimate ——— Confidence Limits

Figure 8. Fed Batch Fermentation: RBF Estimation of Biomass
(with 95% prediction confidence intervals)

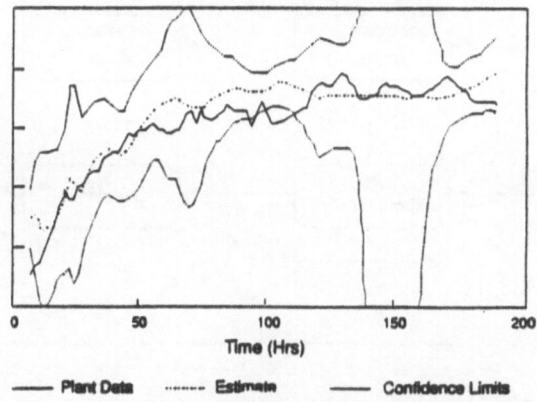

Plant Data ······· Estimate ——— Confidence Limits

Figure 9. Fed Batch Fermentation: RBF Estimation of Biomass
(with 95% prediction confidence intervals)

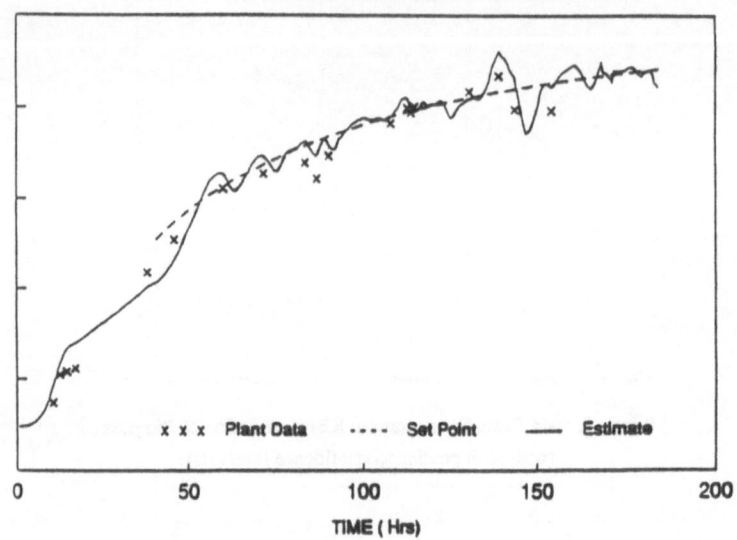

Figure 10. Inferential Control of a Fed Batch Fermentation

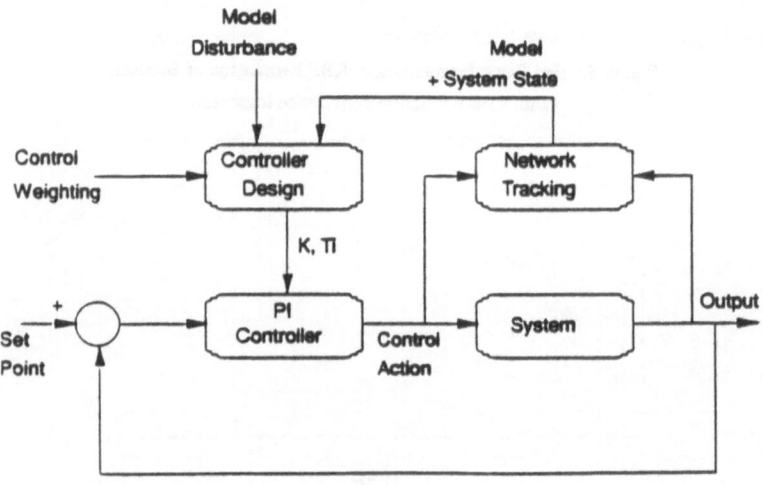

Figure 11. Neural Network Based Auto-tuning Control

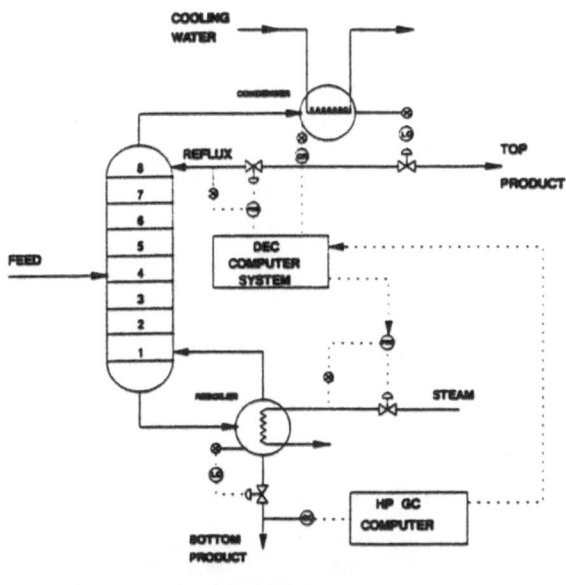

CODE
CR - ANALYZER RECORDER   FRC - FLOW RECORDER/CONTROLLER
GC - GAS CHROMATOGRAPH   LC - LEVEL CONTROLLER

Figure 12. Methanol - Water Distillation Column.

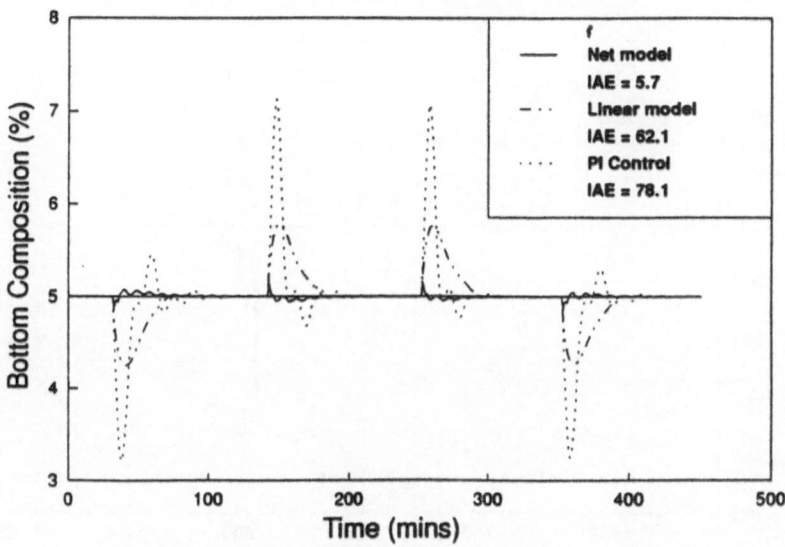

Figure 13. MISO Regulation of Bottom Product Composistion
(P+I, Linear MBPC, Nonlinear MBPC)

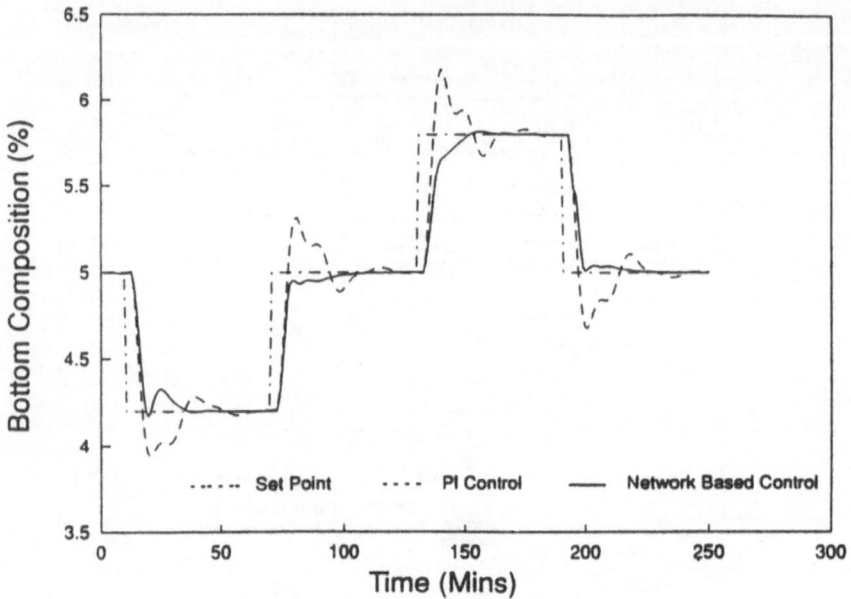

Figure 14. MIMO Nonlinear Predictive Servo Control of Bottom Product Composition.

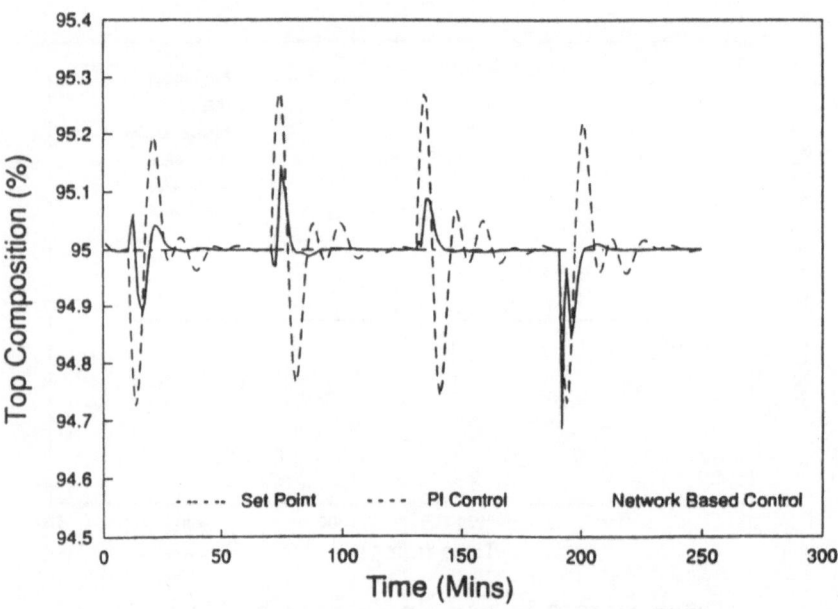

Figure 15. MIMO Nonlinear Predictive Regulatory Control of Top Product Composition.

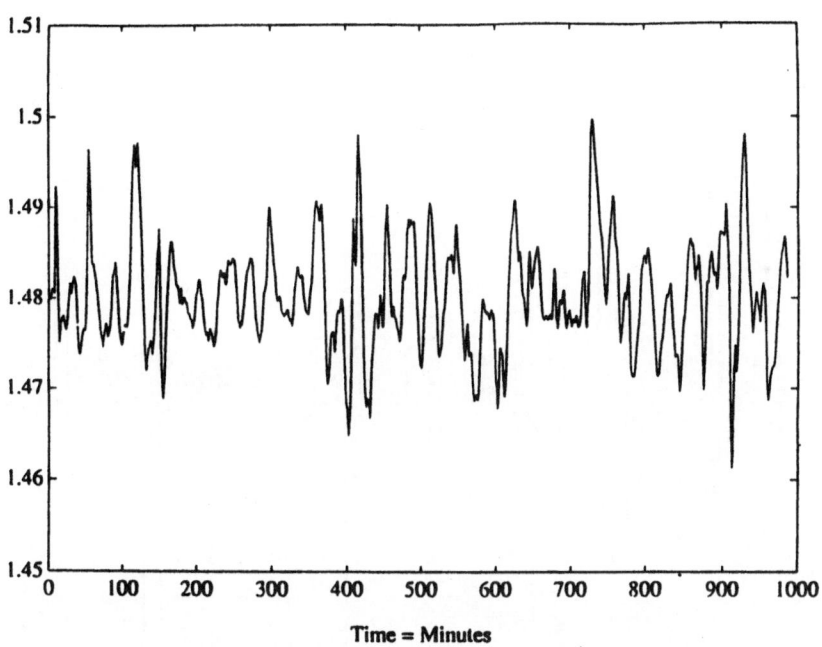

Figure 16. Linear Model Based Predictive Control on a C2 Splitter

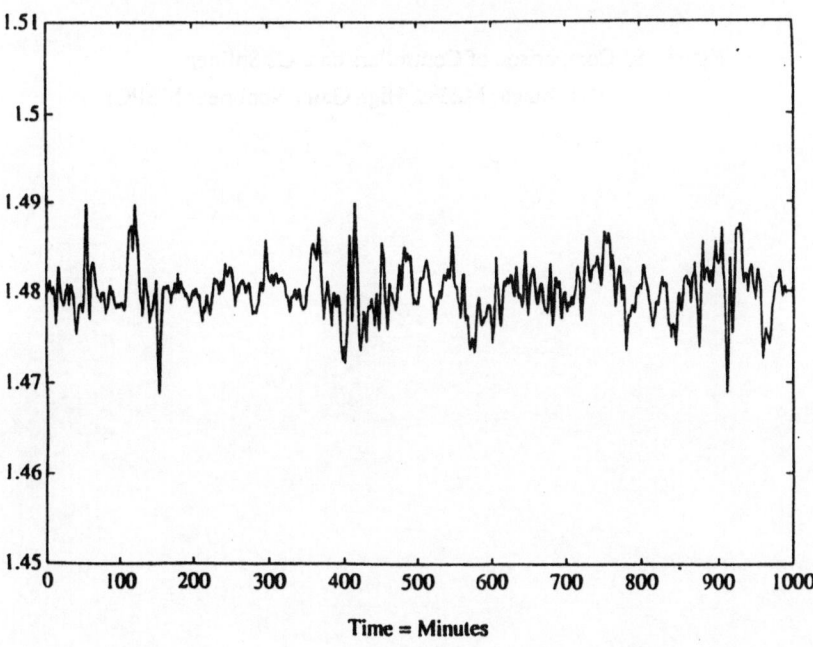

Figure 17. NonLinear Neural Network Model Based Predictive Control on a C2 Splitter

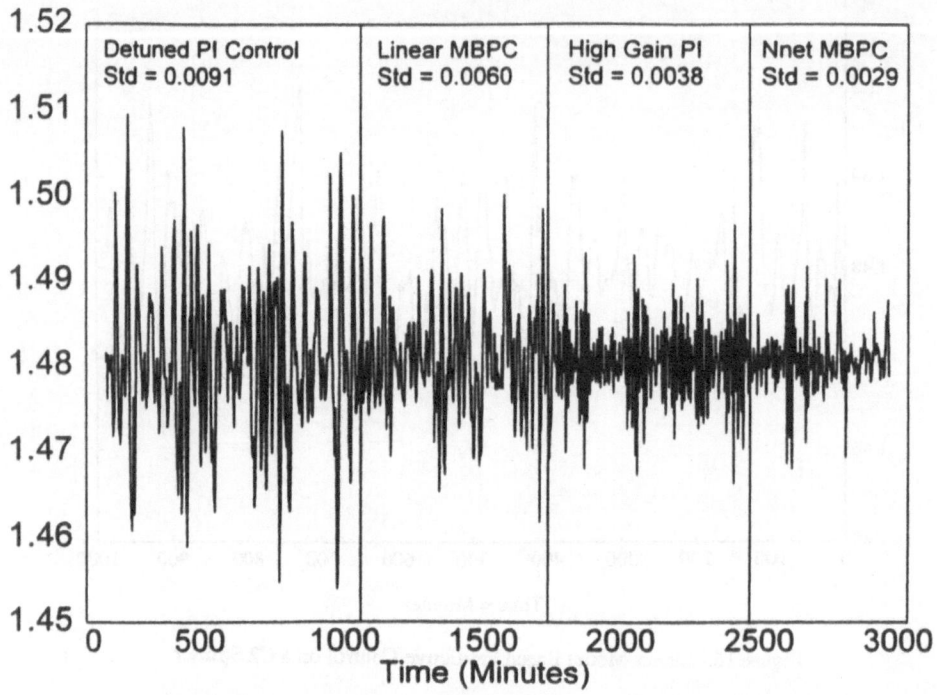

Figure 18. Comparison of Controllers on a C2 Splitter
(P+I, Linear MBPC, High Gain, Nonlinear MBPC)

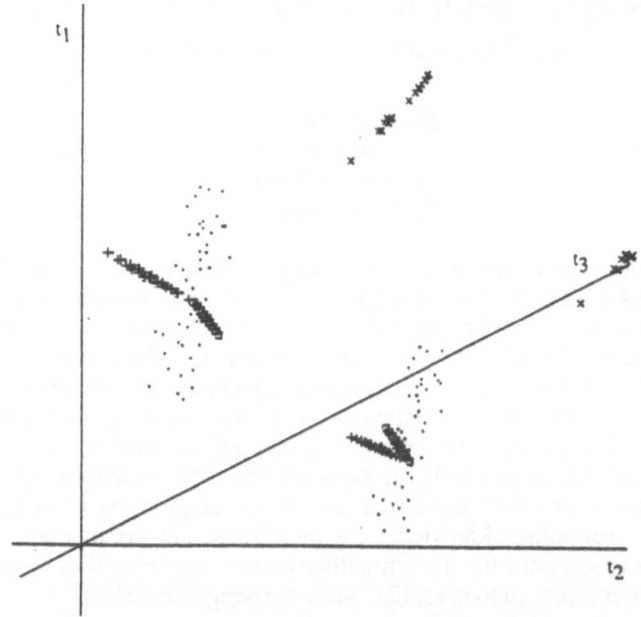

Figure 19. Principal Component Analysis of Continuous Polymer Reactor Data.

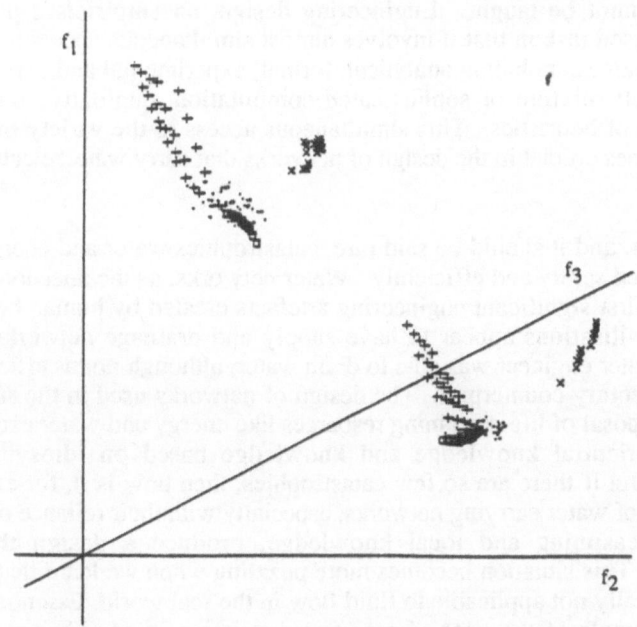

Figure 20. Feature Analysis of Continuous Polymer Reactor Data.

A Knowledge-based Approach to the Safe Design of Distributed Networks

Khurshid Ahmad
AI Group, Department of Mathematical and Computing Sciences
University of Surrey
Guildford, Surrey

Design of complex networks is a knowledge-based task that involves the use of experiential knowledge, the use of simulation models and access to data bases. Safety in the design of such networks is seldom discussed but is, nevertheless, assumed that the knowledge of the expert designer will help in minimising a plethora of risks leading to the failure of the network. Much of the knowledge of designing 'fail-safe' and 'hazard-free' networks is experiential and qualitative. Such knowledge can, in principle, be captured and represented through the use of interview- and text-based knowledge acquisition techniques. This semi-formalised knowledge can then be used in conjunction with simulation software for disseminating knowledge related to the safe design of complex networks, like water carrying networks.

1. Introduction

The design of any artefact, either physical or abstract, is a complex cognitive activity that sometimes defies conventional wisdom and at other times requires skills that cannot be taught. Engineering design, an empiricist's paradise, is a knowledge based task in that it involves almost simultaneous access to a variety of knowledge sources, including analytical, formal, experimental and experiential, and involves a deft mixture of sophisticated computation, qualitative reasoning and extensive use of heuristics. This simultaneous access to the variety of knowledge sources becomes crucial in the design of networks that carry water, electricity, gas or oil.

Barring major, and it should be said rare, catastrophies water and energy networks can be operated safely and efficiently. Water networks, as the anecdote goes, were amongst the first significant engineering artefacts created by human beings and all influential civilisations appear to have supply and drainage networks. The pre-Newtonian water engineer was able to drain water, although not as efficiently as his or her 20th century counterpart. The design of networks used in the supply, usage and waste disposal of life-sustaining resources like energy and water exemplifies the role of experiential knowledge and knowledge based on idiosyncratic local conditions. But if there are so few catastrophies, then how is it, for example, that the designers of water carrying networks, especially with their reliance on heuristics, qualitative reasoning and local knowledge, produce a design that fails so infrequently. This situation becomes more puzzling when we learn that the laws of physics are really not applicable to fluid flow in the real world. Essentially the flow is a good example of the state of a system that is not in equilibrium, is riven by instabilities of one kind or the other.

As we briefly discuss later, the use of simulation programs play an important role in the design of water carrying networks. One could argue that formal proof of the correctness of the program would lead to a safe simulation model. However, we believe that simulation and modelling is carried out throughout the design life-cycle and that it is important to investigate the heuristics the experts use in choosing the so-called parameters of these models. The parameters represent the end-results of a series of physical approximations, and incorporate, albeit indirectly, the operational characteristics of real-world flow for simulation through a model that is based on mathematical and physical idealisations and assumptions. The focus of this paper is on the heuristics used in design, how to verify and validate these heuristics, and make them available on an information system.

Literature on safety critical, high-integrity and dependable computing focuses on formal methods and high technology industry, including VLSI design (Stavridou 1994) on the one hand and prescriptive standards and accident-related reports of control systems on the other (Brazendale and Jeffs 1994).

Bowen and Stavridou (1993) have surveyed the literature on 'safety critical systems, formal methods and standards' in considerable depth and have referred to software safety in industries as diverse as aviation, railways, nuclear power, and health care. These authors have advocated the use of formal methods by showing the use of specification languages and notations, like Vienna Development Method, Oxford Programming Research Group's favourite Z language, and Abrial's B language, hoping that the use of formal methods, notations and languages will lead to increased programmer productivity (*ibid:*193), will help in dealing with real-time kernels (*ibid:*196), and will result in the production of dependable hardware (*ibid:*197).

The issues generally considered in the safety-critical literature deal with a narrow sub-domain of human enterprise, almost a closed-world situation, where the failure of one piece of equipment will lead to some catastrophy. The design of VLSI chips notwithstanding, it is not apparent how a real-world situation is dealt with in detail.

The objective of the project SAFE-*DIS*, is to emphasise the role of Information Systems in making safety-related information available to design engineers in a timely and efficient manner: SAFE-Design of Water Networks through the use of computer-based Information Systems. SAFE-DIS, funded by the EPSRC/DTI Safety Critical Programme, is a three year (1993-96) collaborative project between the University of Surrey and Wallingford Software. Wallingford Software is a commercial software house within Hydraulics Research, developing, marketing and supporting software products to the water and civil engineering industries, and training users. The University of Surrey is the academic partner in the project. The design programs initiated, implemented and marketed by Wallingford Software are used throughout the UK water industry. Many of these design programs will be incorporated in the proposed SAFE-DIS system.

2. Background

2.1 Outline of the Problem
The safe and efficient operation of waste-water drainage networks is essential for human survival. Such an operation depends critically on the design of these networks, and in particular on assumptions about the quantity and quality of water

being discharged through these networks.

Over the last two decades, due to the advances in information technology and, more recently due to mandatory legal and environmental requirements, complex simulation software is used in the design of waste-water networks. The use of this software is replacing manual calculation and error-prone estimates of flow and pressure of water in the networks, together with hand-computation of the costs and benefits of improving an existing network - the 'rehabilitation' of a sewer network - or designing a new network - the 'green-field' sites design. Furthermore, the availability of graphics program packages and the easy access to databases containing topographical details of large conurbations, has increased the scope and frequency of the usage of complex simulation software for designing waste-water networks.

Therefore, in many cases the use of software systems is replacing older manual methods, and in many others completely new applications are undertaken which would not be possible without the waste-water network simulation software. In all cases, however, there is a fundamental problem - that of confidence in the software system itself. The lack of confidence stems in part from the relatively recent use of mathematical modelling techniques in waste-water network design, and in part from the consequent lack of experience with the underlying assumptions and simplifications necessary for expressing complex real-world phenomena, for example, turbulent flow, uneven topology, transients etc, in the idealized language of mathematics including steady flows, smooth topology, etc. Also, the manual calculations and the (error-prone) estimates mentioned above, were cross-checked according to rules and procedures established over the last century by a small hierarchically organized team; the equivalent rules and procedures for computer-based design of waste-water networks are still evolving. Moreover, the calculations and estimates are no longer the exclusive business of public health and civil engineers and contractors, but involve financial experts, environmental bodies, and the increasingly vociferous public at large. The engineers and contractors were largely interested in issues related to the quantity of water flowing through the networks - essentially in the avoidance of flows in urban areas - whereas (the recently enlarged community of) interested parties wants computation related to the overall environmental impact of the performance of a waste-water drainage network.

The increasing complexity of the simulation software, which is geared towards computing water quantity and quality, makes formal proof of correctness even more difficult, despite the recent progress in computer-aided software engineering systems.

The problem related to the lack of confidence in software is relevant to the inexperienced engineer, and can be solved only if the mathematical assumptions are explicated and the software-user has access to the qualitative experiential knowledge of drainage experts. It is important to note that the *drainage experts* have designed a number of safety-operating drainage networks by the use of existing simulation software which contributes to a crisis of confidence for the inexperienced user. Nevertheless, the problem of restoring confidence in the solutions is not unsolvable.

Our main objective for the future, however, is to improve the efficiency of the overall design process and reduce the dependence on human experts, thereby minimising the risk of human error hazarding the operation of a drainage network.

The major inputs to simulation software, particularly for urban drainage, are from computational fluid dynamics, hydraulics, hydrology, applied mathematics and geomorphology for water quantity computations and from microbiology, chemistry and related biological and environmental sciences for water quality. This input is formulated or systematized through the use of differential (or difference) equations and of empirical equations, usually algebraic inequalities. Simulation software for urban drainage network design therefore aims to reproduce physical, chemical and biological phenomena and includes elements of cost-benefit economics. The use of simulation software is not straightforward in that the mathematical model has to be 'calibrated', through the judicious choice of a set of key parameters, to remedy the effects of the necessary simplifying assumptions which are inherent in the development of mathematical models. There is a small number of drainage design experts who can successfully use the existing simulation packages to design (and build) drainage networks. And, as with experts in other fields, these experts are not available on demand and their knowledge is not archived or available for use after they retire, die or leave the industry.

There is no doubt that an incorrect design of a drainage network, and its subsequent implementation, will hazard safety. The water industry has a long and successful history of dealing with safety issues, however the traditional practices, methods and tools are not expected to cope with the ever-growing use of information technology within the industry, and the plethora of legislation governing the entire life-cycle of water use is exerting still greater pressure on the design engineers. There is a need for an information system which will provide guidance to inexperienced engineers in designing safe and cost-efficient drainage networks. This information system will address safety related modelling and software issues, and human factors in design.

1.2 Project Objectives
The principle aim of this project is to specify and prototype an information system which will support a novice (engineer) through the design life-cycle. Such an information system will assure safety, through computer-simulation programs used in conjunction with expert advice, and will increase productivity, through the efficient use of people and their skills. There are two phases of the SAFE-DIS project: definition and initial design (Nov. 1993-Oct. 1994) of the SAFE-DIS IS, followed by the second final design and implementation (Nov. 1994-October 1996). The definition and initial design of the project is first to explore, explicate and archive the largely undocumented knowledge of how effectively to use the existing simulation software, and deriving from the exploration, the requirements for the Safe Design *IS* (SAFE-D*IS*). SAFE-D*IS* will increase the efficiency of an inexperienced drainage engineer and reduce the probability of human error. Second, to examine the efficacy of methods and tools currently available (or under development) in framework design and assess their applicability to the requirements, and establish performance criteria and relevant metrics for the operation of SAFE-D*IS* .

The second phase of the project when completed will lead to the production of a high-level specification for a safety-informed *IS* that could be used to comprehend safety related problems.

2.2 The 'design life-cycle'

The mathematical models of fluid flow used in drainage network simulation and design require the consideration of generic processes - laminar flow, stochastic stability, frictional losses - and a substantial amount of episodic data - rainfall patterns and geomorphological features for instance.

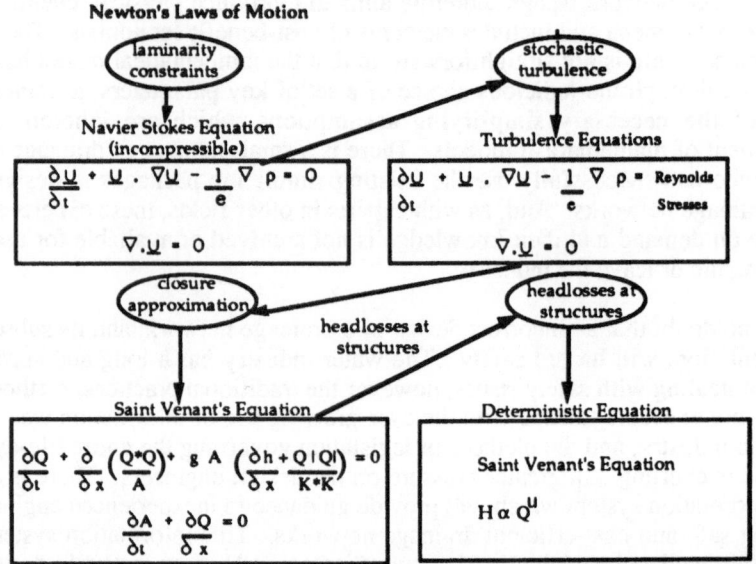

Figure 1. Generic Processes and simplifying assumptions in drainage design computations

Given a complete description of generic processes and access to all relevant episodic data it is, in principle, possible to compute inflows to networks and subsequent discharges through the networks. Although it is possible to describe the generic processes from first principles - i.e., Newton's Laws of Motion - and, with suitable constraints, to outline fluid flow in three-dimensions through the use of Navier Stoke's equation - such a description cannot be computed unless suitable simplifying assumptions are made to the equations. Furthermore, it is not generally possible to have detailed data related to rainfall patterns for a particular (urban) catchment, neither is there data about complex geomorphological features of the catchment.

Over the last four decades, principally in the UK, drainage experts, working together with software engineers, have developed a set of well-tried approximations to adopt pre-programmed models and available data so that network simulation could be effected and a design produced and evaluated. The drainage network design programs, therefore, have a series of parameters, which the end-user has to vary judiciously in order to use the package and simulate networks of users' choice. This is a cyclic process of iterative refinement as the choice of parameters is influenced by the experience and awareness of the end-user and in turn influences the safety of the design. The drainage community terms the design life-cycle *'model validation'*.

The 'model validation' cycle has to be reinitiated for almost every new release of an existing design program or when the software vendors produce a new program product. There are conventional forms for the software failure/bug reports for these design programs and there are informal protocols to record design initiative which have failed - 'bad design'.

2.3 Methods and Tools

The network design life-cycle is sparsely supported by methods and tools thereby encouraging dangerous dependence on human input to ensure design safety and proper use of network simulation software. The methods comprise codes of practice, design guidelines and 'in-house' standards; the tools are essentially flat-file data management systems and simulation-programs of mathematical models (see fig. 1). Safety-related issues are implied in the method-documentation and in the choice of model validation parameters. The novice engineers, with a university degree in engineering, work closely with experts to familiarize themselves with network design. The novices have the opportunity to attend lecture and workshop sessions on design issues and to familiarize themselves with the design software of vendor-organized training/presentation sessions. The teaching and learning of the design life-cycle is largely ad-hoc and leaves much to the individual.

3. Knowledge. Acquisition
3.1 Text-based Knowledge Acquisition

Knowledge acquisition can be regarded as a technique by which a knowledge engineer obtains information from experts, text books, and other authoritative sources for ultimate translation into a machine language and knowledge base. Knowledge acquisition involves extracting, structuring, and organizing domain knowledge with the help of domain experts.

The acquisition of knowledge usually assumes that all the experts in a given domain have similar experiences during their apprenticeship and in applying their knowledge. In most disciplines this may be true. However, there are exceptions and the design of waste-water networks is a good example of this. There are substantial differences of opinion amongst experts, particularly in the design of drainage networks at a geographically well-defined local level. The differences in topography, in the socio-economic makeup of the community, experience of abnormal conditions, and so on, contribute to the divergence in knowledge. This situation is further exacerbated by political considerations: privatised enterprises running these resources versus public ownership of the same. The considerations of safety in the design of the waste-water networks, much of it undocumented, will vary from expert to expert, from region to region and from private sector to the public sector. This variation has to be documented to verify and validate the knowledge of the design process, and through this documentation process, one hopes, hazards to safety of these networks, particularly hazards that may be traced back to the design, can be identified and disseminated to the network design engineers at large.

A systematic approach to the acquisition of knowledge can contribute to the consolidation of policy, to the appraisal of current working practices, and in determining the scope for innovation. Many would claim that all this can be achieved by extracting, analysing and disseminating knowledge in this area by conventional knowledge acquisition techniques and methods. These include psychology-based interviewing techniques, methods related to multidimensional analysis, content analysis and so on. However, as we have pointed out elsewhere much of this knowledge exists in a textual form, or can be rendered in textual form, and a number of knowledge acquisition methods and techniques do not tend to exploit the semantics, especially terminology, and pragmatics of the written text (Ahmad 1993).

Furthermore, in order to achieve consensus, one has to deal with a number of experts and has to extract the similarities and differences in their problem-solving skills and experiential knowledge. *Brainstorming* is a technique used extensively in management, marketing and other related enterprises. This technique appears to be suitable for eliciting opinions of varying shades and in establishing a degree of consensus (Ahmad and Griffin 1994 and Griffin and Ahmad 1994).

3.2 Safety Round Table: Terminology, Texts, Brainstorming, Interviews and Heuristics

In the SAFE-DIS project considerable attention is being paid to identifying and collecting published material on waste-water network design together with documents on safety critical systems. This material - a corpus of texts- has been analysed with a view to creating a terminology data bank of safety in design with particular reference to waste water networks. The contents of this terminology data bank, built on terminological principles and practices adopted by the ISO and refined in various ESPRIT sponsored projects (see Ahmad and Rogers 1992 and forthcoming), will be used in the specification and creation of the object/fact base of the proposed SAFE-DIS information system (see section 4).

The text corpus contains two rather unusual sets of documents. First, a set of interview transcripts were obtained from a video-taped interview of design experts in an earlier (Alvey-sponsored) project (Ahmad et. al 1988). A number of design heuristics have been extracted by analysing the interview texts through the use of System Quirk, a text analysis system (Holmes-Higgin, Ahmad and Abidi 1994). A total of 41 heuristics have been extracted from the interview transcripts. Second, the corpus contains the entire contents of the UK Water Act 1991, a 260-page, 133,268 word document, which is the basis of all laws governing the use and drainage of water, together with a 460-page legal text, comprising 144,189 words, specifically written to interpret other legislation that governs the aquatic environment in the UK. A number of complex objects, like the definition of a watercourse, were extracted from the legal texts. The legal texts have been marked up for use in a hypertext environment and can be used in conjunction with the proposed SAFE-DIS system.

The collection of texts and the building of terminology has been accompanied by the formation of the *Safety Round Table* - a forum of consulting engineers, water companies and (UK) local government organisations responsible for waste water. This forum, which includes experienced network designers, has selected leading experts in the field to be interviewed for the elicitation of the knowledge of network design. The Round Table, working together with knowledge engineers, decided upon a set of key questions during two brainstorming sessions. These questions, together with elaborations on the same, were put by the interviewer to the experts, one from design consultancy and the other working for local government. The one-hour interview was video-taped, and a 10,000 word transcript of the interview was produced.

This transcript served two purposes: **First**, the interview and the transcript were circulated to all the members of the Round Table, including the interviewer and interviewees, and resulted in a number of corrections to the transcripts. These corrections, excluding minor typographical errors, usually involved the experts in qualifying sentences, adding more elaborative notes and deleting certain aspects of the transcripts. **Second**, the transcript was analysed by System Quirk for the extraction of heuristics and for the elaboration of the various domain objects. In total 45 heuristics were identified by the knowledge engineers and are currently being validated by the Round Table.

The interview tape and the transcript were discussed by the Round Table in an all-day brainstorming session in the presence of the experts and the interviewer. The results of this session contributed to the identification of specific areas in drainage network design and helped the knowledge engineers in focusing the expert system's development.

4. SAFE-DIS Environment

Knowledge acquisition without prototyping is a sterile exercise in that it is not possible to indicate to the experts how, and indeed if at all, the knowledge elicited from them can be verified, validated and disseminated. Recall that an important component in the knowledge of drainage network design is the dexterous use of simulation models and the ability to synthesise data from a number of data and knowledge sources. Figure 2 below presents the architecture of a prototype integrated information system that comprises a number of autonomous yet weakly interacting modules.

Figure 2. The user surface of the prototype SAFE-DIS information system.

The architecture of the prototype SAFE-DIS IS has been influenced by design considerations in NASA's Scientists' Intelligent Graphic Modelling System (SIGMA) developed by Keller, Rimon and Das (1994) and in expert systems developed at the University of Surrey for the UK National Water Authority (see Ahmad and Griffin 1994, Ahmad and Griffin 1993).

SIGMA was designed as a rapid prototyping tool for building scientific models and developed as a specialised tool for the scientific modelling domain: the authors claim, and demonstrate how, SIGMA can be used to establish the modelling scope, to specify all quantities associated with any of the 2000 objects SIGMA contains (divided into over 600 classes including planetary modelling and forest ecosystems). Using SIGMA scientists can model complex atmospheric systems and ecosystems. The SIGMA knowledge base contains two kinds of 'scientific knowledge', cross-disciplinary scientific knowledge, that is general scientific background knowledge related to physical quantities, their units and measurements, scientific handbook data and so on, and 'discipline-specific' scientific knowledge, that is quantities and objects that belong one of the biological, physical or ecological sciences. In addition, the SIGMA knowledge base comprises 'problem-specific' knowledge, objects and relations to a specific scientific discipline and 'programming', that is knowledge about numerical programming methods, data structures and control. SIGMA also contains a 'bibliographic citation' module. SIGMA has its own, LISP-based, knowledge representation language called RML.

The W-RAISA (*Water Resources Management Intelligent Assistant*) and ELSIE (*Expert Licensing Support Intelligent Environment*) systems were based on a modular architecture. Both the systems were designed to work on heuristics (obtained from experts through psychological interviewing and brainstorming techniques), simple mathematical computations, access to legal information and to other data that is contained in scientific handbooks or data sheets used in science and engineering. Knowledge in ELSIE (and W-RAISA) was represented using an object-oriented knowledge engineering environment called KAPPA (and KAPPA-PC).

SAFE-DIS IS supplements the textual knowledge by providing access to the corpus of texts that is relevant to the discipline and problem-specific engineering knowledge. Furthermore, the IS users can access one of the major simulation networks in drainage, HYDROWORKS produced by Wallingford Software, and plans are to allow the users to manipulate differential and algebraic equations through a symbol manipulation package. The SAFE-DIS knowledge base is currently being implemented using KAPPA-PC and has 154 rules and 50 objects.

The system will be evaluated by potential end-users, and the Safety Round Table, in the first quarter of next year (1995).

5. Conclusions and Future Work
Conclusions about a project which is only a third of the way through can at best be speculative. However, we remain convinced (and hopeful) that the techniques in artificial intelligence will help in acquiring, representing, and disseminating the safety related knowledge of experts. We believe that an information system which will support the safe design of networks must follow a development strategy

comprising the identification of heuristics used by experts for incorporating safety-features in the design life-cycle, and the identification strategies they adopt to use currently available design packages successfully. Further in-depth interviews have been planned and it is hoped that a substantial body of undocumented knowledge will be collected that will throw some light on problems related to the safe design of water-carrying networks.

The design of a complex artefact, like a drainage network, requires a careful consideration of a number of interdependent knowledge sources, engineering hydraulics and design, environmental sciences, geomorphological and financial data and legal information together with simulation and modelling heuristics. The design engineer works within a community that includes administrators, scientists, lawyers and other engineers, and the experts among the design engineers are fully aware of their limitations and seek to compensate for the limitations by co-operating with the (expert) members of the community. In order to simulate co-operative/collaborative behaviour of experts it is important to look beyond the currently available architecture discussed within conventional AI, because the conventional architectures emphasise the development of an *autonomous* system that does not, and by design should not, seek the help of other systems including its users.

The literature on distributed artificial intelligence contains references to architectures that contain *agents*, knowledge bases with reasoning strategies, that can plan their course of action independently of others yet are still capable of communicating with and supporting each other. Examples of systems based on distributed agents architectures include systems for multi-user project co-ordination (Sathi, Morton and Roth 1986, Klein 1991), and for resource allocation in long haul communication networks (Adler et al 1989). Distributed AI literature claims that multiagent/distributed agent architecture is much closer in operation to the natural systems, for example human societies and insect colonies, in that there is a premium on collaboration and co-operation on the part of the components of these natural systems. The components of a natural system are in themselves well-structured and can behave autonomously and devolve when required.

The SAFE-DIS project is currently exploring the use of multiagent architecture for building a safety-informed knowledge base. For this purpose one of the project team members is exploring the use of the object-oriented distributed AI languages, Oz, developed by the German Centre of Artificial Intelligence (DFKI) (Henz, Smolka and Würtz, 1993). If our explorations yield positive results then the possibility of implementing the SAFE-DIS prototype using distributed agents will be considered.

The benefits of the project are expected to become apparent halfway through Phase 1 (Fourth quarter 1994) particularly through the examination of the experts' knowledge which will be archived and disseminated. The prototypical versions of SAFE-DIS will demonstrate and focus on safety-related issues in the design of drainage networks, during the three year lifetime of the project. We anticipate that it will take another two to four years to develop a production version of SAFE-DIS: the benefits of using the information system during this period.

The prototype SAFE-DIS, and its anticipated production version, will, we hope, provide due motivation for the water industry to adapt a standardized approach to the safe-design of drainage networks. Furthermore, we believe that there will be a substantial number of points of generic interest for other network operators like gas, electricity, etc.

Acknowledgments

The SAFE-DIS project team at the University comprises the author of this paper (KA), Stephen Collingham, who is building the SAFE-DIS surface and the knowledge base, Indrakumar Selveratnam, who is involved in exploring the multiagent architecture, and Andrew Salway, who has recently joined the team and is looking at the 'language' of safety and the possible use of neural networks in network design. The author is grateful to the SAFE-DIS team for stimulating discussions. Caroline McInnes had the tough task of co-ordinating the production of this paper, as usual many thanks. Felix Redmill waited patiently across many deadlines, thanks to him again.

References

Adler, M R., Davis, A. B., Weihmayer, R., & Worrest (1989) 'Conflict Resolution Strategies for Nonhierarchical Distributed Agents'. In (Eds.) L. Gasser and M. N. Huhns. *Distributed Artificial Intelligence*. Vol. 2. London: Pitmans.

Ahmad, Khurshid & Rogers, Margaret A. (1992) Translation and Information Technology: The Translator's Workbench, in *ReCALL*, No. 6:3-9.

Ahmad, Khurshid & Rogers, Margaret A. (Forthcoming) 'The analysis of text corpora for the creation of advanced terminology databases'. In (Eds.) Sue-Ellen Wright and Gerhard Budin The Handbook of Terminology Management. Amsterdam and Philadelphia: John Benjamins.

Ahmad, Khurshid (1993) 'Terminology and Knowledge Acquisition: A Text-based Approach.' In K.-D. Schmidt (Ed.), *Proceedings of the 3rd International Congress on Terminology and Knowledge Engineering*. Frankfurt: Indeks Verlag, pp. 56-70.

Ahmad, Khurshid and Griffin, Stephen M. (1994) NRA ELSIE Final Report 406/1/S.

Ahmad, Khurshid, & Griffin, Stephen M. (1993) 'Intelligent Assistants and Engineering Decision Support: Management of Water Resources'. *Computing Systems in Engineering*. Vol. 4 (Nos. 2 & 3). pp 325-335.

Ahmad, Khurshid, Holmes-Higgin, Paul R., Hornsby, Charles P.W. & Langdon, Andrew J. (1988) 'Expert Systems for Planning and Controlling Complex Physical Networks'. Knowledge-Based Systems Journal Vol. 1 (3) pp 153-165 .

Bowen, Jonathan, and Stavridou, Victoria (1993) 'Safety-critical systems, formal methods and standards'. *Software Engineering Journal*. July 1993. pp 189-209.

Brazendale, J., and Jeffs, A. R. (1994) 'Out of Control: Failures involving Control Systems'. *High Integrity Systems*. Volume 1 (No. 1), pp 67-72.

Griffin, Stephen M. and Ahmad, Khurshid (1994) Archiving Knowledge: Before and After the Interview. In D. Lukose (Ed.), *Proceedings of the Workshop on Knowledge Acquisition Using Conceptual Graph Theory*, International Conference on Conceptual Structures, Maryland, USA.

Henz, Martin, Smolka, Gert and Würtz, Jörg (1993) 'Oz - A programming language for Multi-Agent Systems. IJCAI (International Joint Conference on Artificial Intelligence Chambery, France 28 August - 3 September 1993.

Holmes-Higgin, Paul, Ahmad, Khurshid, & Abidi, Syed Sibte Raza (1994) In (Eds). Martin, Willy, Meijs, Willem, Moerland, Margreet, ten Pas, Elsemiek, van Sterkenburg, Piet & Vossen, Piek 'A Description of Texts in a Corpus: 'Virtual' and 'Real' Corpora.' EURALEX 1994 - Proceedings of the 6th EURALEX International Congress Amsterdam: EURALEX. pp 390-402.

Keller, Richard M., Rimon, Michal & Das, Aseem (1994) 'A Knowledge-Based Prototyping Environment for Construction of Scientific Modelling Software'. *Automated Software Engineering*. Vol. 1 (No.1, March 1994) pp 79-128.

Klein, A. (1991) 'Supporting Conflict Resolution in Co-operative Design Systems'. *IEEE Trans. Sys. Man. Cybrnt.*

Sathi, A., Morton, T. A. & Roth, S. F. (1986) 'Callisto: An Intelligent Project Management System'. *AI Magazine* Vol. 7 (No. 5), pp 34-52.

Stavridou, Victoria (1994) 'Formal Methods and VLSI design'. *The Computer Journal*. Volume 37 (No. 2), pp 96-113.

Where Do Specifications Come From?

Derek Partridge

Department of Computer Science, University of Exeter

Exeter EX4 4PT, UK

Abstract

The notion of 'specification' is generally held to be crucial to the process of software production and <u>the</u> anchor point for the science of computing. It has, however, no simple, well-defined meaning. There are a variety of interpretations and usages of the term. This paper surveys the field of possibilities, both to display this varietal profusion, and to make the point that no one interpretation has a sound claim to be the best.

1   Introduction

From the meeting when the term "software crisis" was coined (or at least accepted as an accurate descriptor of the state of affairs in software engineering) right up to the present time, some 25 years on, the notion of 'specification' has been thought by many to be the key to our problems. The use of this key is often presented as an upfront, complete, unambiguous, etc. specification of the problem *before* the final design and implementation are undertaken. With a subsequent validation of the implemented system in terms of how true it is to the specification. In the extreme, some wish to formally prove that the actual implementation is a 'correct' implementation of the specification, and to present this as the main plank upon which to build a proper software science.

In this paper I wish to raise the general level of awareness of the fact that the term 'specification' has multiple and not mutually consistent meanings, that the favoured ones are not indisputably the best ones, and that no one interpretation is totally satisfactory. Part of the confusion stems from the fact that software persons use the term rather casually on the (mistaken) understanding that it has a clear meaning and we all agree (more or less) on what it is. This is not true.

Blame, if that is the right word, for the wealth of conflicting interpretations can be attributed to the multiplicity of problem types, system priorities and personal philosophies to be found in the full spectrum of activities centred on the process of getting a computer to compute something — loosely, the world of software system construction.

One way to explore this crucial notion is through a consideration of the origin and provenance of the different manifestations of it. Hence, the title of this paper "Where do specifications come from?"

Where <u>do</u> they come from? The sources are many and their routes are no less.

2 Divinely inspired

The slightly flippant heading for this first section has a serious purpose: it is to make the point that specifications can be seen as the characterisation of the problem at hand. It is true that not many persons could be found who would want to credit a General Omniscient Designer (GOD) with the construction of software specifications, but many are quite prepared to treat the specification *as if* it had the good housekeeping stamp of divine approval.

From this viewpoint, the specification is the problem, and the task of the software person is 'simply' to implement it correctly. It is not the software person's concern to second guess why various aspects of the specification are as they are. He or she has no business looking 'behind', 'underneath' or 'through' the specification. The software engineering (SE) task is to translate what's given into a machine-executable form.

The best rationale for this particular viewpoint is, as Dijkstra (amongst others) would argue: software development is a very complex process and so the sensible way to tackle it is to break it into more manageable, less complex chunks and deal with each separately — divide and conquer by a separation of concerns. The essential separation underlying this viewpoint is that between "the 'pleasantness problem,' i.e., the question of whether an engine meeting the specification is the engine we would like to have" and "the 'correctness problem,' i.e., the question of how to design an engine meeting the specification." [Dij89] p. 1414.

In more everyday terminology Dijkstra is arguing for a separation between requirements analysis, which terminates with a specification, and software design and implementation, which starts with a specification. He wants a "logical firewall between [these] two different concerns". And wouldn't it be a great thing to have, but then so would summer all year round, a major pay rise for academics, and a mango tree growing in my back garden. But at some point we must face reality, some nice things are just not going to happen either because the world doesn't work that way, or it would be a breach of government guidelines, or it's a logical impossibility, etc.

A very different rationale for this particular separation of concerns comes from the world of commercial software production: a specification serves admirably as (the core of) the contract between supplier and client. From this viewpoint it is again a 'fixed point', but as a commercial practical expedient.

Can we really expect to break our software problems at the specification stage? Some we can and some we can't. This takes us on to the notion of types of software problem.

3 Unequal opportunities

Just like people, all software problems are not equal. Coming from an AI background, I would first suggest that AI problems are not at all like conventional SE problems. Here's a succinct tabulation of the differences (from [Par92]):

AI problems	Conventional SE problems
1. incomplete, performance-mode specifications	fairly complete, abstract specifications
2. solutions adequate or inadequate	solutions testably correct or incorrect
3. poor static approximations	quite good static approximations
4. resistant to modular approximation	quite good modular approximation
5. poorly circumscribable	accurately circumscribable

As point 1 makes explicit, the nature of the specification for AI problems and for conventional SE problems is fundamentally different. The response from the software engineer might be that this acceptance of such poorly specified problems is the reason why AI persons have such difficulty constructing robust and reliable implementations. "If only the AI fraternity would insist on a prior, proper specification of their problems, then they would have far less trouble", the software engineer might explain.

However, problems precede specifications. One simple answer to the opening question, but one that makes a forgotten point, is that the specification comes from the problem. Now consider the nature of AI problems.

Take natural-language processing (NLP) as an example of an AI problem. The problem then is the human phenomenon of communication in, say, English (to pick a natural language at random). What defines this problem? Ultimately, it is some sort of consensus of what native speakers actually do when communicating in their mother tongue. Observed behaviour defines the problem. Hence, we have, in effect, a *performance-mode specification* which gives us a *data-defined* problem.

The NLP problem as manifest in individuals varies greatly from one person to another, and in addition, it is changing with time at varying rates and in varying ways. There is a core of the problem that is perhaps invariant with time and fairly constant over the full range, but even an abstract characterization (i.e., specification) of this core would necessarily be an incomplete specification of the NLP problem.

Given this sort of characterization of what the AI person, who hopes to build an NLP system, is tackling, a sensible response might be that NLP is not a suitable problem for computerization. And, although there is some merit in this dismissal, it is the case that nearly all humans are competent natural-language processors. So, presumably there is a good implementation of the NLP problem in each of our heads. Why not build a similar one, even a vastly inferior one, into a computer? Its uses would be legion.

4   In the beginning there were specifications

In fact it is not true that problems always precede specifications, and, although the exceptional situation is somewhat specialized, it has had more than its fair share of influence on computing. The exception to this general rule is abstract, technical problems — mathematics, to put in succinctly and without great loss of accuracy. The reason for the undue influence on computing exerted by this rather special type of problem is that the introductory examples in computer science are usually of this type.

Mathematical problems are fundamentally abstractions, and they are complete and well-defined. A mathematical problem is, or can be, almost a specification. Certainly, it is quite likely that no major transformation of the problem statement is required to generate a specification which the software engineer will find acceptable.

Once you have stated that the problem is to compute the greatest common divisor (GCD) of two integers and you've formally defined the notion GCD, you have a specification of the problem. And, moreover, the specification is, more or less, the problem.

Contrast this problem with one of designing and implementing an inventory management system. It is true that there will be a core of well-defined mathematics pertaining to the addition and substraction of items from the inventory. Such well-defined elements are however but a small part of the overall system and probably not where the design and implementation difficulties will occur.

The necessary specification must be developed from the real-world problem — e.g. the customer wants to computerize inventory management with all the attendent problems of efficiency, user interfaces, the nature of the goods inventoried, etc. No longer is the specification more or less the problem. The problem exists in the real world quite independent of the specification, and the questions is how well does the specification actually specify the what the customer really wants (whether he or she knows it or not)? The end product should meet expectations rather than specifications — the customer is likely to assert.

Again we can tabulate problem differences, this time between conventional SE problems and the abstract mathematical ones — the model computer science problems.

Conventional SE problems	Model computer science problems
1. fairly complete abstract specification	complete abstract specification
2. solutions testably correct or incorrect	solutions provably correct or incorrect
3. quite good static approximations	perfect static approximations
4. quite good modular approximations	perfect modular approximations
5. accurately circumscribable	completely circumscribable

A question that arises from such tabulations is: are SE problems more like AI problems than model computer science problems? There is, of course, no simple answer, but the point of raising this issue is to make it clear that the latter alternative is not (as much discussion of specification issues assumes) 'obviously' the right answer.

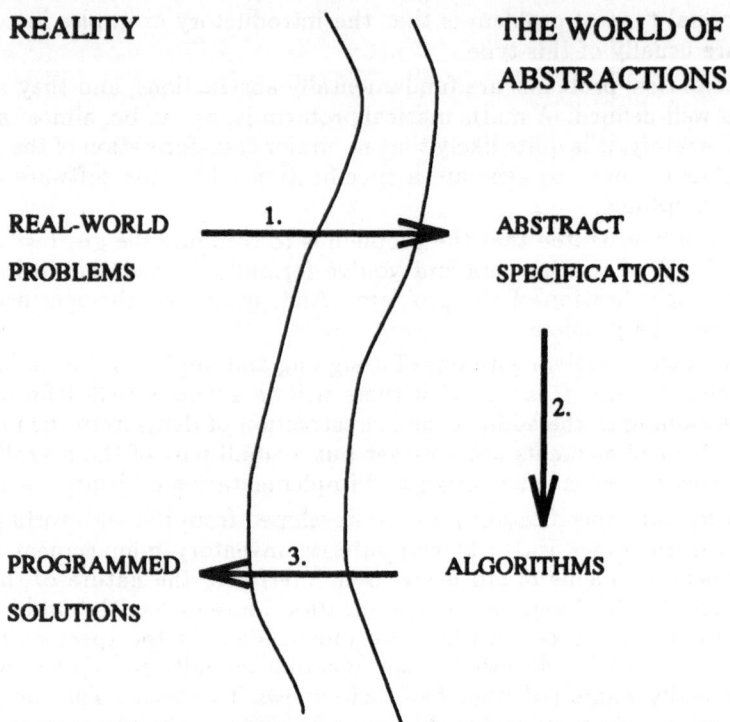

REALITY

THE WORLD OF
ABSTRACTIONS

REAL-WORLD PROBLEMS — 1. → ABSTRACT SPECIFICATIONS

2.

PROGRAMMED SOLUTIONS ← 3. — ALGORITHMS

Figure 1: an idealised map of the software engineering world

5 The bigger picture: earth and water, but no fire

In order to obtain a proper perspective on specifications we need to consider the big picture. It is instructive to view the abstract world and the concrete world as two distinct worlds separated, by, say, water — a river Styx if you like.

Problems to which we might want to apply computer technology occur in the concrete world, the real world, and similarly the computer system that is constructed to address a given problem must also function in reality. However, it has proved immensely useful to travel from the problem in the real world to its solution also in the real world by detouring through the abstract world. The benefits conferred by abstraction and symbol manipulation within the abstract world cannot be underestimated. But in order to reap the benefits that abstract conceptions carry with them it is necessary that the software developer crosses this body of separating water not once, but twice. Figure 1 will put you in the picture.

As you will see it is quite clear where specifications come from — the other side of the river. The modern river Styx separates the infernal regions of reality from the celestial perfection of the world of abstraction, and over which pure

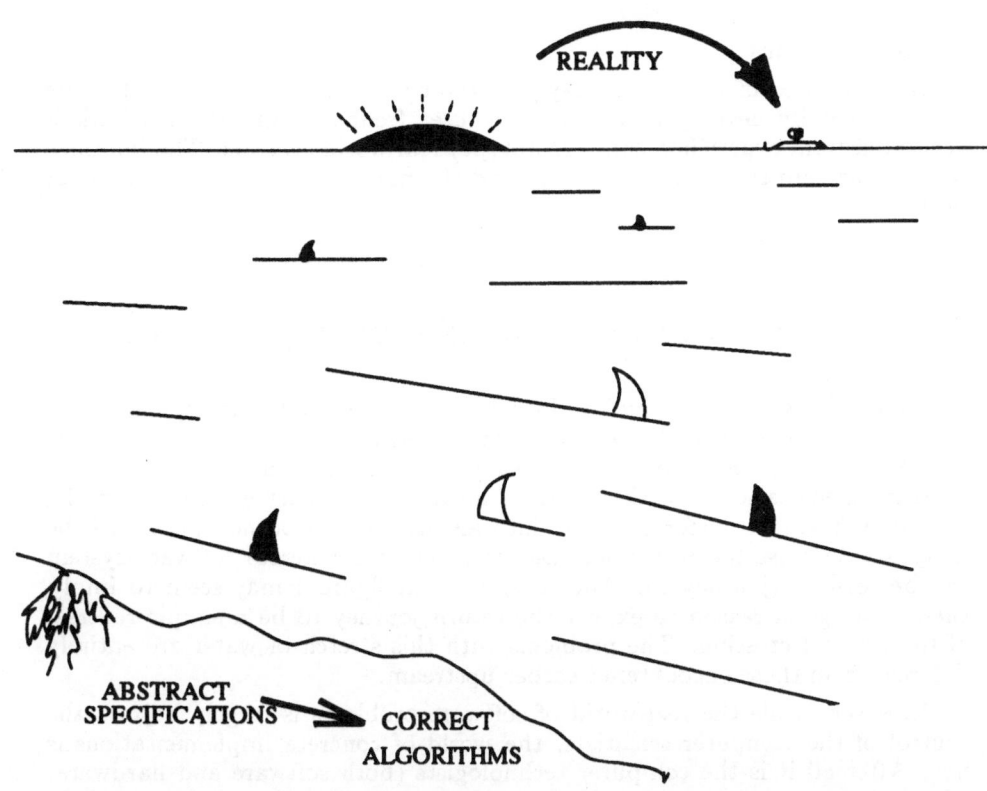

REALITY

ABSTRACT
SPECIFICATIONS ➡ CORRECT
ALGORITHMS

Figure 2: a real picture of most of the software engineering world

specifications are ferried by software engineers. The question is how easy is it to prise them out and drag them across without undue distortion? And how reasonable is it to expect that once you have a complete, precise and unambiguous specification settled as an abstraction, you can sever the connections — blow up the ferryboat, or, less destructively, erect a "logical firewall" as Dijkstra would prefer?

Again the answer depends on what sort of problem you are dealing with. If it's a model computer science problem, then the separation is likely to be sensible and effective. The river is narrow and calm — not a Styx at all. If it's an AI problem then the separation is likely to be fatal to the process of developing an adequate programmed solution, a practically useful piece of software. The water crossing again might not be Stygian — it might be worse, as figure 2 conjectures.

But what if it's a conventional software engineering problem? It can fall anywhere within the span from near-separability (e.g., an accounting package) to non-separability (e.g., a human-computer interface system). In the latter case, one that typically elicits a prototyping methodology for software development, specifications are iteratively developed subsequent to user-feedback on successive versions.

In the context of an evolutionary prototyping methodology, the specification is incrementally developed from specific user feedback, and the feedback is generated from experience with (prototype) real-world systems. The boatmen will be busy crossing back and forth, and the firewall had better have a door in it.

6 The unexpected cost of return journeys

Unlike the railway system, and many other terrestrial transport organizations, a return journey in the real software world is likely to cost at least twice as much as a one-way trip. For having struggled to extract and transport an accurate abstract specification into the world of abstractions (journey 1 in Figure 1), and after having transformed that into an algorithm (journey 2), it will be necessary to cross the water one last time before a concrete software system can be realised (journey 3). And, despite what figure 1 may seem to imply, there is no good reason to expect the return journey to be a simple reversal of the original crossing. The problems with this stretch of water are entirely different from those encountered earlier upstream.

However, while the real-world of software problems is largely beyond the control of the computer scientists, the world of concrete implementations is not. After all it is the computer technologists (both software and hardware) who design and construct actual computer systems. Therefore they can largely determine the structure of reality for computer programs, and hence the nature of the return trip.

On the strength of this sort of observation we can manufacture further good justification for concentrating maximum effort within the world of abstractions. The argument is that with perfect implementations the river is effectively dried up. There will be no water to cross once the hardware and software technologists have properly sorted out their problems. There are, however, good reasons to believe that the river will remain wide, deep and dangerous as a result of all conceivable hardware and software advances.

Nelson, for example, has argued this point at length [Nel92]. He argues, for example, for the fundamental mismatch between the assumed continuity of arithmetic underlying program 'proofs' and the non-continuity of mathematical functions as pieces of software. "In a digital computer, the continuum of real numbers is 'replaced' by the representations of a finite sub-set of the rational numbers that is not uniformly dense and many of whose members are finite approximations" p. 295. In summary he states that "the transformation of an algorithm into an executable program is a *wrenching metamorphosis* that changes a mathematical abstraction into a prescription for concrete actions to be taken by real computers" page 283 (my emphasis).

It is, of course, always possible that time together with technological advances will reduce the degree of wrenching change necessarily wrought on an algorithm on crossing back across the water, but it seems that there are some very good reasons for believing that the river will in fact never run dry.

7 By its fruit it will be known

If you take the stance that either you believe that the river will never dry up, or that because the software is wanted last week you cannot afford to wait any longer and must therefore make the best of the current situation, you might be forgiven for thinking that the best source of specifications is implementations. For if the abstract algorithm does undergo a "wrenching metamorphosis" when it is brought back across the water, then it is wise (if not essential) to check what it is that this new object actually implements. The specification should be a succinct characterization of what the software actually does, not what it would have done had the world been different.

There are a number of different reasons why the source of the specification may be the implementation.

The maintenance programmer when faced with a large and undocumented (or worse, wrongly documented) piece of *legacy software* has absolutely no choice — a meaningful specification must be derived from the existing pile of code. This resort to the implementation may be seen as an unfortunate practical expedient which, although frequent and widespread, is just another symptom of bad practice in software engineering. It is not a 'good' reason for viewing an implementation as the source of a specification.

So here's an excellent reason for taking this view of specifications: inductive programming techniques, ranging from induction of decision trees (see Michie [Mic91] for an eloquent argument for the effectiveness of this particular approach to software production) to training of neural networks (see [Gal88] and [PS94] for two totally different neural computing approaches), being quite close to fully automatic programming deliver an implementation from a set of initial conditions.

Data-defined problems (the SE problems closely akin to performance-mode specified problems of AI) can be particularly appropriate for these techniques.

The available data is the best characterization of the problem. An abstract specification may be hard to extract, likely to be inaccurate, and above all unnecessary for an inductive programming approach. Some abstract characterization of the problem is useful as it helps to guide the set up of initial conditions for the automatic induction process, but no great weight needs to be put upon it.

8 In the end

This paper is loosely derived from another [PG94] that lays out the full range of interpretations and resultant conflicts that surround the idea of 'specification' in the software world. It also provides full references to the many earlier discussions of the various roles of specifications in the the software development process.

The crucial notion of specification is important in almost all of the various manifestations of the software development enterprise. It is not however a simple unitary idea that can be simply applied in exactly the same way whatever the problem type, software development paradigm or particular practical situation. It must be applied with proper regard for the surrounding context and the overall goals of the exercise. It is not merely bad practice and general sloppiness that dictates interpretations other than the computing science ideal. There are a variety of entirely sound and sensible reasons for rejecting this much-touted view and its implied separation of concerns.

References

[Dij89] Edsger W Dijkstra. On the cruelty of really teaching computing science. *Communications of the ACM*, 32(12):1398–1414, 1989.

[Gal88] S. I. Gallant. Connectionist expert systems. *Communications of the ACM*, 31(2):152–169, 1988.

[Mic91] Donald Michie. Methodologies from machine learning in data analysis and software. *The Computer Journal*, 34(6):559–565, 1991.

[Nel92] David A. Nelson. Deductive program verification (a practitioner's commentary). *Minds and Machines*, 2(3):283–307, 1992.

[Par92] D. Partridge. *Engineering artificial intelligence software*. Intellect:Oxford, UK, 1992.

[PG94] D. Partridge and A. Galton. The specification of 'specification'. *Minds and Machines*, 1994. in press.

[PS94] D. Partridge and N. E. Sharkey. Use of neural computing in multi-version software reliability. In F. Redmill and T. Anderson, editors, *Technology and Assessment of Safety-Critical Systems*, pages 224–235. Springer-Verlag, London, 1994.

Formalising Fault Trees

Janusz Górski and Andrzej Wardziński
Franco-Polish School of New Information
and Communication Technologies
Poznań, Poland

Abstract

The paper presents a systematic approach to formalisation of events in Fault Trees. The formal meaning of events is given in terms of the Extended CSDM model. The paper gives an algorithm which guides through the process of event formalisation together with some classification of the typical event classes. The approach is illustrated by a number of examples which present formalisation of events derived from real Fault Trees.

1 Introduction

Formal Methods is one of the technologies which may help in improving safety analysis. The motivations behind efforts to extend presently used safety analysis techniques with formal semantics have been presented in [Górski 94]. In [BCG 91] and [Górski 94] the Common Safety Description Model (CSDM) has been introduced as a candidate base for Fault Tree formalisation. In those papers the focus was mainly on defining the precise meaning of gates with particular stress on the causal and temporal dependencies among events.

This paper focuses on the problem of how to formally express the events of a tree. As the event descriptions are given with reference to the system structure (physical, functional, etc.) some means of representing this structure are necessary. We propose to use the state modelling facilities of VDM (see e.g. [Jones 90] for this purpose). The paper is structured as follows. First we recall the definition of the CSDM model and extend it with the VDM-like state notation. Then we give an algorithm which guides the user through the process of passing from an informal to a formal event representation. We also provide some useful classification of events and give a specification scheme for each class. Then the application of the algorithm is illustrated by a number of case studies. The case studies are based on the examples derived from Fault Trees published in the literature. They present different aspects of the formalisation task. In the Appendix a more complete Fault Tree is formalised with respect to both its events and gates.

2 Event Modelling

The CSDM (*Common Safety Description Model*) formalism [BCG'91, Górski'93] is used as a formal base for modelling. The basic underlying concept is that of *event*. A linear time model is assumed, i.e. time is treated as an infinite set of time moments, linearly ordered. E*vents* may occur in time. The same event can have many occurrences - this is distinguished by giving each occurrence its unique *label*. We distinguish between two classes of happenings: *timed events* (*actions*) and *instantaneous events* (*transitions*). Transition take no time, i.e. its occurrence in time is specified by a single time moment. An occurrence of an action x is specified by giving two associated transitions: x_s and x_e which denote the start and the end of x. One of the basic relations among events which is sought for during safety analysis is *causality*. It tells how events contribute to occurrences of other events. Another important concern of safety analysis is *non determinism* which provides for specification of possible alternative paths of causal relationships among events.

CSDM comprises the following:

E - a set of *events*, its elements denoted by X, Y, Z.

L - a set of *labels* which are used to uniquely identify event occurrences, with individual labels denoted by l, m, n.

A - a set of *actions* - labelled events plus a distinguished *silent action* $\perp$, $\mathbf{A} = (\mathbf{L} \times \mathbf{E}) \cup \{\perp\}$; individual actions denoted by x, y, z.

T - a set of *transitions* with its elements denoted by w; a transition is essentially an action name with a subscript s or e; **T** is related to **A** in the following sense $\mathbf{T} = \{x_s | x \in \mathbf{A}\} \cup \{x_e | x \in \mathbf{A}\}$; intuitively, for an action x, x_s denotes the *start* transition and x_e denotes the *end* transition of x.

$\prec_c$ - a *causality* relation on $(\mathbf{T} \times \mathbf{T})$, a partial order that is irreflexive and transitive.

$=_c$ - a *causality equivalence* on $(\mathbf{T} \times \mathbf{T})$, representing transitions that have exactly the same causes; in particular, for any transition w, $(\perp_s \prec_c w \vee \perp_s =_c w)$ and $(w \prec_c \perp_e \vee w =_c \perp_e)$ hold.

R - the set of *real numbers*, and its elements denoted by r; the notation $[r_1, r_2]$ denotes a closed interval of real numbers.

Time ($\in \mathbf{T} \to \mathbf{R}$) a partial function which assigns real time to transitions. *Time* can also be interpreted as a set of pairs $((l,X), r)$, where $r \in \mathbf{R}$, $(l, X) \in \mathbf{T}$ and $Time((l, X)) = r$. Elements of *Time* we will denote by t, u. *Time* is restricted by $\prec_c$ and $=_c$ in the following sense, assuming $\{w, w'\} \in \mathbf{dom}$ *Time* (domain of the function *Time*), $w \prec_c w' \Rightarrow Time(w) < Time(w')$ and $w =_c w' \Rightarrow Time(w) = Time(w')$, and furthermore we demand that $\perp_s \in \mathbf{dom}$ *Time* $\wedge \perp_e \in \mathbf{dom}$ *Time*.

The set of possible *Time* functions is called the *behaviour space* of the modelled system and its elements are called *possible behaviours*. It is the task of the analyst to distinguish *actual behaviours* from this set and to reason about their properties.

start ($\in \mathbf{A} \to \mathbf{R}$) a partial function that, given x, returns r if $(x_s, r) \in$ *Time*, otherwise *start*(x) is undefined.

end ($\in \mathbf{A} \to \mathbf{R}$) a partial function that, given x, returns r if $(x_e, r) \in$ *Time*, otherwise *end*(x) is undefined.

For any action x, $x_s \prec x_e$ and $(x_s, r) \in Time \Rightarrow \exists r' \cdot ((x_e, r') \in Time \wedge (r < r'))$ and $(x_e, r) \in Time \Rightarrow \exists r' \cdot ((x_e, r') \in Time \wedge (r < r'))$. From that we see that an action always has a duration in time and once started it always ends.

The *temporal ordering*, $\prec_t$, and *temporal equality*, $=_t$ relations ($\subseteq Time \times Time$) are defined as follows. Let $t = (w, r)$ and $t' = (w', r')$; we define $t \prec_t t' \Leftrightarrow w \prec_c w' \wedge r < r'$ and $t =_t t' \Leftrightarrow w =_c w' \wedge r = r'$. Note that for those transitions which occur in *Time*, the ordering $\prec_t$ is implied by $\prec_c$ and similarly $=_t$ is implied by $=_c$

Now we come to a relation about actions and their starting and ending time moments - a relation $\approx (\subseteq \mathbf{A} \times (Time \times Time))$ which is defined as follows: for each $x \in \mathbf{A}$, $x \approx ((x_s,r),(x_e,r')) \Leftrightarrow (x_s,r), (x_e,r') \in Time$.

Causal relations for actions are defined as follows. Let x and y be actions, $x \approx (t_s, t_e)$ and $y \approx (u_s, u_e)$. The relations $\prec_i$ and $\prec_h$ on $\mathbf{A} \times \mathbf{A}$ are defined based on the $\prec_t$ between transitions of actions: $x \prec_i y \Leftrightarrow t_e \prec_t u_s$ and $x \prec_h y \Leftrightarrow t_s \prec_t u_s$. If $x \prec_i y$, we say that x is *interior causal* of y; if $x \prec_h y$, we say that x is *head causal* of y.

Observe that in the definitions above, *start, end,* $\prec_t, =_t, \prec_h, \prec_i$ and $\approx$ are *Time* specific, whereas $\prec_c$ and $=_c$ do not depend on *Time*. Different *Time* functions represent different behaviour in our model. Intuitively, when *Time* is undefined for a transition w, we understand that w never occurs in the corresponding behaviour. Note that the range of the function *Time* falls in the interval [*start*($\perp$), *end*($\perp$)].

For notational convenience, we introduce the following abbreviations.

- predicate *occur*: $\mathbf{A} \rightarrow \mathbf{B}$ is defined as $occur(x) \Leftrightarrow (x_s \in \textbf{dom } Time)$.

- function ϕ (of type $\mathbf{E} \rightarrow \rho(\mathbf{A})$, where $\rho(\mathbf{A})$ denotes the powerset of $\mathbf{A}$) applied to every event to obtain a set of actions such that $\phi(X) = \{(l,X)|l \in \mathbf{L}\}$.

- predicate *overlap*: $\mathbf{A} \times \mathbf{A} \rightarrow \mathbf{B}$ is defined as follows $overlap(x, y) \Leftrightarrow \exists r \cdot (start(x) < r < end(x) \wedge start(y) < r < end(y))$.

Events in Fault Trees are always defined in relation to the real system structure. Typically, they refer to the state of some elements of the system, e.g. 'engine on'. To refer to this structure we need some way of system state representation. In our approach we use a VDM-like [Jones 90] approach to system state modelling. Thus, we can have the following definition of state components:

> **state** SYSTEM **of**
> ENGINE   : { ON, OFF }
> ENGINE_SPEED   : REAL

An event can be defined by giving its *characteristic predicate* which imposes restrictions on the system state. We interpret such predicates in the time domain. It is assumed that the event occurs at a given time moment t iff its characteristic predicate holds. The CSDM model with the VDM notation for state definition is refered to as the Extended CSDM (ECSDM). In ECSDM, an event E can be formally expressed in the following way:

$$E(Time, t) \equiv PE/(Time, t)$$

where E is an event, PE is its characteristic predicate and Time is a behaviour of the system.. The above definition reads as follows:

> *E occurs in behaviour Time at time moment t iff in the behaviour Time PE is true at the time moment t.*

Note that such definition does not distinguish between different *actions* (event occurrences). It identifies all occurrences of the event in a given behaviour of the system of interest.

3 The Algorithm of Event Formalisation

Formalisation of events is a basic step in the process of transition from an informal to a formal Fault Tree representation. This step is implemented by the algorithm given below.

For a given Fault Tree, the algorithm traverses all nodes of the tree, starting with the top event. For each node (event description) the following steps are performed.

STEP 1: Identification of the State Space

a) Identification of State Elements
From an informal description given in this node, identify all the system elements which contribute to the event. This can be done by picking up all the nouns denoting real objects (like "pump", "relay", "valve", etc.) or value-objects like "limit", "boundary", etc.

b) Domains Definition
Distinguish between **state** elements and constant **values**. Identify and write down the domains of all state elements. Use additional knowledge about the system available from other sources of information from the system documentation. Do not relay exclusively on the event description from the tree.

c) State Space Definition
Associate state elements with the corresponding domains. If necessary introduce a structure among state elements. Identify any restrictions on the resulting state space which come from the physical laws and/or can be derived from the application domain knowledge. Express such restrictions as state invariants.

STEP 2: Event Definition

Build the *characteristic predicate* which captures meaning of the event. In a simple case this predicate refers to a single state element and has a form

> *simple predicate:* S *rel* V

where S is a state element, *rel* denotes a relation and V denotes a value from the domain of S. In some cases however, an event refers to more than one state element. In this more complex situation the characteristic predicate comprises several simple predicates related one to another by some relationships.

a) Identification of simple predicates
From the event description, identify simple predicates and make a list out of them. Each such predicate defines a (simple) event which contributes to the event under consideration.

b) Identification of relationships
In the case of a complex event, identify the sort of relationships among the contributing simple events. Identify additional temporal restrictions on the participating events, if any.

314

c) Formal event definition

In this step the event is expressed in terms of the ECSDM model. Below we give definitions relating to the three most frequently occuring situations:

- E is a *simple event* and PE is its characteristic predicate. Then the formal definition has a form

    $$E(\ Time,\ t\ ) \equiv PE/(\ Time,\ t\ )$$

    which reads *E occurs at time moment t in a behaviour Time iff in the behaviour Time PE holds at time t.*

- E is a *composed* event given by an expression 'E1 when E2' where E1 and E2 are simple events and their corresponding state predicates are PE1 and PE2. Then the definition of E has the form

    $$E(\ Time,\ t\ ) \equiv (PE1 \wedge PE2)/(\ Time,\ t\ )$$

    which reads *E occurs at time moment t in a behaviour Time iff in the behaviour Time PE1 and PE2 hold at time t*

- E is a *timed* event (i.e. its informal definition includes an explicit reference to time).

    Let PE1 be a state predicate characterising E and PT be a predicate giving the time constraint on a given occurence of the event. Then the definition has the form

    $$E'(\ Time,\ t\ ) \equiv PE1/(\ Time,\ t\ )$$
    $$E(\ Time,\ t\ ) \equiv \exists\ e \in \phi_{Time}(E') \cdot PT(\ e,\ t\ )$$

    where PT may refer to some characteristics of the action e, like its *start, end, duration,* etc. This definition reads as follows: *E occurs at time moment t in a behaviour Time iff E' occurs at t and this particular action satisfies PT.*

STEP 3: Validation

In this step the validity of the formal definition is being checked. We have to make sure that the formal description adequately captures the information content of the original Fault Tree. The event definition is validated against the original text with possible help of a domain specialist and a safety analyst. In particular, if there is the possibility of different interpretations of the original tree, the meaning chosen for the formal counterpart should be carefully checked and justified.

4 Case Studies in Event Formalisation

In this section the applicability of the Algorithm is illustrated through examples.

4.1 Simple Events

An example of a simple event formalisation is given below.
Example 1: Switch manipulation.

E | switch pressed

Step1: Identification of state space.

a) State elements:

SWITCH - is a monostable switch which after being depressed returns automatically to the initial position.

b) Domains:

SWITCH may assume the following positions:
- ON - if it is pressed
- OFF - if it is depressed

c) State Space definition:

state SYSTEM **of**
SWITCH: { ON, OFF }

Step 2: Event Definition.

a) Identification of state predicates:
SWITCH=ON

b) Identification of relationships:
No relationships, E is a simple event

c) Formal event definition:

$$E(\ Time,\ t\ ) \equiv (\ SWITCH = ON\ )/(\ Time,\ t\ )$$

Step 3. Validation:

The interpretation of the above definition is that event E occurs in behaviour *Time* at time moment t iff SWITCH assumes position ON. E stands for all such occurrences.

Example 2: Pump failure [BA 93].

E
```
pump
stuck on
```

1a)   PUMP - a physical object

1b)   Additionally to the normal "on" and "off" states we explicitly
       distinguish the failure state of PUMP. Thus possible states are:
       ON, OFF, STUCK-ON

1c)
       state SYSTEM **of**
       PUMP: { ON, OFF, STUCK-ON }

2a)
       PUMP=STUCK-ON

2b)   No relationships, E is a simple event

2c) $E(\ Time,\ t\ ) \equiv (\ PUMP = STUCK\text{-}ON\ )/(\ Time,\ t\ )$

3)   E occurs iff, for a given behaviour *Time* and at a given time moment t,
       PUMP assumes value STUCK-ON.

4.2 Composed Events

Composed events refer to two or more simple events. In most cases a composed event is specified as a coincidence of two simple events, by joining them using the word "when".

Example 3: In the following tree E1, E2, E4 i E6 are composed events [BA 93]:

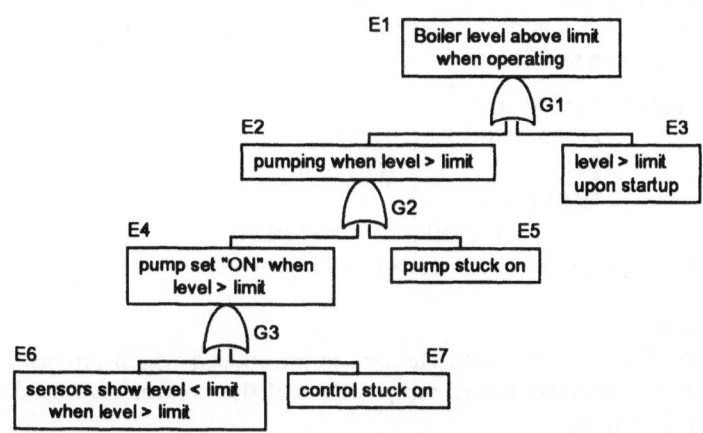

Formalisation of E1:

1a)
BOILER	a physical object
LEVEL	gives current amount of water in the BOILER
LIMIT	a constant specifying the safe level of water
SYSTEM	the system of interest (an implicit object)
OPERATING	characterises a mode of SYSTEM

1b)
LIMIT	assumes a fixed value **max**
BOILER	is a composed object and LEVEL is one of its elements which gives the value of current level of water
OPERATING	relates to the whole system and assumes boolean values.

1c)

values LIMIT = **max**

state SYSTEM **of**

OPERATING   : BOOL

BOILER ::

LEVEL   : REAL

2a) E1 is composed of two simple events:

E1.1: water level above limit

its characteristic predicate is PE1.1 ≡ LEVEL > LIMIT

E1.2: system is operating

and its characteristic predicate PE1.2 ≡ OPERATING=TRUE

2b) E1: E1.1 "when" E1.2

2c)

E1(Time, t) ≡ ((LEVEL > LIMIT) ∧

(OPERATING = TRUE))/(Time, t)

317

3)   In the formal specification given above OPERATING relates to the whole
        system. Note that it was not clear from the Fault Tree itself. This
        interpretation has to be validated against an additional documentation of
        the system. "When" from the informal description is interpreted as the
        coincidence of the two simple events.

Formalisation of E2:
Formalisation of E2 adds new components to the formal model: the resulting
specification looks as follows.

> **state** SYSTEM **of**
> > OPERATING   : BOOL
> > PUMP   :{ON,OFF}
> > BOILER ::
> > > LEVEL   : REAL
>
> E2(Time, t) ≡ ((LEVEL > LIMIT) ∧ (PUMP = ON))/(Time, t)

Formalisation of E4:
The PUMP_COMMAND state element is introduced which provides for making
distinction between the actual pump state and the control command issued with
respect to the pump.

> **state** SYSTEM **of**
> > BOILER ::
> > > LEVEL   : REAL
> > > OPERATING   : BOOL
> > > PUMP   :{ON,OFF}
> > > PUMP-COMMAND   :{ON,OFF}
>
> E4(Time, t) ≡ ((LEVEL > LIMIT) ∧ (PUMP-COMMAND = ON))/(Time, t)

Formalisation of E6:
The MEASURED-LEVEL state component is added to distinguish between the
actual level of water and the level measured by the sensor:

> **state** SYSTEM **of**
> > BOILER ::
> > > LEVEL   : REAL
> > > MEASURED-LEVEL   : REAL
>
> E6(Time, t) ≡ ((LEVEL > LIMIT) ∧ (MEASURED-LEVEL < LIMIT))/(Time, t)

4.3 Timed Events

Events occur in time. For each such occurrence we can identify its beginning and
its end (in our model we assume a finite observation interval for a system). This
fact is usually not specified in a Fault Tree - the time duration of events is
understood implicitly. In some cases however, an event definition includes
explicit reference to time. It is interpreted as a restriction which relates to all
possible occurrences of this event.

Example 4: Restriction of duration time:

E | motor runs
for t > 60 sec.

1a)   MOTOR
1b)   ON, OFF
1c)   **state** SYSTEM **of**
             MOTOR: { ON, OFF }
2a)   MOTOR=ON
2b)   duration time greater than 60sec.
2c) $F(\text{Time}, t) \equiv (\text{MOTOR=ON})/(\text{Time}, t)$
    $E(\text{Time}, t) \equiv \exists\, f \in \phi_{\text{Time}}(F) \cdot duration(f) > 60s \wedge start(f) < t < end(f)$

Another example demonstrates formalisation of an event which is defined in a rather unclear way.

Example 5: [Vesely 81].

E | EMF (electromotive force)
applied to K2 relay coil
for t > 60 sec

To understand this description we have to recall the physical model of the system:

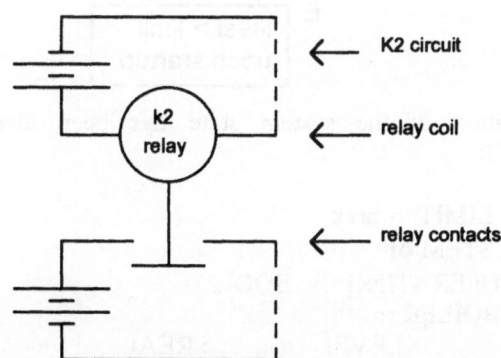

In the scheme shown above, if the K2 relay works reliably and the current flows through the K2 circuit then the K2 relay closes the second circuit. Thus, after studying this additional documentation, we can realise that the phrase "EMF applied to relay coil" denotes that the electric circuit which includes the K2 relay coil is closed and that electric current is presently flowing though it.

1a)   CIRCUIT   the electric circuit of interest
            K2   the relay in the circuit
            CURRENT   the electric current in the CIRCUIT

1b)   K2   relay contacts may be OPEN or CLOSED
         CURRENT   ON (the current flows),
                             OFF (no current in the circuit)
1c)

state SYSTEM **of**
              CIRCUIT ::
                    K2   : {OPEN,CLOSED}
                    CURRENT   : {ON,OFF}
         inv (CIRCUIT.CURRENT = ON) ⇒ (CIRCUIT.K2 = CLOSED)

The above invariant establishes that K2 is closed on the condition that the current flows through the CIRCUIT.

2a)   CIRCUIT.CURRENT = ON
2b)   minimal duration of the event is limited to 60 sec
2c)

         · E'(Time, t) ≡ (CIRCUIT.CURRENT = ON)/(Time, t)
           E(Time, t) ≡ ∃ e ∈ φ$_{Time}$(E') · *duration*(e) > 60s ∧ *start*(e) < t < *end*(e)

3)   The meaning of the formal definition is that electric current flows through the circuit which includes K2 relay coil and this situation stays for at least 60 sec.

4.4 Synchronisation Dependencies

Let us consider the following event:
Example 6: [BA 93].

E | level > limit
    | upon startup

1)   formalisation of the system state has been already considered in example 3:

         values LIMIT = **max**
         state SYSTEM **of**
                    OPERATING   : BOOL
                    BOILER ::
                            LEVEL   : REAL

2a)   E1: LEVEL > LIMIT
         E2: OPERATING=ON
2b)   The dependencies between the simple events E1, E2 are shown below:

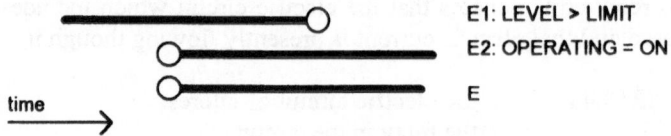

E is an event which is synchronised by the occurrence of E1.

2c)   E1(Time, t) ≡ (LEVEL > LIMIT)/(Time, t)
        E2(Time, t) ≡ (OPERATING = ON)/(Time, t)
        E(Time, t) ≡ ∃ e1 ∈ ϕ_{Time}(E1), e2 ∈ ϕ_{Time}(E2) ·
                start(e2) ≥ start(e1) ∧ overlap(e1,e2) ∧ start(e2) < t

3)   the definition says that the system has been started in a situation where the
        water level in the tank is higher than the maximal limit.

4.5 'Measurable' Events.

Example 7: The pump again.
Let us come back to the system from the Example 2. This time we are interested in
the 'pump stuck-off' situation.

E ┌─────────────┐
  │   pump   │
  │ stuck off │
  └─────────────┘

        Previously, we have introduced the STUCK-ON value which explicitly
represents the pump failure. At the first glance it seems that the 'stuck-off' situation
could be handled in a similar way - by doing this we would directly refer to the
'pump stuck-off' failure. However, it is not clear how the 'stuck-off' situation is to
be detected in the system: the event would be defined in an 'unmeasurable' way.
One of possible ways of approaching this problem is to 'translate' the event
expressing it in terms of 'measurable' attributes, for instance: *the pump is stuck-off
if, after issuing the 'pump start' command, the pump does not start to operate.*

1a)   PUMP
        COMMAND send to the pump
1b)   PUMP may be ON, OFF or STUCK-ON
        COMMANDs are START and STOP
1c)
        state SYSTEM **of**
                PUMP   : { ON, OFF, STUCK-ON }
                COMMAND   : { START, STOP }

2a)   COMMAND = START
        PUMP = OFF
2b)   The event occurrs iff (COMMAND = START) and (PUMP = OFF)
        hold together
2c)
        E(Time,t)≡ (COMMAND=START) ∧ (PUMP=OFF)/(Time, t).

3)   While validating the above definition we can discover the following
        problem, however. Because of its mechanical inertia the pump will not
        turn on at the moment of issuing the START command. Instead, the ON
        state of the pump can be unambiguously detected only after some time
        delay, say *d*. Consequently, the definition of the 'pump stuck-off' situation
        should be modified to:
        E1(Time,t)≡ (COMMAND=START)/(Time, t)

E2(Time,t)≡ (PUMP=OFF)/(Time, t)
E(Time,t)≡ ∃ e1∈ φTime(E1),e2∈ φTime(E2) ·
　　　　overlap(e1, e2) ∧ *end*(e1) > *start*(e2) > *start*(e1) + d ∧
　　　　t >=*start*(e2)

A graphical illustration is given below:

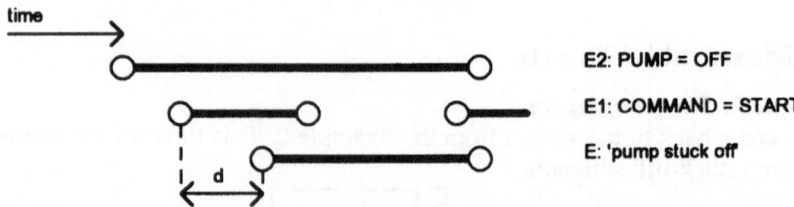

The event E holds as long as the event E2 (i.e. as long as the pump is OFF). It terminates when the pump switches to the ON or STUCK-ON state.

5 Putting Events and Gates Together

Having specified the events of a Fault Tree we can proceed with the specification of gates. The basic structures of the CSDM based specification of Fault Tree gates can be found in [Gorski 94, BCG 91]. We will not repeat this material here. Instead, we illustrate the possibilities by presenting an example OR-gate. An OR-gate is the *Generalisation OR* if the output event coincides with the input events. In such case there is no time delay between the input and the output event. The events coincide in time and their causes are identical. To illustrate this concept let us consider an electric circuit with two independent switches, as shown below:

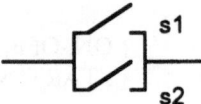

If electric current flows through this circuit then at least one of the switches must have been closed. This is illustrated by the following tree:

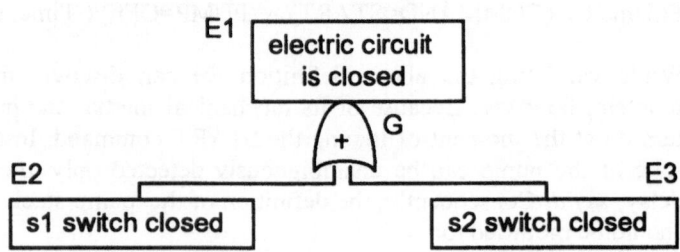

322

1a) State elements:

  electric CIRCUIT,
  switch S1,
  switch S2.

1b) Domains:

  CIRCUIT may be OPEN or CLOSED
  switch S1 may be OPEN or CLOSED
  switch S2 may be OPEN or CLOSED

1c) State space definition:

  state SYSTEM **of**

CIRCUIT	: { OPEN, CLOSED }
S1	: { OPEN, CLOSED }
S2	: { OPEN, CLOSED }

2a) Identification of state predicates:

  E1: CIRCUIT = CLOSED
  E2: S1 = CLOSED
  E3: S2 = CLOSED

2b) Identification of relationships:

  Events E1, E2 and E3 are simple events

2c) Formal events definitions:

  E1(Time,t) ≡ (CIRCUIT = CLOSED)/(Time,t)
  E2(Time,t) ≡ (S1 = CLOSED)/(Time,t)
  E3(Time,t) ≡ (S2 = CLOSED)/(Time,t)

  The semantics of the gate G is given below:

  $\mathcal{M}$(G(E2, E3, E1)) ≡
      $\forall e \in \phi(E1) \cdot \exists f \in \phi(E2) \cup \phi(E3) \cdot$
      $(occur(e) \Rightarrow occur(f) \wedge (start(e) =_c start(f)) \wedge$
      $(end(e) =_c end(f)))$

A possible relation in time between the input events E2, E3 and their *generalisation* E1 is shown below.

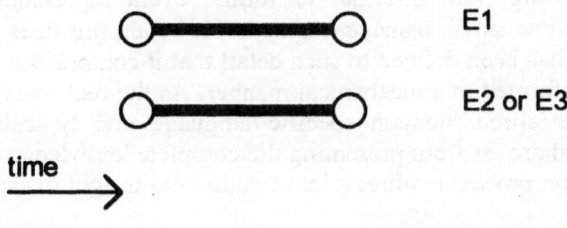

E1

E2 or E3

time

323

Let us note that the semantics given above excludes the following situation:

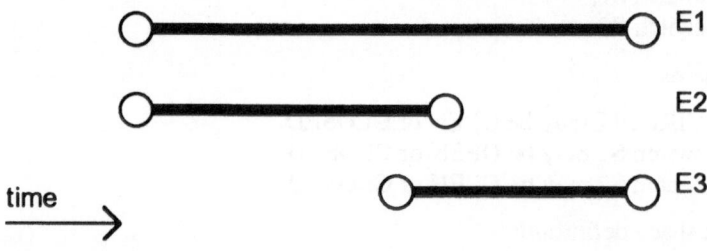

which should be clearly included in the case of our circuit. This can be achieved by the following modification:

$$\mathcal{M}(G(\ E2,\ E3,\ E1\ )) \equiv$$
$$\forall\, e \in \phi(E1)\cdot$$
$$(\exists\, f \in \phi(E2) \cup \phi(E3) \cdot \forall\, h \in \phi(E2) \cup \phi(E3)\ \cdot$$
$$\mathit{occur}(e) \Rightarrow \mathit{occur}(f) \wedge \neg \mathit{overlap}(f,h)\ \wedge$$
$$\mathit{start}(e) =_c \mathit{start}(f) \wedge \mathit{end}(e) =_c \mathit{end}(f))$$
$$\vee\, (\exists\, f \in \phi(E2),\ h \in \phi(E3)\ \cdot$$
$$\mathit{occur}(e) \Rightarrow \mathit{occur}(f) \wedge \mathit{occur}(h)\ \wedge$$
$$\mathit{overlap}(f,h)\ \wedge$$
$$\mathit{start}(e) =_c \min(\mathit{start}(f),\mathit{start}(h))\ \wedge$$
$$\mathit{end}(e) =_c \max(\mathit{end}(f),\mathit{end}(h))).$$

The above example shows how the semantics of the gate is adjusted to a specific context of a particular application. Although the specifier may be provided with a catalogue of typical schemes giving the semantics to gates, his choice will always depend on the application the gate refers to and therefore should be carefully validated against the application domain knowledge.

6 Conclusions

The paper presented a systematic approach to formalisation of Fault Trees. The formalisation is accomplished in two steps: Events Formalisation and Gates Formalisation. As the Gates Formalisation step has been already discussed in the previous papers [BCG 91, Górski 93], the main focus of this presentation was on formalisation of events. The events of a tree were expressed in terms of the ECSDM model. The paper gave an algorithm which guides the user through the process of passing from informal to formal event representation. The whole process is split into small, manageable steps. However, this does not mean that the whole process has been defined to such detail that it comprises a sequence of steps that can be performed in a mechanical manner. As the real trees are often defined in a very specialised, domain specific language and typically contain many ambiguities and are far from presenting the complete knowledge about the system, the formalisation process requires a lot of additional insight to capture the intended meaning of a tree. Therefore a full algorithmisation of this process it is very difficult if not impossible. The approach presented here should be therefore understood as a guide which gives a systematic way of proceeding from informal to formal rather than a complete algorithm. Also the variety of different 'event definitions' encountered in real trees is very wide. We have introduced here some classification of events, for each class giving a suitable formalisation scheme. This classification is based on a review of trees published in the literature as well as on data extracted from the documentation of some real systems. The further work in this area will include the following

- collecting more 'real' trees to validate and refine our approach to event formalisation,
- more work towards establishing the formal semantics of the whole tree, resulting from the formal descriptions of events and gates,
- looking into possible analyses which can be done with respect to formalised tree,
- relating the formal description to other representations of the system and its components (e.g. software).

7 Acknowledgement

The work presented in this paper has been partially supported by the Associated Contract with the SHIP (Safety of Hazardous Industrial Processes) Project within the Environment EC research Programme.

References

[BM 89]   L. Bass, D. L. Martin, *Cost-Effective Software Safety Analysis,* Proceedings of Annual Reliability and Maintainability Symposium, IEEE 1989

[BCG 91]   R. E. Bloomfield, J. H. Cheng, J. Górski, *Towards A Common Safety Description Model,* Proceedings of Safecomp'91, Pergamon Press, 1991

[BA 93]   G. Bruns, S. Anderson, *Validating Safety Models With Fault Trees,* in: Safecomp'93, (J. Górski, Ed.), Springer-Verlag, 1993

[Górski 94] J. Górski, *Extending Safety Analysis Techniques With Formal Semantics,* In Technology and Assessment of Safety Critical Systems, (F.J. Redmill, Ed.), Springer-Verlag, 1994

[HSE 87]   Health and Safety Executive, *Programable Electronic Systems In Safety Related Applications, General Technical Guidelines,* Her Majesty's Stationery Office, London, 1987

[Jones 90] C. B. Jones, *Systematic Software Development using VDM,* Prentice Hall Int., 1990

[Vesely 81] W. E. Vesely et el., *Fault Tree Handbook,* Nureg 0492, US Nuclear Regulatory Commission, 1981

Appendix: Formalisation of a simple Fault Tree

A partial Fault Tree of a robot system is shown below [BM89]:

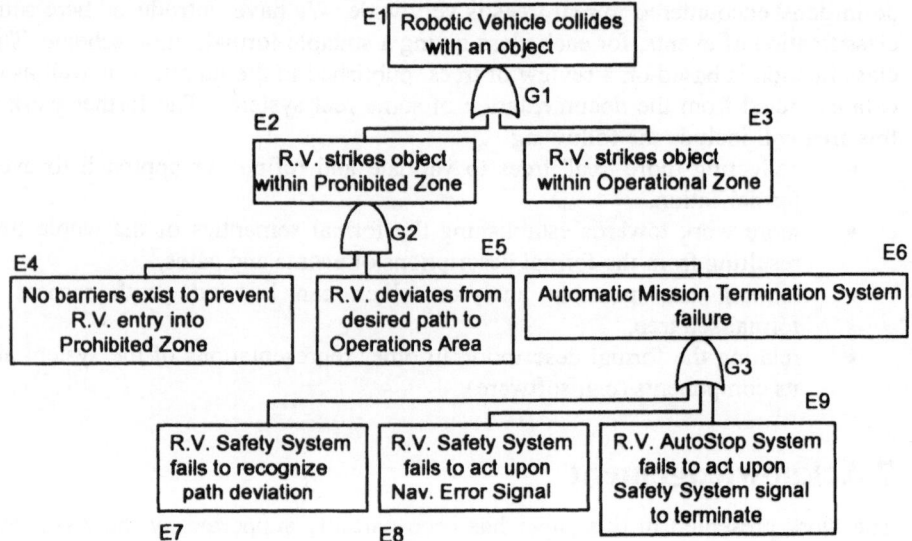

Formalisation of events:

1a) State elements:

       Robotic Vehicle (RV)
       RV Safety System
       RV AutoStop System
       Barriers
       Path
       Zone
       navigation error signal

1b) Domains:

       RV may collide or be OK,
       RV may be in the operational zone or in the prohibited zone,
       the RV path may be OK or deviated from the desired one,
       the RV Safety System may detect the path deviation (and signal *'nav error'*
            to the AutoStop System),
       the barriers can be present or not (this depends on the physical construction)
       the navigation error signal (from Safety System to AutoStop System) may
            be active (UP) or inactive (DOWN)
       the RV AutoStop System may stop the RV or just do nothing.

1c) System definition

values BARRIERS-PRESENT= *here we have a BOOL value depending on the physical construction of the system*

state SYSTEM **of**

RV-STATUS	: { OK, COLLISION }
RV-ZONE	: { OPERATIONAL, PROHIBITED }
RV-PATH	: { CORRECT, DEVIATED }
RV-SAFETY-SYSTEM	: { OK, PATH-DEVIATION-DETECTED}
NAV-ERROR-SIGNAL	: { DOWN, UP }
RV-AUTOSTOP	: { OK, STOP }

2a) Identification of state predicates:

E1: RV-STATUS = COLLISION
E2: (RV-STATUS = COLLISION) and
     (RV-ZONE = PROHIBITED)
E3: (RV-STATUS = COLLISION) and
     (RV-ZONE = OPERATIONAL)
E4: BARRIRES-PRESENT
E5: RV-PATH = DEVIATED
E6: (RV-PATH = DEVIATED) and
     (RV-AUTOSTOP ≠ STOP)
E7: (RV-PATH = DEVIATED) and
     (RV-SAFETY-SYSTEM ≠ PATH-DEVIATION-DETECTED)
E8: (RV-SAFETY-SYSTEM = PATH-DEVIATION-DETECTED)
     and (NAV-ERROR-SIGNAL ≠ UP)
E9: (NAV-ERROR-SIGNAL = UP) and
     (RV-AUTOSTOP ≠ STOP)

2b) Identification of relationships:

E2, E3, E6, E7, E8 and E9 are composed events. They occur during the coincidence of the contributing simple events.

2c) Formal events definitions:

E1(Time, t) ≡ (RV-STATUS = COLLISION)/(Time, t)
E2(Time, t) ≡ ((RV-STATUS = COLLISION) ∧
          (RV-ZONE = PROHIBITED))/(Time, t)
E3(Time, t) ≡ ((RV-STATUS = COLLISION) ∧
          (RV-ZONE = OPERATIONAL))/(Time, t)
E4(Time, t) ≡ BARRIERS-PRESENT
E5(Time, t) ≡ (RV-PATH = DEVIATED)/(Time, t)

AUTHOR INDEX